This book gives an account of the modern view of the global circulation of the atmosphere. It brings the observed nature of the circulation together with theories and simple models of the mechanisms which drive it. Early chapters concentrate on the classical view of the global circulation, on the processes which generate atmospheric motions and on the dynamical constraints which modify them. Later chapters develop more recent themes including low frequency variability and the circulation of other planetary atmospheres.

The book will be of interest to advanced students and researchers who wish for an introduction to the subject before engaging with the original scientific literature. The book is copiously illustrated, and includes many results of diagnostic and modelling studies. Each chapter includes a set of problems and bibliographical notes.

Cambridge atmospheric and space science series

Introduction to circulating atmospheres

Cambridge atmospheric and space science series

Editors
Alexander J. Dessler
John T. Houghton
Michael J. Rycroft

Titles in print in this series

M. H. Rees, *Physics and chemistry of the upper atmosphere*
Roger Daley, *Atmospheric data analysis*
Ya. L. Al'pert, *Space plasma, Volumes 1 and 2*
J. R. Garratt, *The atmospheric boundary layer*
J. K. Hargreaves, *The solar–terrestrial environment*
Sergei Sazhin, *Whistler-mode waves in a hot plasma*
S. Peter Gary, *Theory of space plasma microinstabilities*
Martin Walt, *Introduction to geomagnetically trapped radiation*
Tamas I. Gombosi, *Gaskinetic theory*
Boris A. Kagan, *Ocean–atmosphere interaction and climate modelling*

Introduction to circulating atmospheres

Ian N. James

University of Reading

CAMBRIDGE
UNIVERSITY PRESS

Published by the Press Syndicate of the University of Cambridge
The Pitt Building, Trumpington Street, Cambridge CB2 1RP
40 West 20th Street, New York, NY 10011–4211, USA
10 Stamford Road, Oakleigh, Melbourne 3166, Australia

© Cambridge University Press 1994

First published 1994
First paperback edition (with corrections) 1995

A catalogue record for this book is available from the British Library

Library of Congress cataloguing in publication data available

ISBN 0 521 41895 X hardback
ISBN 0 521 42935 8 paperback

Transferred to digital printing 2003

TAG

Contents

Preface

The wind blows where it wills; you hear the sound of it, but you do not know where it comes from, or where it is going. (John, 3, 8)

In the ancient world, the question of where the wind came from and where it was going to was one of life's unanswerable puzzles. Indeed, so unpredictable and unknowable was the wind, that the evangelist used it as an elaborate pun on the inscrutable purposes of God's Spirit (in the Greek of the New Testament, the same word means 'wind' and 'spirit'). Meteorology of any sort must have been a most frustrating study. Hints of regular patterns emerged, only to vanish on closer inspection. Wise saws about the weather were wrong as often as they helped. And if you were a farmer, a sailor or a campaigning soldier, guessing the winds or the weather wrongly could lead to disaster. Truly, through the weather and the winds, the gods played with man, teased and tormented him, and confirmed their authority.

And here matters more or less rested until the Newtonian revolution of the seventeenth century. By then, many different aspects of the natural world had been reduced to reproducible laws. Kepler showed how the motions of the planets were governed by strict rules, although he did not quite succeed in explaining why his laws of planetary motion had the forms they did. Galileo showed how the motions of simple bodies, such as heavy metal spheres on inclined planes, or pendulum bobs, were orderly and predictable. Newton with his three Laws of Motion and single Law of Universal Gravitation unified all these various studies and showed how the observations of earlier scientists could be predicted by his set of four laws.

It was not long before the problem of atmospheric motions was addressed using the new Newtonian mechanics. Halley, in 1687, advanced a theory of the Trade Winds, which he suggested were caused by the rising of heated air on the equator sucking in air from higher latitudes. Hadley, in 1720, advanced much the same ideas, suggesting that there must be poleward flow aloft over the surface trade winds. Thus the earliest attempts to demystify the winds and the large scale circulation of the atmosphere began. In the latter part of the eighteenth century, the motions of a fluid on the Earth were put on a more rigorous mathematical basis, and with the formulation of the 'Coriolis force', the crucial role of the Earth's rotation began to be appreciated. Nevertheless, these attempts were hampered by lack of both theoretical tools and, especially, of observations.

The dream of Newton's disciples was to describe the atmospheric circulation as a predictable mechanism, an objective to be achieved by the systematic application of Newton's laws of motion to the air. It is ironic that this dream appeared to become more of a possibility after the First World War, just as quantum mechanics and the theory of relativity were undermining the simple deterministic Newtonian account of the dynamics of systems in other branches of physics. As aviation developed, the need for detailed meteorological information, and the capability of gathering, and more particularly collating, the necessary data, both expanded. With the development of the Norwegian frontal cyclone model by members of the Bergen school, and their clear objective of quantifying that model in terms of Newtonian fluid mechanics, the scene was set for the great flowering of meteorology in the wake of the Second World War. The stability analyses of Charney and Eady accounted for the development of cyclones in terms of fluid instabilities. The development of the computer made numerical weather prediction possible, based upon the integration of the intractible nonlinear equations of motion. And as a global network of observing stations and telecommunications links was established, a truly global view of the atmospheric circulation began to emerge.

To my mind, in the computer the meteorologist found a most diverting toy. The effort to develop realistic models, first for short range weather prediction, and more recently for climate simulation and prediction, took precedence over attempts to understand the dynamical nature of the large scale atmospheric circulation. Since the governing equations are complex and nonlinear, it seemed to many that their numerical solutions would afford the only feasible approach to understanding their implications; and so many of the most ingenious brains in the business devoted themselves to refining numerical schemes and improving the representation of various unresolved but important processes. The deep physical insights into the nature of the atmospheric circulation gained by such men as Rossby and Ertel were all but forgotten by a whole generation of atmospheric scientists. A few academic meteorologists continued to be drawn to the attempt to reduce the atmosphere to its simplest possible essentials, and out of that effort came a deeper appreciation of the way in which nonlinearity manifests itself in the atmosphere. The work of Lorenz and others led to the development of so-called 'chaos' theory, which has become so fashionable in many fields of modern science. This revealed that nonlinearity can act to generate great complexity out of really rather simple equations. The language of chaos now illuminates our studies of the large scale atmospheric circulation.

The Newtonian dream was severely compromised with the realization that unpredictability is a fundamental property of such systems as the atmosphere, and not just a technical limitation. Unpredictability is most obvious in the weather forecasting problem: two nearly identical approximations to the atmospheric circulation at a particular moment quickly diverge as the forecast proceeds. The attempt to carry out long range forecasts by analogy with the history of similar atmospheric states is clearly doomed. Intimately related to the limited validity of weather forecasts is the difficulty of identifying discrete 'parcels' of air. As the chaotic motions in the atmosphere advect any such parcel around, it becomes distorted; long, thin streamers of material are pulled off it, and eventually it is completely dispersed through the rest of the atmosphere. In the process, it loses any identity.

After the naïve optimism of the Newtonian era, we now realize that indeed, St John's observation is quite right; where the wind comes from and where it is going to are, in a sense, profoundly and essentially unknowable. From the static concept of a precise and definable global circulation, we now understand the atmosphere as an infinitely varied, evolving system, with complex structure on all space and timescales. By dint of much time and spatial averaging, we can arrive at a static view of the global circulation of the atmosphere which is built up from the combined effects of smaller subsystems or 'building blocks', in the classic Newtonian reductionist tradition. But when we look more closely, the same building blocks are capable of building entirely different edifices. The building itself modifies the building blocks. We conclude that these simple pictures always have limited validity and are severely limited when we try to extend them to different situations. That is partly why predicting the climate of the high carbon dioxide world that mankind is rapidly creating is so difficult and controversial. It is also why recent attempts to survey the atmospheric circulations of the other planets have led to unexpected surprises on every front.

Much of the earlier part of this book is taken up with a description of the modern elaboration of the deterministic account of the global atmospheric circulation. The approach will be based on physical principles, relating the thermodynamical processes which drive the circulation to the mechanical processes which determine the resulting motions of the atmosphere. I have tried to present both observations of the actual circulation as well as simple models based on the laws of mechanics and thermodynamics. These models are approximate and limited, but they do provide a vocabulary, a set of conceptual models, which enable the researcher to interpret observations of the real atmosphere on one hand or the results of immensely complex 'global circulation models' on the other. Towards the end of the book, I will survey some more recent developments in the field which point to the complexity of the real circulation, and the limitations of our simple models.

Chapters 1–6 concentrate essentially upon the deterministic view of the global circulation, developed over the last 50 years or so. The basic physical laws, expressed in suitably quantitative language, will be given in Chapter 1. Chapter 2 discusses both observations of the large scale atmosphere and the weather prediction and global circulation models which between them provide our modern tools for the study of the global circulation. The thermodynamic imperative for that circulation is provided by the inhomogeneous heating of the atmosphere. This will be addressed in Chapter 3. The dynamics of the resulting motion is strongly constrained by the Earth's rotation and other mechanical factors. The atmosphere exhibits considerable ingenuity in circumventing these constraints in order to transport heat. The various responses of the atmosphere to differential heating will be outlined in Chapters 4, 5 and 6.

The remaining chapters focus on some more recent developments in our understanding of the global circulation. The traditional view of the circulation as a meridional overturning is of course a great oversimplification. Variations in the zonal direction have been studied more intensively in recent years, with an emphasis on the maintainance of the midlatitude storm tracks and on those processes which generate the long wavelength, semi-permanent troughs and ridges in the midlatitude westerlies. Zonal asymmetries in the tropics have assumed greater importance, not least because they seem to offer some limited degree of predictability of weather regimes at higher latitudes on the seasonal timescale. The underlying theme in these chapters is the interaction between different scales of motions in the atmosphere. The end result is that the large scale structure of the circulation is, to a considerable extent, organized by smaller scale, transient features, and not the reverse. These topics form the subject of Chapter 7.

Similar results hold in the time domain. The timescales predicted by the work of the earlier chapters are all fairly short, not generally exceeding a few days or, at most, weeks. Yet the atmospheric circulation fluctuates on much longer periods. 'Low frequency' variability, that is, variability with timescales of ten days and longer, is considered in Chapter 8. While the preceding material is of some help in understanding some of the basic mechanisms for low frequency variability, the modern developments in the field of chaotic behaviour assume explicit importance.

Routine satellite observations, and greatly improved satellite instrumentation have led to a considerable upsurge of interest in the stratosphere and the circulation of the 'middle atmosphere', that is, of the stratosphere and mesosphere. Such research has gained considerable urgency as the interactions between ozone chemistry and mass circulations in the middle atmosphere are seen as central to attempts to understand human influences on ozone. The appearance of the 'ozone hole' in the southern hemisphere spring has clearly demonstrated that such human influences are present. The increasing suspicion that ozone depletions on a more global scale are beginning means that there is an urgent need to understand the circulation of the stratosphere. The transport of trace consituents such as ozone and the various trace chemicals which interact with it again brings into sharp focus the chaotic and unpredictable nature of the atmospheric circulation. Chapter 9 is devoted to an introduction to the stratospheric circulation and its influence on ozone distribution.

Finally, in Chapter 10, we step back from a myopic consideration of the Earth's atmosphere in isolation. Properly, we may understand the circulation of the Earth's atmosphere as a particular realization of a great range of possibilities, some of which we see exemplified in the other planets of the solar system. The study of the global circulation has perhaps always suffered from being the study of a single system. Now at least a small core of data is available for virtually all the bodies with substantial atmospheres in the solar system. Meteorology can join with the other earth sciences, and embark on the *comparative* study of the range of possible global circulations.

The level of the exposition is intended to be suitable for advanced undergraduate and for masters level students. I hope it will also be found useful by researchers who wish to acquire enough background knowledge of the global circulation to tackle the original literature confidently. The text is an expansion of courses that I have taught to our MSc and BSc students over the past few years. Indeed, an important motivation for the text was the need to recommend suitable reading to accompany these courses. The standard texts on dynamical meteorology devote only a small proportion of their pages to the global circulation. Other more general texts are very descriptive and scarcely enable the student to get to grips with the physical processes underlying the global circulation at any quantitative level. So this book deliberately sets out to address, not the whole of dynamical meteorology, but the application of dynamical meteorology to the large scale circulation. It is presumed that the reader will have studied the basics of atmospheric dynamics, though in order to make the text self-contained and to establish the notation that I will use, the most important results are quoted without rigorous derivation in the opening chapter. The various processes which drive the circulation, such as absorption and emission of electromagnetic radiation, microphysical cloud processes and turbulent transports in the atmospheric boundary layer (a group of processes often miscalled 'atmospheric physics', as if Newton's laws are *not* physics!) will not be developed to any great extent; once more, a few important results will be quoted, and readers requiring more information will be referred to other texts. What this book will concentrate on is the transport of heat by large scale atmospheric motions, and the way in which these transports release more energy to drive the circulation. Transport of heat inevitably implies transport of other quantities, such as moisture or momentum. It is the transport of these quantities which form the central theme of our studies.

Each chapter finishes with a short section of problems. These serve a number of purposes. One is to encourage the student to develop a quantitative, as well as a qualitative, understanding of the subject. I hope the problems will lead the student to develop the habit of carrying out little calculations and estimates whenever data and equations occur in proximity. In this way, an appreciation of the relative importance of different effects and of the circumstances in which different sorts of balances can occur will be built up. Another function of the problems is to suggest some lines of thought or to introduce some additional results which have had to be dropped from the main text. These extensions are mainly fairly straightforward. It is worth spending time with them; only when they can be tackled confidently does one truly understand the material of the text. Of course, staring at an insoluble problem for hours is a depressing experience of limited educational value. For this reason, I have included outline answers to the problems. Brevity and clarity do not always go together in such a context, and I hope that what I have written will at least contain enough hints to help the student produce a more detailed solution.

The bibliography gives further reading for each chapter, and in some cases for each section. The sources of many of my data and figures are given there. However, in the main, the bibliography is deliberately pedagogical in intent; I have not attempted to make this complete or slavishly sought out first attributions. Rather, the books and papers which are mentioned are chosen to be helpful to the student who requires more detail or a fuller explanation of particular topics. One of the results is that there is a preponderance of references to papers by my own colleagues. This parochial bias is not meant as a criticism of the work of those who are omitted; rather, it is a consequence of the huge amount of global circulation related research which has been published in recent years.

Many people have helped me in the preparation of this book. Some have helped unknowingly, by their teaching, writing and conversation about matters relating to my themes. In particular, I would like to record my gratitude to Dr Raymond Hide, who introduced me to the study of atmospheric circulations, and to Professors Brian Hoskins and Robert Pearce, who provided a stimulating environment at Reading, where I learned about, and explored, the Earth's global circulation, with the aid of both data and simple models. I would like to record my grateful thanks to a host of colleagues who generously helped me with advice and who spent considerable time and effort attempting to satisfy my insatiable appetite for data and figures. Among them, I would mention Paul Berrisford, James Dodd, Paul James, David Jackson, Adrian Matthews, Michiko Masutani, Fay Nortley, Mark Ringer, Keith Shine, Julia Slingo, and Paul Valdes. Michael Blackburn deserves special thanks for his patience in helping me to write the computer programs needed to prepare plots of the ECMWF climatology. Many colleagues in other institutes have not only given me permission to reproduce their diagrams, but have generously supplied original artwork. I would like to mention X. Cheng, R. Hide, D. A. Johnson, L. V. Lyjak, B. Naujokat, A. O'Neill, R. A. Madden, A. Scaife, J. M. Wallace, and G. P. Williams. Several of my colleagues have read and commented upon early drafts of various sections of the text. Among them are M. Collins, M. Juckes, F. Nortley, R. Pearce and S. Rosier.

A large number of the diagrams in this book have been prepared from climatologies based on operational meteorological analyses carried out at the European Centre for Medium Range Weather Forecasts (ECMWF). The climatologies have been built up over several years as part of a joint project with the UK Meteorological Office and the University of Reading. I must acknowledge a debt to all my colleagues who have worked on this project over the last ten years or so; among them are Paul Berrisford, Huang Hsu, Michiko Masutani, Sarah Raper, Prashant Sardeshmukh and Glenn White. Special thanks are due to ECMWF for their interest and patient support for this diagnostics project and to the UK Meteorological Office who afforded access to the data.

It is also traditional for authors to thank their long suffering families at this point, and I am no exception. I owe a tremendous debt to their patience and understanding. I trust their resigned indulgence can now be rewarded with a positive orgy of gardening, decorating and model making.

But above all, this book owes its existence to my students at Reading. Over the years their interest in my lectures, and their determined efforts to learn about the global circulation, have inspired me to refine my material and write this book. I am also conscious of many able students around the world who would like to attend universities in the West to learn about the atmosphere, but for whom raising the necessary finance to travel is impossible. Perhaps books like this will help them to develop their studies. While it is all too imperfect and partial, I hope that all these people and their successors will find my account of the atmospheric circulation helpful.

Ian James
Bracknell
February 1993

Notation

A	Wave action density, Eq. (6.32).
a	Radius of planet (6.371×10^6 m for Earth).
c_p	Specific heat at constant pressure.
c_v	Specific heat at constant volume.
c_D	Drag coefficient.
c_{gx}, c_{gy}	Components of group velocity.
$\mathbf{c}_g$	Group velocity vector.
c_0	Phase speed of external gravity wave for equivalent depth h_0, see Eq. (7.7).
D	Horizontal divergence.
$\mathscr{D}$	Form drag, see Eq. (8.15).
D_p/Dt	Rate of change following a wave packet, Eq. (6.22).
E	Rate of evaporation.
$\mathbf{E}$	'E-vector', Eqs. (7.20) and (7.26).
e	Vapour pressure.
e_s	Saturated vapour pressure.
e_{s0}	Saturated vapour pressure at reference temperature; for water, taken to be 611 Pa at 273 K.
F	Friction tendency of the vertical component of the relative vorticity.
$\mathbf{F}$	Eliassen–Palm flux, Eq. (6.68).
f	Coriolis parameter.
f_0	Coriolis parameter at some reference latitude.
$\mathscr{F}$	Friction force per unit mass.
g	Acceleration due to gravity (9.81 m s^{-2} for Earth).
H	Pressure scale height, RT/g.
$h(p)$	Ratio of $\partial\Phi/\partial p$ and θ, see Eqs. (1.54) and (5.23).
h_0	Equivalent depth, see for example Eq. (7.4).
i	$(-1)^{1/2}$.
K	1 – Diffusion coefficient. 2 – Kinetic energy per unit mass. 3 – Total wavenumber, $(k^2 + l^2)^{1/2}$.
K_R	L_R^{-1}.
K_s	Total stationary wavenumber.

k Zonal wavenumber.

$\mathbf{k}$ Vertical unit vector.

k_β Rhines wavenumber, see Eq. (10.25).

L 1 - Characteristic horizontal length scale.

    2 - Latent heat of condensation of water vapour.

L_R Rossby radius of deformation, NH/f.

l Meridional wavenumber.

M $(\overline{v'^2} - \overline{u'^2})/2$.

m Vertical wavenumber.

N 1 – Brunt–Väisälä frequency.

    2 – $\overline{u'v'}$.

P Rate of precipitation.

p Pressure

p_R Standard reference pressure (usually taken as 100 kPa for Earth).

p_s Surface pressure.

Q 1 – Heat added per unit mass to an air parcel.

    2 – Arbitrary scalar variable.

$\mathcal{Q}$ $D\theta/Dt$ due to heating.

$\mathbf{Q}$ Forcing of omega equation, see Eqs. (1.77) and (1.78).

q Quasi-geostrophic potential vorticity, Eq. (1.75).

q_E Ertel's potential vorticity, Eq. (1.79).

R Gas constant ($287 \, \text{J kg}^{-1} \, \text{K}^{-1}$ for dry air).

R^* Universal gas constant ($8314 \, \text{J (kg mole)}^{-1} \, \text{K}^{-1}$).

R_d Gas constant for dry air.

R_v Gas constant for water vapour ($461.5 \, \text{J kg}^{-1} \, \text{K}^{-1}$).

r Humidity mixing ratio.

r_s Saturated humidity mixing ratio.

S 1 – Heating rate, dQ/dt.

    2 – Flux of solar radiation.

s 1 – Static stability parameter, Eq. (1.72).

    2 - Specific entropy, Eq. (3.13).

T Temperature.

t Time.

U 1 – Internal energy per unit mass.

    2 – Characteristic velocity scale.

u Zonal component of wind.

u_a Ageostrophic zonal component of wind.

u_g Geostrophic zonal component of wind.

$\mathbf{u}$ Vector velocity (u, v, w).

v Meridional component of wind.

v_a Ageostrophic meridional component of wind.

v_g Geostrophic meridional component of wind.

$\mathbf{v}$ Horizontal component of velocity vector $(u, v, 0)$.

w Vertical component of wind.

W Work per unit mass done by an air parcel.

x Zonal coordinate.

Y Width of Hadley cell in Held–Hou theory, see Eq. (4.12).

y Meridional coordinate.

z Height above Earth's surface.

z' Pseudo-height, $-H \ln(p/p_R)$.

Z	Geopotential height.
α	1 – Specific volume.
	2 – $u\Delta t/\Delta x$.
	3 – Angle between Rossby ray and zonal direction.
	4 – Coefficient of thermal expansion.
β	$\partial f/\partial y$.
$\hat{\beta}$	$\beta - \partial^2[u]/\partial y^2$.
γ	Dimensionless parameter in theory of baroclinic regime, Eq. (10.19).
δ	Phase difference between two waves.
κ	R/c_p.
θ	Potential temperature.
θ_E	Radiative equilibrium potential temperature.
χ	Velocity potential.
η	Meridional particle displacement.
ρ	Density.
$\rho_R(z)$	Density of standard atmosphere.
Ψ	Streamfunction wave amplitude.
ψ	Streamfunction.
ψ_g	Geostrophic streamfunction.
Φ	Geopotential.
ϕ	Latitude.
λ	1 – Longitude.
	2 – Wavelength.
ξ	Vertical component of relative vorticity.
ξ_g	Geostrophic vorticity.
$\boldsymbol{\xi}$	Relative vector vorticity.
ζ	Vertical component of absolute vorticity.
$\boldsymbol{\zeta}$	Absolute vector vorticity.
σ	1 – Vertical co-ordinate, p/p_s.
	2 – Stefan–Boltzmann constant, $5.67 \times 10^{-8}\,\mathrm{W\,m^{-2}\,K^{-4}}$.
	3 – Growth rate of unstable wave.
τ	Period of a time average.
τ_D	Drag or 'spinup' timescale.
τ_E	Radiative equilibrium timescale.
$\boldsymbol{\tau}$	Stress vector.
ω	1 – Pressure vertical velocity.
	2 – Circular frequency of wave.
Ω	Rotation rate of planet ($7.292 \times 10^{-5}\,\mathrm{rad\,s^{-1}}$ for Earth).

1

The governing physical laws

The aim of this chapter is to introduce the basic physical laws which govern the circulation of the atmosphere and to express them in convenient mathematical forms. No attempt is made at either completeness or rigour beyond the requirements of the later chapters. Those who wish for a more detailed discussion are referred to one of the many excellent texts on dynamical meteorology which are now available. Those by Holton (1992) and by Gill (1982) are particularly recommended.

1.1 The first law of thermodynamics

The first law may be stated simply in its qualitative form: heat is a form of energy. The transformation of heat energy into various forms of mechanical energy is the process which drives the global circulation of the atmosphere and which is responsible for the formation of the weather systems whose cumulative effects define the climate of a particular region. These transformations will be discussed in more detail in Chapter 3. In this section, the first law will be expressed in mathematical terms. But, first, it will be necessary to consider the thermodynamic properties of the air which makes up the atmosphere.

The 'thermodynamic state' of a parcel of air is defined by specifying its composition, pressure, density, temperature, and so on. In fact, these properties are not independent of one another, but are related through the 'equation of state' of the air.

For our purposes, only one constituent of the air varies significantly, and that is water vapour. The remaining gases which make up the bulk of the atmosphere are present in constant proportions, at least up to very great heights. These are principally nitrogen and oxygen, with smaller concentrations of argon and carbon dioxide. Other gases are present in very

Table 1.1. *Composition of dry air*

Gas	Volume mixing ratio
Nitrogen (N_2)	0.780 83
Oxygen (O_2)	0.209 47
Argon (Ar)	0.009 34
Carbon dioxide (CO_2)	0.000 33

small amounts; some are important in determining the transparency of the atmosphere to various frequencies of electromagnetic radiation, and some play a crucial role in the chemistry of the atmosphere. But for our purposes, we may ignore them. Table 1.1 summarizes the normal composition of dry air.

We will return to water vapour shortly. If we consider 'dry air', then its pressure p, temperature T and density ρ are related by the 'ideal gas law':

$$p = \rho R T. \tag{1.1}$$

This equation of state needs modification at very high pressures and low temperatures. But over the range of temperature and pressure encountered in the atmosphere, it is perfectly adequate. The gas constant R is related to the universal gas constant R^* by

$$R = R^*/m. \tag{1.2}$$

where m is the mean (by volume) molecular weight of the cocktail of gases comprising dry air. The equation of state, Eq. (1.1), means that it is only necessary to know any two of p, T or ρ to specify the thermodynamic state of the air completely. It is sometimes more convenient to work with 'specific volume' $\alpha = 1/\rho$ (i.e. the volume occupied by a unit mass of air) rather than with ρ.

The temperature of the air is simply a measure of the 'internal energy' of the air, that is, of the energy which is associated with the random motion of the molecules and possibly with their rotation and internal vibration. If two masses of gas are brought into intimate contact, this internal energy is rapidly shared between them and their temperatures become equal. When their temperatures are unequal, a flow of heat from the hotter to the colder mass takes place. An infinitesimal change of the internal energy U of a unit mass of dry air is related to the temperature change by:

$$dU = c_v dT, \tag{1.3}$$

where c_v is the 'specific heat at constant volume'.

If an infinitesimal quantity of heat dQ is added to an element of air, it may contribute to an increase in its internal energy, or it may be converted into mechanical energy, or a combination of the two. But the change of internal energy plus the mechanical work done must balance the heat added. This is the mathematical statement of the first law of thermodynamics:

$$dQ = dU + dW. \qquad (1.4)$$

Typically, work is done by the air parcel when it expands against the pressure exerted by the surrounding gas. Assuming that the pressure of the element of gas is equal to the pressure of its surroundings (always true if the expansion is gentle), the work done is related to the change of volume:

$$dW = p d\alpha. \qquad (1.5)$$

Thus a working form of the first law of thermodynamics can be written:

$$dW = c_v dT + p d\alpha. \qquad (1.6)$$

A more convenient form is obtained using the equation of state, Eq. (1.1):

$$dQ = c_p dT - \alpha dp, \qquad (1.7)$$

where $c_p = c_v + R$ is the 'specific heat at constant pressure'. This form is useful since many atmospheric processes occur more nearly at constant pressure than at constant volume.

A 'thermodynamic process' is a slow change of the thermodynamic state of an element of air; it may be described by a curve on a 'thermodynamic diagram' on which any two of the state variables are plotted. A particularly important class of thermodynamic processes is the 'adiabatic' process, in which no heat enters or leaves the element. From Eq. (1.7),

$$c_p dT = \alpha dp, \qquad (1.8)$$

during an adiabatic process, or, integrating,

$$T = \theta(p/p_R)^\kappa, \quad \kappa = R/c_p, \qquad (1.9)$$

where θ is a constant of integration which may be interpreted simply as the temperature at pressure p_R during the adiabatic process; θ is generally called the 'potential temperature' and p_R is usually (but arbitrarily) taken to be 100 kPa. Indeed, the potential temperature may be regarded as a new thermodynamic variable and Eq. (1.9) as an alternative equation of state.

Yet another form of the first law is obtained if Eq. (1.7) is written in terms of potential temperature:

$$dQ = c_p T \, d(\ln \theta). \tag{1.10}$$

Finally, if the heat is added over a time dt, the rate of change of potential temperature of the element is:

$$\frac{d\theta}{dt} = \frac{1}{c_p}(p/p_R)^{-\kappa}\frac{dQ}{dt} = \mathcal{Q}. \tag{1.11}$$

The term dQ/dt is sometimes called the 'diabatic warming rate'; $\mathcal{Q}$ denotes the rate of change of θ due to heating. This is the rate of change of θ when a particular fluid element is followed, and is more usually written $D\theta/Dt$, the 'Lagrangian derivative'. This differs from the 'Eulerian derivative', which measures the rate of change at a fixed point in space. If the gradient of θ at any instant is $\nabla\theta$, then the difference between the Eulerian and Lagrangian derivatives is simply the rate of change due to advection, $-\mathbf{u} \cdot \nabla\theta$. Thus:

$$\frac{\partial\theta}{\partial t} + \mathbf{u} \cdot \nabla\theta = \mathcal{Q} \tag{1.12}$$

The quantity of moisture in the air may be measured by the mass mixing ratio of water vapour $r = \rho_v/\rho_d$, ρ_v being the mass of water vapour in a unit volume and ρ_d the mass of dry air in the same volume. The saturation mixing ratio r_s is a function of temperature and pressure of the air, and may be as large as 0.030 in the warmest parts of the tropics. Generally, it is much less, with a typical value of r_s of 0.010 at the surface. For an average atmospheric temperature of 255 K and pressure of 50 kPa, $r_s = 0.005$. The equation of state of moist air is obtained by writing the total pressure as the sum of the vapour pressure and the partial pressure of the dry air, the ideal gas equation applying to both components separately with a suitable gas constant. The result can be written:

$$p = R_d \frac{(1 + (R_v/R_d)r)}{(1 + r)}\rho T. \tag{1.13}$$

In fact, for most of the atmosphere, the difference between the equation of state for moist air and that for dry air is not very large, and may frequently be ignored when discussing the large scale circulation. The primary importance of the variable moisture content of air is the huge latent heat of condensation of water vapour, larger than that of any other common substance, which means that very large amounts of heat are released when water condenses. Equally, large amounts of heat must be supplied when water evaporates. A quantity of heat

$$dQ = -L\,dr \qquad (1.14)$$

is released when the mixing ratio is reduced by condensation, where L is the latent heat of condensation. Thus if 10 mm of rain falls during a 24 hour period, the release of latent heat amounts to 289 W m^2, which is comparable to the typical insolation per unit area.

An equation describing the evolution of the humidity mixing ratio is analogous to the equation of conservation of energy. It is simply based on the hypothesis that any change of the moisture content of an air parcel is due to a rate of evaporation E into the parcel, or of condensation P taking water vapour out of the parcel. Small amounts of water are created or destroyed by chemical reactions, but these can generally be neglected. For our purposes, it is often enough to suppose that any condensed water falls out of the air immediately as rain, though some sophisticated models carry the suspended liquid and solid water content of the air as separate variables. Then

$$\frac{\partial r}{\partial t} + \mathbf{u} \cdot \nabla r = E - P. \qquad (1.15)$$

The Lagrangian rate of change of water mixing ratio leads to an important contribution to the heating rate:

$$S = -L\frac{Dr}{Dt} = L(P - E). \qquad (1.16)$$

This term is frequently dominant in the Earth's atmosphere, particularly in localized regions of persistent rainfall.

1.2 Conservation of matter

Consider some fixed volume of space V, enclosed by a surface A. The mass of air enclosed in this volume is:

$$m = \int_V \rho\,d\tau. \qquad (1.17)$$

Any change in this mass must be accomplished by a flux of mass into or out of the volume, so that

$$\frac{\partial}{\partial t}\int_V \rho\,d\tau = -\int_A \rho\mathbf{u} \cdot \mathbf{n}\,dA = -\int_V \nabla \cdot \rho\mathbf{u}\,d\tau, \qquad (1.18)$$

where the divergence theorem has been used. Since this must apply to any arbitrary volume, the two integrands in the volume integrals must be equal, so that:

$$\frac{\partial \rho}{\partial t} + \nabla \cdot (\rho \mathbf{u}) = 0. \tag{1.19}$$

This is the full form of the 'equation of continuity'. It may be simplified further if the density is broken into a reference profile ρ_R, which represents the mean density at any height and depends only on height, and the departure ρ_A from this reference density. For flow in planetary atmospheres, the variation of density in the vertical is very much larger than any horizontal fluctuations. Then scale analysis shows that:

$$\left| \frac{\partial \rho}{\partial t} \right| \ll |\nabla \cdot (\rho \mathbf{u})|, \tag{1.20}$$

so that the continuity equation can be reduced to:

$$\nabla \cdot \mathbf{v} + \frac{1}{\rho_R} \frac{\partial \rho_R w}{\partial z} = 0. \tag{1.21}$$

This result would become invalid if the flow speed approached the sound speed, in which case the full continuity equation, Eq. (1.19), must be used.

1.3 Newton's second law of motion

Newton's second law of motion is used to calculate how the motion of the atmosphere evolves. It states that the acceleration of a parcel of air of unit mass is equal to the vector sum of the forces acting upon it, that is,

$$\frac{D\mathbf{u}}{Dt} = \sum_i \mathbf{F}_i, \tag{1.22}$$

This is frequently called the 'equation of motion' or the 'momentum equation'. The forces which we need to consider in the case of atmospheric motion are:

(i)   The gravitational force. We consider this to be a constant vector $\mathbf{g}$ directed towards the centre of the Earth. It can be written as the gradient of a 'gravitational potential' $\nabla \Phi$.

(ii) The pressure gradient force. Figure 1.1 shows two surfaces of constant pressure a distance Δs apart. Consider a small volume of air, cross sectional area ΔA between them. The mass of air in the volume is $\rho \Delta A \Delta s$ and the net force due to the pressure of the surrounding air is

$$p_0 \Delta A - \left(p_0 + \left| \frac{\partial p}{\partial s} \right| \Delta s \right) \delta A = - \left| \frac{\partial p}{\partial s} \right| \Delta s \Delta A. \tag{1.23}$$

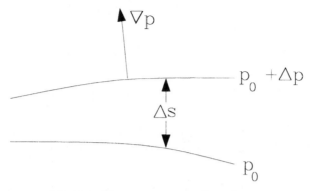

Fig. 1.1. The pressure gradient force.

The pressure gradient force per unit mass is therefore:

$$\mathbf{F}_p = -\frac{1}{\rho}\nabla p. \tag{1.24}$$

(iii) The friction force. Friction is generally a result of turbulent exchanges of momentum between the Earth's surface and the overlying layers of air. Accurate simple formulae for this transfer do not exist, and rather complex empirical relationships have to be employed in global circulation models. Generally, we will simply call the friction force $\mathscr{F}$, and note that it will usually act in such a direction as to reduce the wind towards rest. A very approximate linear parametrization of friction will be used on occasions where an analytical expression for friction is needed:

$$\mathscr{F} = -\frac{\mathbf{v}}{\tau_D}, \tag{1.25}$$

where τ_D is a drag or 'spin up' timescale. Such a term represents an exponential decay of the velocity towards zero in the absence of other forces. It is sometimes called 'Rayleigh friction'. A typical global mean spin up time for the Earth's atmosphere is around five days.

Equation (1.22) describes the acceleration of a parcel of air in an inertial frame of reference, that is, in a frame of reference which is not accelerating and which is therefore not rotating. It is usual to describe motion in the atmosphere relative to a noninertial frame of reference which is embedded in the rotating Earth. The relationship between acceleration in an inertial frame of reference, denoted I, and in a uniformly rotating frame of reference, denoted R, is derived, for example, in Pedlosky, page 17; the result is

$$\left(\frac{D\mathbf{u}_I}{Dt}\right)_I = \left(\frac{D\mathbf{u}_R}{Dt}\right)_R - \nabla\left(\frac{|\mathbf{\Omega}\times\mathbf{r}|^2}{2}\right) + 2\mathbf{\Omega}\times\mathbf{u}_R. \tag{1.26}$$

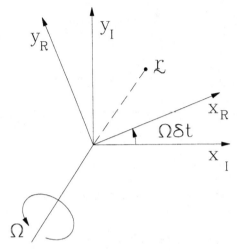

Fig. 1.2. A uniformly rotating frame of reference

Figure 1.2 illustrates the notation. $\mathbf{u}_I$ is the velocity in an inertial frame and $\mathbf{u}_R$ is the velocity in a rotating frame. From now on, the velocities and derivatives without any such subscript will be assumed to refer to a frame which is rotating with the solid Earth.

The second term on the right hand side of Eq. (1.26) is the centripetal acceleration. Since it is the gradient of a scalar, it introduces no structural change to the equation of motion; it can be absorbed into the definition of gravitational potential. The centripetal acceleration makes a very small correction to the gravitational acceleration, which is largest at the equator.

Thus, Newton's second law may be written:

$$\frac{\partial \mathbf{u}}{\partial t} + \mathbf{u} \cdot \nabla \mathbf{u} = 2\mathbf{\Omega} \times \mathbf{u} - \frac{1}{\rho}\nabla p + \mathscr{F}. \qquad (1.27)$$

This has now been written in terms of the Eulerian rate of change of velocity. The first term on the left hand side arises from the rotation of the frame of reference and is a most important term for global scale circulations. It is sometimes called the 'Coriolis force'. Strictly, it should be regarded as a 'pseudo-force', that is, a mental construct which is designed to make it appear that Newton's second law is holding despite the rotation of the frame of reference. Note that since the Coriolis force always acts at right angles to the fluid motion, it can do no work. Acting in isolation from other forces, it will cause parcel trajectories to be circular, with radius $|\mathbf{u}|/(2|\mathbf{\Omega}|)$. Such motion is termed 'inertial flow'.

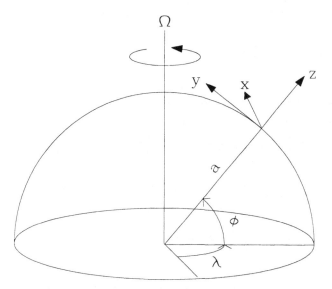

Fig. 1.3. Coordinates used to describe atmospheric motions relative to a spherical planet.

1.4 Coordinate systems

Up to now, the equations governing the atmospheric circulation have been expressed in a general vector notation. But for practical purposes, they must be written in terms of the components of velocity, etc., in orthogonal directions. This will lead us to consider the very marked asymmetry between the vertical and horizontal directions, and thereby to simplify the equations, obtaining a usable set for computation.

The Earth is nearly a sphere, and so it is natural to employ spherical coordinates ϕ (latitude), λ (longitude) and r (distance from the centre of the Earth). In fact, it is possible (though not trivial) to show that the slightly oblate shape of the Earth can be ignored, and that its effect can be represented by small variations of **g**, the acceleration due to gravity, with latitude if necessary. Recognizing that the depth of the atmosphere is very small compared to the radius of the Earth a, we write:

$$r = a + z, \text{ with } z \ll a, \qquad (1.28)$$

where z is the height above mean sea level. The three components of velocity are denoted u (zonal), v (meridional) and w (vertical). Figure 1.3 defines the notation. The equations of motion are derived in general curvilinear coordinates in standard texts on fluid dynamics such as Batchelor (1967). The results are quoted here for reference:

Equations of motion:

$$\frac{\partial u}{\partial t} + \frac{u}{a\cos\phi}\frac{\partial u}{\partial\lambda} + \frac{v}{a}\frac{\partial u}{\partial\phi} + w\frac{\partial u}{\partial z} + \frac{uv}{a}\tan\phi + \frac{uw}{a}$$

$$= 2\Omega v\sin\phi - 2\Omega w\cos\phi - \frac{1}{\rho a\cos\phi}\frac{\partial p}{\partial\lambda} + \mathcal{F}_1, \qquad (1.29a)$$

$$\frac{\partial v}{\partial t} + \frac{u}{a\cos\phi}\frac{\partial v}{\partial\lambda} + \frac{v}{a}\frac{\partial v}{\partial\phi} + w\frac{\partial v}{\partial z} + \frac{u^2}{a}\tan\phi + \frac{vw}{a}$$

$$= -2\Omega u\sin\phi - \frac{1}{\rho a}\frac{\partial p}{\partial\phi} + \mathcal{F}_2, \qquad (1.29b)$$

$$\frac{\partial w}{\partial t} + \frac{u}{a\cos\phi}\frac{\partial w}{\partial\lambda} + \frac{v}{a}\frac{\partial w}{\partial\phi} + w\frac{\partial w}{\partial z} - \frac{(u^2+v^2)}{a}$$

$$= 2\Omega u\cos\phi - g - \frac{1}{\rho}\frac{\partial p}{\partial z} + \mathcal{F}_3. \qquad (1.29c)$$

Equation of continuity:

$$\frac{1}{a\cos\phi}\frac{\partial u}{\partial\lambda} + \frac{1}{a\cos\phi}\frac{\partial(v\cos\phi)}{\partial\phi} + \frac{1}{\rho_R}\frac{\partial(\rho_R w)}{\partial z} = 0. \qquad (1.30)$$

Thermodynamic equation:

$$\frac{\partial\theta}{\partial t} + \frac{u}{a\cos\phi}\frac{\partial\theta}{\partial\lambda} + \frac{v}{a}\frac{\partial\theta}{\partial\phi} + w\frac{\partial\theta}{\partial z} = \mathcal{Q}. \qquad (1.31)$$

If the meridional extent of the motion is limited, it is often advantageous to use a local Cartesian set of coordinates (x, y, z), where $y = a(\phi - \phi_0)$ is the distance poleward from some reference latitude and $x = a\lambda\cos\phi$ is the eastward distance along the latitude circle. Such a coordinate system, neglecting many of the curvature terms in Eqs. (1.29a) – (1.31), simplifies the equations without removing any of the primary physical processes they represent. While it is inadequate for exact work, such as constructing numerical models of the global circulation or constructing budgets of global or zonal mean quantities, it often very helpful for expository purposes and will be used frequently in later chapters.

1.5 Hydrostatic balance and its implications

The vertical component of the momentum equation is dominated by the vertical pressure gradient term and the acceleration due to gravity. These are many orders of magnitude larger than any of the other terms in the equation. Hence the atmosphere is very close to a state of hydrostatic balance, in which:

$$\frac{\partial p}{\partial z} = -\rho g. \tag{1.32}$$

This balance only breaks down for small scale phenomena, such as thunderstorm updrafts and flow in the vicinity of very rugged mountain surfaces. On scales greater than around 10 km, hydrostatic balance is usually valid.

The contrast between the vertical scale of the global atmosphere, which can be taken as 7 – 10 km, and its horizontal scale, of around 6000 km, means that the vertical component of velocity is very much smaller than either of the horizontal components. The stable stratification of the atmosphere and the rotation of the system further inhibit vertical motion. This means that several terms involving w in the governing equations, such as the $2\Omega w \cos \phi$ term in the zonal momentum equation, Eq. (1.29a), can be neglected. The result is the so-called 'primitive equation set', which is widely used for numerical weather prediction and global circulation models. The primitive equations on a spherical planet of radius a are set out in Table 1.2 for easy reference. The quantity $f = 2\Omega \sin \phi$ is twice the component of the Earth's angular velocity parallel to the local vertical, known as the 'Coriolis parameter'.

Table 1.2. *The 'primitive' equations*

Equations of motion:

$$\frac{\partial u}{\partial t} + \frac{u}{a \cos \phi} \frac{\partial u}{\partial \lambda} + \frac{v}{a} \frac{\partial u}{\partial \phi} + w \frac{\partial u}{\partial z} + \frac{uv}{a} \tan \phi = fv - \frac{1}{\rho a \cos \phi} \frac{\partial p}{\partial \lambda} + \mathscr{F}_1, \tag{1.33a}$$

$$\frac{\partial v}{\partial t} + \frac{u}{a \cos \phi} \frac{\partial v}{\partial \lambda} + \frac{v}{a} \frac{\partial v}{\partial \phi} + w \frac{\partial v}{\partial z} + \frac{u^2}{a} \tan \phi = -fu - \frac{1}{\rho a} \frac{\partial p}{\partial \phi} + \mathscr{F}_2, \tag{1.33b}$$

Hydrostatic equation:

$$\frac{\partial p}{\partial z} = -\rho g \tag{1.34}$$

Equation of continuity:

$$\frac{1}{a \cos \phi} \frac{\partial u}{\partial \lambda} + \frac{1}{a \cos \phi} \frac{\partial (v \cos \phi)}{\partial \phi} + \frac{1}{\rho_R} \frac{\partial (\rho_R w)}{\partial z} = 0. \tag{1.35}$$

Thermodynamic equation:

$$\frac{\partial \theta}{\partial t} + \frac{u}{a \cos \phi} \frac{\partial \theta}{\partial \lambda} + \frac{v}{a} \frac{\partial \theta}{\partial \phi} + w \frac{\partial \theta}{\partial z} = \mathscr{Q}. \tag{1.36}$$

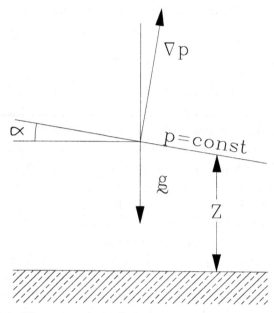

Fig. 1.4. The pressure gradient term in pressure coordinates.

Returning to the hydrostatic equation, Eq. (1.32), the right hand side of the equation is always negative, so that pressure always decreases with height. In fact, integrating from height z to infinity (where $p = 0$):

$$p(z) = \int_z^\infty \rho g \, dz. \tag{1.37}$$

That is, the pressure at any level in the atmosphere is simply equal to the weight of the overlying layers of air. The monotonic decrease of pressure with height means that pressure (or indeed, any single valued, monotonic function of pressure) can be used as a vertical coordinate just as well as can height z. The advantage of this procedure is that the equations of motion and the continuity equation are simplified. The disadvantage is that the lower boundary condition is rendered more complicated.

The principal simplification arises in the pressure gradient terms. Consider Fig. 1.4. An isobaric surface will be nearly, though not exactly, horizontal; denote the angle between the normal to the pressure surface and the vertical by α, and denote the magnitude of ∇p by $\partial p/\partial s$; α is typically less than 10^{-3}. From hydrostatic balance,

$$\frac{\partial p}{\partial s} \cos \alpha = \rho g, \tag{1.38}$$

so that the horizontal components of the pressure gradient become

$$-\frac{1}{\rho}\frac{\partial p}{\partial s}\sin\alpha = -g\tan\alpha. \tag{1.39}$$

But $\tan\alpha$ is simply the slope of the isobaric surface $|(\partial Z/\partial x, \partial Z/\partial y)|$, where Z denotes the height of the isobaric surface. It follows that in pressure coordinates, the horizontal components of the pressure gradient force can be written:

$$-\frac{1}{\rho}\frac{\partial p}{\partial x} = -g\frac{\partial Z}{\partial x}, \quad -\frac{1}{\rho}\frac{\partial p}{\partial y} = -g\frac{\partial Z}{\partial y}. \tag{1.40}$$

Using pressure as a vertical coordinate, vertical advection terms such as $w\,\partial Q/\partial z$ transform to

$$w\frac{\partial Q}{\partial z} = \omega\frac{\partial Q}{\partial p}, \tag{1.41}$$

where $\omega = Dp/Dt$ is the pressure coordinate vertical velocity. The pressure vertical velocity is approximately related to the geometric vertical velocity by

$$\omega \approx -\rho g w. \tag{1.42}$$

Similarly, using the hydrostatic relationship, the continuity equation transforms to

$$\nabla \cdot \mathbf{v} + \frac{\partial \omega}{\partial p} = 0. \tag{1.43}$$

Here, the vector $\mathbf{v}$ denotes the *horizontal* component of the velocity vector, $(u, v, 0)$.

The lower boundary condition, which in geometrical coordinates is simply $w = \mathbf{v} \cdot \nabla h$, h being the height of the surface, is considerably less straightforward in pressure coordinates. In the first place, the pressure at the ground fluctuates, so that the boundary moves. In the second, the surface of the Earth is not a coordinate surface. It is sometimes enough to apply a boundary condition $\omega = 0$ at $p = p_R$, but this is certainly not adequate for numerical modelling purposes.

The 'sigma coordinate' system is widely used in numerical models, and combines the simple form of the pressure gradient force in pressure coordinates with the straightforward lower boundary condition of geometric coordinates. Define the vertical coordinate as

$$\sigma = p/p_s, \tag{1.44}$$

where p_s is the actual surface pressure. The Earth's surface is therefore the surface $\sigma = 1$. The vertical advection terms can be written:

$$w\frac{\partial Q}{\partial z} = \dot{\sigma}\frac{\partial Q}{\partial \sigma}, \tag{1.45}$$

where $\dot{\sigma} = D\sigma/Dt$ is the equivalent of vertical velocity. The boundary conditions are simply $\dot{\sigma} = 0$ at $\sigma = 0$ and $\sigma = 1$. The continuity equation is rendered more complicated; it becomes a prognostic equation for surface pressure:

$$\frac{\partial p_s}{\partial t} + \nabla \cdot (p_s\mathbf{v}) + p_s\frac{\partial\dot{\sigma}}{\partial\sigma} = 0. \tag{1.46}$$

This rather strange equation relates the rate of change of surface pressure to the divergence at an arbitrary level in the atmosphere. The vertical advection term relates the flow at the chosen level to that at other levels. The equation may be integrated with respect to σ; making use of the boundary conditions yields:

$$\frac{\partial p_s}{\partial t} + \int_0^1 \nabla \cdot (p_s\mathbf{v})\mathrm{d}\sigma = 0. \tag{1.47}$$

For analytical work, the extra complications of the sigma coordinate system makes it impractical. It is usually reserved for numerical integration.

It is sometimes helpful, especially for work in the stratosphere, to introduce a 'pseudo-height', proportional to $\ln(p)$:

$$z' = -H\ln(p), \tag{1.48}$$

where H is a constant called the 'pressure scale height'. Integration of the hydrostatic relation shows that this is equal to geometric height for the special case of an atmosphere whose temperature does not change with height. The temperature in the lower stratosphere varies only weakly with height. An equation set based on $\ln(p)$ retains the advantages of a simple pressure gradient term, but expands the rarefied upper levels of the atmosphere.

Other, more complicated, vertical coordinates, such as the potential temperature θ ('isentropic coordinates'), as well as hybrid coordinates which are defined to be sigma coordinates near the ground and pressure coordinates at higher levels, will not be discussed further here. The interested reader is referred to texts on dynamical meteorology for a discussion of generalized vertical coordinates.

1.6 Vorticity

Taking the curl of the momentum equation gives a vorticity equation where relative vorticity $\xi = \nabla \times \mathbf{u}$. In some ways, the vorticity equation is a

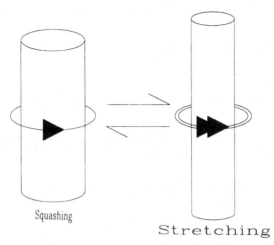

Squashing

Stretching

Fig. 1.5. Vortex stretching mechanism for generating relative vorticity.

more convenient way of expressing atmospheric dynamics since, in pressure coordinates, it involves no explicit reference to the pressure field. After some manipulation, the full vorticity equation may be written:

$$\frac{\partial \boldsymbol{\xi}}{\partial t} + \mathbf{u} \cdot \nabla \boldsymbol{\xi} = (2\boldsymbol{\Omega} + \boldsymbol{\xi}) \cdot \nabla \mathbf{u} - (2\boldsymbol{\Omega} + \boldsymbol{\xi}) \nabla \cdot \mathbf{u} + \frac{\nabla \rho \times \nabla p}{\rho^2} + \nabla \times \mathscr{F}. \quad (1.49)$$

In fact, the large scale dynamics of the atmosphere are determined by the vertical component of the vorticity. The numerically larger horizontal components play a less active role in determining the evolution of the meteorological flow. The vertical component of Eq. (1.49) is most simply written in pressure coordinates. The third term on the right hand side is zero since we are concerned with the component of vorticity perpendicular to pressure surfaces. The second term on the right hand side is zero by virtue of continuity, Eq. (1.35), and so the result is:

$$\frac{\partial \zeta}{\partial t} + \mathbf{u} \cdot \nabla \zeta = (f + \zeta)\frac{\partial \omega}{\partial p} + \mathbf{k} \cdot (\nabla \times \mathscr{F}). \quad (1.50)$$

Meteorologists frequently refer to this vertical component of the relative vorticity simply as 'vorticity'. The Coriolis parameter f is sometimes referred to as the 'planetary vorticity'. The absolute vorticity, $(f + \zeta)$, has a simple physical interpretation. It is simply twice the angular velocity of the air parcel about a vertical axis. On a rapidly rotating planet such as the Earth, one might suspect that this rotation is generally dominated by the rotation of the planet itself. This turns out to be the case, a fact which will be used in Section 1.7 to derive the 'quasi-geostrophic' approximation.

The crucial term in governing changes to the vorticity is the first remaining term on the right hand side of Eq. (1.50), which represents the generation of vorticity by stretching of vortices as a result of vertical motions. If the vertical velocity stretches a column of air, the column will assume a smaller radius and will rotate more rapidly about its vertical axis, that is, its vorticity will increase. Conversely, squashing of the column will reduce its vorticity. Figure 1.5 illustrates vortex stretching.

Friction generally acts to reduce the relative vorticity towards zero. Newtonian friction can be written in terms of vorticity as

$$\mathbf{k} \cdot (\nabla \times \mathscr{F}) = -\frac{\xi}{\tau_D},$$ (1.51)

a term which represents an exponential decay of the relative vorticity towards zero in the absence of other processes. In fact, Ekman layers, which result from laminar flow over a solid rotating boundary, give rise to precisely such a dissipative term, which arises because the friction near the surface induces small vertical velocities at the top of the boundary layer. The magnitude of these vertical motions is proportional to the relative vorticity just above the boundary layer, and their direction is such as to induce vortex squashing when the interior relative vorticity is positive, and vice versa. Pedlosky gives a good account of this 'spin up' process. The structure of the Ekman layer is a poor approximation to the observed structure of the atmospheric boundary layer, but this effect of spinning up of the interior vorticity is a helpful qualitative model of the effect that boundary layer friction has on vorticity.

1.7 The quasi-geostrophic approximation

Away from the equator, the large scale meteorological flow is close to a state of geostrophic balance. That is, the dominant terms in the horizontal momentum equations, Eqs. (1.33a, b), are the Coriolis terms and the pressure gradient terms. Thus, the 'geostrophic' velocity field is determined by the gradients of the geopotential height:

$$u_g = -\frac{g}{f}\frac{\partial Z}{\partial y}, \quad v_g = \frac{g}{f}\frac{\partial Z}{\partial x}.$$ (1.52)

Differentiating these relationships and making use of the hydrostatic equation, Eq. (1.32) and the definition of potential temperature, the vertical variations of u_g and v_g are related to horizontal variations of potential temperature:

$$\frac{\partial u_g}{\partial p} = \frac{h}{f}\frac{\partial \theta}{\partial y}, \quad \frac{\partial v_g}{\partial p} = -\frac{h}{f}\frac{\partial \theta}{\partial x}$$ (1.53)

where

$$h(p) = \frac{R}{p} \left(\frac{p}{p_R} \right)^{\kappa}. \tag{1.54}$$

Equation (1.53) shows that the geostrophic wind and temperature field are not independent, but are related in a state of 'thermal wind balance'. A crucial aspect of any process which (for instance) changes the temperature field is that there must be a compensating adjustment of the wind field in order to preserve thermal wind balance. Examples of this adjustment process will be discussed in Chapters 4 and 5. As an alternative to Eq. (1.53), the variation of geostrophic vorticity with height can be written:

$$\frac{\partial \xi_g}{\partial p} = -\frac{h}{f} \left(\frac{\partial^2 \theta}{\partial x^2} + \frac{\partial^2 \theta}{\partial y^2} \right). \tag{1.55}$$

These relationships between the geostrophic velocity (or vorticity) fields and the geopotential height and temperature fields can be used to simplify the governing equations, giving an approximate set which is called the 'quasi-geostrophic' equation set. This has now fallen out of favour as an equation set for modelling the atmospheric circulation since it is not uniformly valid as one approaches the equator; but it remains of great value in diagnosing, and gaining insight, into the dominant dynamical processes in the midlatitude and subtropical regions.

Although Eq. (1.52) represents the dominant terms in the momentum equations, it is of little use for predicting the evolution of the flow. The time derivative terms have been dropped, and so the approximated equations are simply diagnostic, relating the velocity to the pressure field. To determine the evolution of these fields, some ageostrophic effects must be retained. The approach in this section will be via the vorticity equation. First, it is necessary to examine the conditions for which geostrophic balance will hold.

Suppose that a typical horizontal velocity has magnitude U and that it varies over a characteristic length scale L. In the Earth's midlatitudes, U is around 10 m s^{-1} and L might be of the order of 10^6 m. Then the typical magnitude of the horizontal advection terms in the momentum equations will be U^2/L. The magnitude of the Coriolis term will be fU. The ratio of the two is called the 'Rossby number' Ro:

$$\frac{U^2/L}{fU} = \frac{U}{fL} = Ro. \tag{1.56}$$

A necessary condition for geostrophic balance to be achieved is that the Rossby number be small. Other conditions are that the friction term be small, and that the trajectories of fluid elements be only gently curved. For typical terrestrial midlatitude flows, $Ro \sim 0.1$, which meets this criterion. When f becomes small in the tropics, or for mesoscale systems where L is small, the Rossby number becomes comparable to unity or larger, and the quasi-geostrophic approximation ceases to be meaningful.

Now consider the vorticity equation, Eq. (1.50), and estimate the magnitude of the various terms. The quasi-geostrophic vorticity equation involves dropping those terms whose typical magnitude is $O(Ro)$ times the magnitude of the dominant terms. The vortex stretching term is made up of two terms: the stretching of planetary vorticity and the stretching of relative vorticity. The typical magnitude of the relative vorticity is simply U/L, so the relative magnitude of these two terms is simply

$$\frac{|\xi|}{f} \sim \frac{U}{fL} = Ro. \tag{1.57}$$

We shall therefore retain only the stretching of planetary vorticity. With this result, we can now estimate the typical vertical velocity W in terms of the typical horizontal velocity U. Assuming that horizontal advection roughly balances vortex stretching gives:

$$\frac{U^2}{L^2} \sim f\frac{W}{\Delta p},$$

i.e.,

$$W \sim \left(\frac{\Delta p}{L}\right)\left(\frac{U}{fL}\right)U. \tag{1.58}$$

The first term is simply a geometrical factor (or 'aspect ratio') based on the differing vertical and horizontal scales of the atmosphere. A simple scaling argument based on the continuity equation would suggest that this term alone would be the ratio of vertical to horizontal winds. The second factor is the small Rossby number. Our scale analysis has revealed a fundamental property of rapidly rotating fluid systems: rapid rotation (i.e. small Rossby number) implies that vertical motion is suppressed. For present purposes, this indicates that we can neglect the vertical advection of vorticity compared to the horizontal advection terms.

The horizontal advection terms comprise two components: the advection of relative vorticity and the advection of planetary vorticity. To estimate the relative importance of these two terms, it is necessary to estimate the gradient of planetary vorticity. It is frequently adequate to approximate f by a linear function of poleward distance (or latitude):

$$f = f_0 + \beta y, \tag{1.59}$$

where $\beta = O(f/a)$. More precisely,

$$\beta = \frac{2\Omega}{a} \cos \phi. \tag{1.60}$$

This 'β-plane' approximation is very useful for our purposes, since it represents the most important effects of the Earth's curvature without having to write the equations in spherical coordinates, a cumbersome procedure at best. The ratio of the two horizontal advection terms is therefore:

$$\frac{\beta U}{U^2/L^2} = \frac{\beta L^2}{U}. \tag{1.61}$$

For the Earth's midlatitudes, $\beta L^2/U$ is roughly unity. Both the horizontal advection terms must therefore be retained.

Friction can generally be ignored away from the Earth's surface, but may be large in the boundary layer. We will retain the friction term in a schematic way by writing it in terms of a Rayleigh friction. The drag timescale τ_D may in general be a complicated function of the flow and temperature fields.

Finally, we obtain a quasi-geostrophic form of the vorticity equation:

$$\frac{\partial \xi_g}{\partial t} + \mathbf{v}_g \cdot \nabla \xi_g + \beta v_g = f_0 \frac{\partial \omega}{\partial p} - \frac{\xi_g}{\tau_D}. \tag{1.62}$$

In fact, this equation involves only two variables. From Eq. (1.52, the geostrophic velocities can be written in terms of a 'geostrophic streamfunction'

$$\psi_g = \frac{gZ}{f}, \tag{1.63}$$

so that

$$u_g = -\frac{\partial \psi_g}{\partial y}, \quad v_g = \frac{\partial \psi_g}{\partial x}. \tag{1.64}$$

The vorticity equation involves the geostrophic stream function and the vertical velocity only.

For some purposes, we will use the momentum equations in a form which is consistent with the quasi-geostrophic approximation. It can be verified by differentiation that the pair:

$$\frac{\partial u_g}{\partial t} + \mathbf{v}_g \cdot \nabla u_g = f_0 v_a + \beta y v_g, \tag{1.65a}$$

$$\frac{\partial v_g}{\partial t} + \mathbf{v}_g \cdot \nabla v_g = -f_0 u_a - \beta y u_g, \tag{1.65b}$$

lead to the quasi-geostrophic vorticity equation. The vector $\mathbf{v}_a = (u_a, v_a, 0)$ is the 'ageostrophic velocity':

$$\mathbf{v}_a = \mathbf{v} - \mathbf{v}_g, \tag{1.66}$$

which, since the geostrophic velocity is purely rotational, is related directly to the vertical velocity from the continuity equation, Eq. (1.35):

$$\frac{\partial \omega}{\partial p} = -\nabla \cdot \mathbf{v}_a \tag{1.67}$$

It follows that the ageostrophic velocity components have magnitude $O(Ro)$ U. Note that there are no vertical advection terms in Eqs. (1.65a, b).

To obtain a complete and consistent equation set, it is necessary to approximate the thermodynamic equation also. But first note that the potential temperature can be written in terms of the geostrophic streamfunction. From the hydrostatic relation, Eq. (1.32), the definition of potential temperature, Eq. (1.9), and using Eq. (1.54), we find:

$$\frac{\partial \psi_g}{\partial p} = -\frac{h(p)}{f_0} \theta. \tag{1.68}$$

The thermodynamic equation is simplified by writing the potential temperature as the sum of a standard reference potential temperature which depends only upon pressure and a departure from this reference atmosphere:

$$\theta = \theta_R(p) + \theta_A(x, y, p, t). \tag{1.69}$$

The reference profile θ_R may be taken to be the global mean potential temperature at pressure p. The stable stratification of the atmosphere means that θ_R decreases markedly with pressure. The quasi-geostrophic approximation of the thermodynamic equation involves the assumption that

$$\left| \frac{\partial \theta_R}{\partial p} \right| \gg \left| \frac{\partial \theta_A}{\partial p} \right|, \tag{1.70}$$

so that only the vertical advection of the reference potential temperature is retained. The quasi-geostrophic form of the thermodynamic equation is therefore:

$$\frac{\partial \theta}{\partial t} + \mathbf{v}_g \cdot \nabla \theta = -\frac{\partial \theta_R}{\partial p} \omega + \mathcal{Q}. \tag{1.71}$$

Since $\partial \theta_R / \partial p$ is always negative for a stably stratified atmosphere, we will introduce a (positive) stratification parameter which can be related to the Brunt–Väisälä frequency N:

$$s^2 = -h(p) \frac{\partial \theta_R}{\partial p}. \tag{1.72}$$

The assumption that vertical advection of θ_A is negligible is certainly the weakest tenet of quasi-geostrophic theory. In fact, the stratification of the atmosphere at a given pressure varies substantially across the globe, and can fluctuate appreciably even within individual weather systems. Nevertheless, quasi-geostrophic theory provides a helpful basis for understanding many aspects of the global circulation, and for the diagnosis of dynamical processes from observations or model output. It will be used repeatedly in later chapters.

Eqs. (1.62) and (1.71) involve just two unknown dependent variables. These are the geostrophic streamfunction and the vertical velocity. The quasi-geostrophic equations therefore constitute a complete set of equations for the motion of the atmosphere, although of course, they will become invalid near the equator.

1.8 Potential vorticity and the omega equation

It is straightforward to eliminate the vertical velocity between Eqs. (1.62) and (1.71), obtaining a single prognostic equation for the geostrophic stream function. Differentiating Eq. (1.71) with respect to p and eliminating $\partial\omega/\partial p$ between the thermodynamic and vorticity equations gives:

$$\frac{D_g}{Dt}\left\{f_0 + \beta y + \xi - \frac{\partial}{\partial p}\left(\frac{f_0 h\theta}{s^2}\right)\right\} = F - \frac{f_0 h\mathscr{Q}}{s^2}$$

or

$$\frac{D_g}{Dt}\left\{f_0 + \beta y + \nabla^2\psi + \frac{\partial}{\partial p}\left(\frac{f_0^2}{s^2}\frac{\partial\psi}{\partial p}\right)\right\} = F - \frac{f_0 h\mathscr{Q}}{s^2}, \qquad (1.73)$$

where the operator D_g/Dt indicates a rate of change following the geostrophic wind. More compactly, Eq. (1.73) can be written:

$$\frac{D_g q}{Dt} = F - \frac{f_0 h\mathscr{Q}}{s^2}, \qquad (1.74)$$

where

$$q = \left\{f_0 + \beta y + \nabla^2\psi + \frac{\partial}{\partial p}\left(\frac{f_0^2}{s^2}\frac{\partial\psi}{\partial p}\right)\right\} \qquad (1.75)$$

is the 'quasi-geostrophic potential vorticity'. Note that this is conserved following the geostrophic wind if friction and heating can be neglected. Furthermore, Eq. (1.75) may be regarded as an elliptic equation relating the geostrophic streamfunction to the potential vorticity. If the distribution of potential vorticity is known, and boundary conditions on ψ are specified, then, in principle, Eq. (1.75) can be inverted to yield $\mathbf{v}_g$ and θ everywhere.

An alternative is to eliminate the time derivatives between Eqs (1.62) and (1.71). When this is done, one obtains an elliptic diagnostic equation for the vertical velocity:

$$\nabla_H^2 \omega + \frac{f_0^2}{s^2}\frac{\partial^2 \omega}{\partial p^2} = \frac{f_0}{s^2}\frac{\partial}{\partial p}\left\{\mathbf{v}_g \cdot \nabla \xi + \beta v + \frac{\xi_g}{\tau_D}\right\} + \frac{h}{s^2}\nabla^2\{\mathbf{v}_g \cdot \nabla\theta - \mathscr{Q}\}. \quad (1.76)$$

Eq. (1.76) is frequently called the 'omega equation'. Again, this equation may be inverted, given suitable boundary conditions on ω, so that the vertical velocity can be deduced from the geostrophic streamfunction. If the 'source terms' on the right hand side are positive, then ω will tend to be negative, that is, there will be ascent. Conversely, descent will be associated with a minimum of the source terms. The right hand side of Eq. (1.76) consists of two terms. The first includes the vertical gradient of the advection of absolute vorticity, together with the vorticity tendency due to friction. The second includes the Laplacian of the advection of potential temperature, and the heating term. In many common meteorological situations, the friction and heating terms are small compared to the advection terms; the advection terms tend to cancel out, and so a qualitative determination of the vertical velocity field can be difficult.

Finally, it may be noted that an alternative formulation of the right hand side of the omega equation exists which avoids problems of cancellation. It is written as the divergence of a vector $\mathbf{Q}$ where:

$$\nabla^2 \omega + \frac{f_0^2}{s^2}\frac{\partial^2 \omega}{\partial p^2} = 2\nabla \cdot \mathbf{Q} \qquad (1.77)$$

where

$$\mathbf{Q} = \mathbf{k} \times \frac{\partial \mathbf{v}_g}{\partial s}\frac{\partial \theta}{\partial n}. \qquad (1.78)$$

Here, s denotes a coordinate parallel to the local θ contour, and n is a coordinate at right angles to the θ contour. The $\mathbf{Q}$-vectors converge on to regions of ascent and diverge from regions of descent. In the vertical plane, continuity implies that there is an anticlockwise ageostrophic circulation around the $\mathbf{Q}$-vector. Figure 1.6 illustrates the relationship between $\mathbf{Q}$ and the ageostrophic circulation. In a localized region, across which variations of f can be neglected, a simple recipe can be deduced for determining the direction of the $\mathbf{Q}$-vector. The recipe is: take the vector change of the wind along a θ contour (moving with cold air on the left) and rotate the vector clockwise through 90°. Figure 1.7 illustrates the method for a jet exit region.

To summarize, and to provide a handy reference for later chapters,

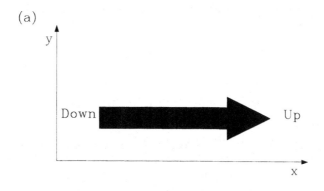

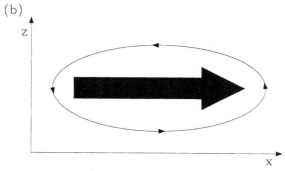

Fig. 1.6. Illustrating the relationship between the **Q**-vector and (a) the vertical velocity and (b) the meridional ageostrophic circulation.

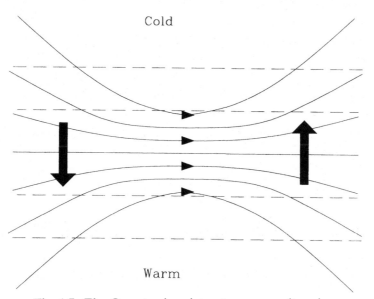

Fig. 1.7. The **Q**-vector in a jet entrance or exit region.

Table 1.3 draws together the complete set of quasi-gesotrophic equations, together with the definitions of the notation used.

1.9 Ertel's potential vorticity

The quasi-geostrophic potential vorticity cannot be applied near the equator where the geostrophic approximation breaks down, nor is it valid to apply it in regions where the static stability changes sharply on a pressure surface. A more general potential vorticity can be defined which avoids this problem. Sometimes called 'Ertel's potential vorticity', it is defined as

$$q_E = \frac{(2\mathbf{\Omega} + \xi)}{\rho} \cdot \nabla\theta. \tag{1.79}$$

If friction and heating are negligible, then q_E is conserved following the full three-dimensional motion, that is,

$$\frac{\partial q_E}{\partial t} + \mathbf{u} \cdot \nabla q_E = 0. \tag{1.80}$$

As in the case of the quasi-geostrophic potential vorticity, the distribution of q_E can be 'inverted' to yield both the velocity field and the potential temperature field. This requires two conditions to be met. First, suitable boundary conditions must be specified. Second, there must be a 'balance condition' relating the temperature and velocity fields to each other. Thermal wind balance is the simplest such condition.

A physical interpretation of the Ertel potential vorticity is shown in Fig. 1.8. If the θ surfaces move apart, then conservation of θ means that a column of fluid between these surfaces must remain bounded by them and so must stretch. Conservation of angular momentum (friction is presumed zero), then, means that the relative vorticity must become more cyclonic. Conversely, if the θ surfaces move together, the column must become more anticyclonic.

Scale analysis shows that q_E is dominated by the contribution of the vertical terms, so that to a good approximation:

$$q_E = \frac{(f + \xi)}{\rho} \frac{\partial \theta}{\partial z}. \tag{1.81}$$

(An equivalent expression in pressure coordinates may be preferred.) It is this dominance of the vertical contribution to the Ertel potential vorticity which justifies ignoring the horizontal components of the vorticity equation in Section 1.6. The quasi-geostrophic potential vorticity can be recovered from this relationship. First, note that $|f| \gg |\xi|$ for geostrophic conditions.

Table 1.3. *A summary of quasi-geostrophic relationships and definitions.*

Geostrophic streamfunction:

$$\psi_g = \frac{gZ}{f} \qquad (1.61)$$

Geostrophic winds:

$$u_g = -\frac{g}{f}\frac{\partial Z}{\partial y} \qquad (1.52a)$$

$$v_g = \frac{g}{f}\frac{\partial Z}{\partial x} \qquad (1.52b)$$

Thermal wind relationships:

$$\frac{\partial u_g}{\partial p} = \frac{h}{f}\frac{\partial \theta}{\partial y}$$

$$\frac{\partial v_g}{\partial p} = -\frac{h}{f}\frac{\partial \theta}{\partial x} \qquad (1.53)$$

where

$$h(p) = \frac{R}{p}\left(\frac{p}{p_R}\right)^{\kappa} \qquad (1.54)$$

Thermodynamic equation:

$$\frac{\partial \theta}{\partial t} + \mathbf{v}_g \cdot \nabla\theta = -\frac{s^2}{h}\omega + \mathscr{Q} \qquad (1.71)$$

where

$$s^2 = -h(p)\frac{\partial \theta_R}{\partial p} \qquad (1.72)$$

Vorticity equation:

$$\frac{\partial \xi_g}{\partial t} + \mathbf{v}_g \cdot \nabla\xi_g + \beta v_g = f_0\frac{\partial \omega}{\partial p} - \frac{\xi_g}{\tau_D} \qquad (1.62)$$

Momentum equations:

$$\frac{\partial u_g}{\partial t} + \mathbf{v}_g \cdot \nabla u_g = (f_0 + \beta y)v_a - \frac{u_g}{\tau_D} \qquad (1.65a)$$

$$\frac{\partial v_g}{\partial t} + \mathbf{v}_g \cdot \nabla v_g = -(f_0 + \beta y)u_a - \frac{v_g}{\tau_D} \qquad (1.65b)$$

Potential vorticity equation:

$$\frac{D_g}{Dt}\left\{ f_0 + \beta y + \nabla^2\psi + \frac{\partial}{\partial p}\left(\frac{f_0^2}{s^2}\frac{\partial \psi}{\partial p}\right) \right\} = F - \frac{f_0 h\mathscr{Q}}{s^2} \qquad (1.73)$$

Omega equation:

$$\nabla_H^2\omega + \frac{f_0^2}{s^2}\frac{\partial^2\omega}{\partial p^2} = \frac{f_0}{s^2}\frac{\partial}{\partial p}\left\{ \mathbf{v}_g \cdot \nabla\xi + \beta v + \frac{\xi_g}{\tau_D} \right\} + \frac{h}{s^2}\nabla^2\left\{ \mathbf{v}_g \cdot \nabla\theta - \mathscr{Q} \right\} \qquad (1.76)$$

Definition of **Q**-vector:

$$\mathbf{Q} = \mathbf{k} \times \frac{\partial \mathbf{v}_g}{\partial s}\frac{\partial \theta}{\partial n} \qquad (1.78)$$

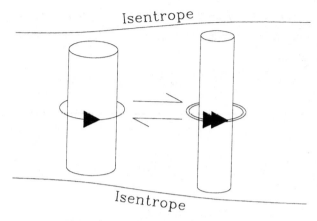

Fig. 1.8. Ertel's potential vorticity.

Second, divide θ into a reference profile and a small anomaly. Then, multiplying out the terms in Eq. (1.81) and dropping the products of small terms, the quasi-geostrophic potential vorticity can be recovered.

The Ertel potential vorticity combines the dynamics and thermodynamics of the atmosphere into a single equation. It is the most compact, and arguably the most fundamental description of the atmospheric flow. An ideal goal is to describe the global circulation entirely in terms of the sources, sinks and transports of potential vorticity. This is a goal which has not yet been attained, not least because accurate observations of q_E have only recently become available. There are also conceptual difficulties. There is every reason to suppose that the structure of the potential vorticity field contains large fluctuations on the smallest scales that one cares to resolve. Modelling studies show that initially smooth distributions of q_E rapidly develop extremely fine structure. Implicit in any description of the global circulation of q_E is some sort of large scale smoothing; whether a rational basis for such smoothing exists has yet to be demonstrated.

1.10 Problems

1.1 Deduce an expression relating the mass mixing ratio of atmospheric constituents to their volume mixing ratio, and hence translate Table 1.1 into mass mixing ratios.

1.2 A parcel of air, temperature 10 °C, is lifted slowly from the Earth's surface to a height of 3 km, such that it remains in thermodynamic equilibrium with its surroundings. Assuming that the Brunt–Väisälä frequency

$N = \sqrt{g/\theta)\partial\theta/\partial z}$ is a constant, equal to $10^{-2}\,\mathrm{s}^{-1}$, calculate the amount of heat gained or lost by the parcel.

1.3 A parcel of fluid at latitude ϕ on a rotating planet is observed moving at a speed U. If only the Coriolis force acts upon it, show that it moves in a circular trajectory, and calculate the time taken to execute each circle.

1.4 Use the hydrostatic relation to derive a correction of the observed surface pressure to obtain mean sea level pressure, assuming that the lower atmosphere has a constant lapse rate Γ. A station situated 150 m above mean sea level observes a pressure of 98.5 kPa, a temperature of 15 °C. Assuming a lapse rate of $6\,\mathrm{K\,km}^{-1}$, estimate the mean sea level pressure. What error would be made in the mean sea level pressure correction if the atmosphere was assumed to be isothermal?

1.5 Starting from Eqs. (1.65a,b), deduce the ageostrophic wind in a steady jet exit, in which the 25 kPa wind decreases from $50\,\mathrm{m\,s}^{-1}$ to $25\,\mathrm{m\,s}^{-1}$ over a distance of 1500 km at a latitude of 40 °N. Estimate the strength of the vertical velocities in the jet exit assuming the jet width is 1000 km.

1.6 Show that the quasi-geostrophic potential vorticity, using height as a vertical coordinate, may be written:

$$q = \left(f_0 + \beta y + \nabla^2 \psi + \frac{1}{\rho_R} \frac{\partial}{\partial z} \left(\rho_R \frac{f^2}{N^2} \frac{\partial \psi}{\partial z} \right) \right).$$

Show that under suitable conditions, Ertel's potential vorticity is proportional to this expression, and deduce the constant of proportionality.

2

Observing and modelling global circulations

2.1 Averaging the atmosphere

Strictly speaking, describing global atmospheric circulations requires the specification of the evolving three-dimensional fields of meteorological variables. Such a data compilation would be indigestible, and our description of the global circulation generally implies that some kind of averaging has been carried out. The flow is thought of as consisting of a 'mean' part, and a fluctuating or 'eddy' part. It is assumed that the details of any individual eddies are unimportant, though the average properties of eddies may well affect the mean fields. There are a number of different ways of averaging atmospheric data, the most frequently used of which are the average with respect to longitude or 'zonal average', and the average with respect to time. The concept of an ensemble average is also of importance.

The earliest studies of the global circulation were concerned with the zonal average. It is easy to understand why. Most atmospheric variables change much less in the zonal direction than they do in the vertical or in the meridional directions. Indeed, the latitude of an observing site on the Earth's surface is probably the most important single factor in determining its climate. The zonal average of any scalar quantity Q is denoted $[Q]$, and may be defined as:

$$[Q] = \frac{1}{2\pi} \int_0^{2\pi} Q \mathrm{d}\lambda. \tag{2.1}$$

Alternatively, in terms of the distance x along the latitude circle,

$$[Q] = \frac{1}{L} \int_0^L Q \mathrm{d}x. \tag{2.2}$$

Note that, by definition, $[Q]$ is independent of longitude. The local value of Q will generally be different from $[Q]$. This deviation is variously called 'the

28

eddy part' or the 'zonal anomaly' of Q, and is denoted Q^*:

$$Q^* = Q - [Q]. \tag{2.3}$$

It follows immediately that

$$[[Q]] = [Q] \text{ and } [Q^*] = 0. \tag{2.4}$$

Furthermore, if Q is a continuous function of latitude:

$$\left[\frac{\partial Q}{\partial x}\right] = 0. \tag{2.5}$$

Very similar results follow for time averaging. Denote the time mean of Q over some time τ as $\bar{Q}$, where

$$\bar{Q} = \frac{1}{\tau} \int_0^\tau Q \, \mathrm{d}t. \tag{2.6}$$

The 'transient' part of Q is denoted Q' where

$$Q' = Q - \bar{Q}. \tag{2.7}$$

Provided that τ is sufficiently long, the time mean value of Q will be independent of τ. 'Sufficiently long' generally means 'greater than the typical lifetime of weather systems', and for the midlatitudes, most mean quantities are roughly independent of τ for τ greater than 15 or 20 days. In the tropics, the necessary time is perhaps somewhat shorter. The global circulation changes significantly as the seasonal cycle progresses, and so a popular averaging period is a three-month 'season', consisting of around 91 or 92 days. The usual seasons chosen are December, January and February, denoted DJF, and referring to northern hemisphere winter or southern hemisphere summer, and June, July and August, denoted JJA. The equinoctial seasons March, April and May (MAM) and September, October and November (SON) are less frequently studied, not least because there is a large and systematic trend in many fundamental meteorological variables through the equinoctial periods. In fact, there is some evidence that important features of the seasonal cycle have rather different phases in different locations. Nevertheless, this simple scheme of four equal seasons will be adequate for our purposes in this book.

Although many general characteristics of the circulation are reproduced year after year, there is also an element of interannual variability. Consequently, wherever possible, we will use 'ensemble' means, in which a number of (say) DJF seasons are averaged together. The ensemble average is denoted

$\hat{\bar{Q}}$ where

$$\hat{\bar{Q}} = \sum_{i} \bar{Q}_i. \qquad (2.8)$$

However, this notation is becoming cumbersome, and the ensemble average will generally be assumed rather than stated explicitly. The number of seasons averaged together in this way is usually determined by practical rather than scientific considerations, since suitable global observations of the atmosphere (especially at levels away from the surface) have only recently become available. In any case, we must recognize that the ideal average winter circulation is a myth. Historical and paleontological studies amply demonstrate that the global circulation exhibits fluctuations on all timescales, up to the longest encompassed by the geological record.

The mean level of eddy activity is measured by the 'variance' of a given quantity, either with respect to time or to longitude. The variance is defined as $[Q^{*2}]$ or $\overline{Q'^2}$, and will generally be nonzero. Sometimes the variance may be partitioned into the contributions from different ranges of space scale or frequency.

Similarly, the covariance of two independent quantities is often of interest. Suppose a second scalar is R; then the covariance of Q and R is $[Q^*R^*]$ or $\overline{Q'R'}$. Again, spatial or temporal filtering can be used to partition the covariance into contributions from different scales or frequencies. The covariance of two quantities is closely related to whether or not they fluctuate in phase. To illustrate this, suppose Q^* and R^* both vary sinusoidally in the zonal direction, but with a phase difference δ:

$$Q^* = Q_0 \sin(kx), \quad R^* = R_0 \sin(kx + \delta). \qquad (2.9)$$

Then elementary trigonometry can be used to show that

$$[Q^*R^*] = \frac{1}{2} Q_0 R_0 \cos(\delta). \qquad (2.10)$$

The covariance is a maximum when $\delta = 0$, and is zero when $\delta = \pi/2$. Particularly important covariances are between various meteorological quantities and the velocity components. These are called the 'eddy fluxes' of the quantity involved. When there is a systematic tendency for large values of Q and large values of poleward velocity to occur at the same location, then the poleward eddy flux of Q, $[v^*Q^*]$, will be positive, that is, the eddies are systematically advecting Q polewards.

As an example of these various circulation statistics, consider the 'transport equation' for a scalar Q in pressure coordinates:

$$\frac{\partial Q}{\partial t} + u\frac{\partial Q}{\partial x} + v\frac{\partial Q}{\partial y} + \omega\frac{\partial Q}{\partial p} = S. \tag{2.11}$$

Here, S is a 'source term', describing sources and sinks of Q following the parcel motion. The continuity equation, Eq. (1.43) enables the transport equation to be written in 'flux form':

$$\frac{\partial Q}{\partial t} + \frac{\partial}{\partial x}(uQ) + \frac{\partial}{\partial y}(vQ) + \frac{\partial}{\partial p}(\omega Q) = S. \tag{2.12}$$

Now apply the zonal averaging operator to this equation. Noting that $[[v]Q^*]$ and similar terms are identically zero, the evolution of $[Q]$ is given by:

$$\frac{\partial}{\partial t}[Q] = -[v]\frac{\partial[Q]}{\partial y} - [\omega]\frac{\partial[Q]}{\partial p} - \frac{\partial}{\partial y}[v^*Q^*] - \frac{\partial}{\partial p}[\omega^*Q^*] + [S]. \tag{2.13}$$

The first two terms on the right hand side represent the advection of $[Q]$ by the mean meridional flow. The second pair of terms represents the convergence of the eddy fluxes of Q, and demonstrates how the eddies might play a crucial role in determining the mean distribution of $[Q]$, even though Q^* itself averages to zero. In the climatological mean, $\partial[Q]/\partial t$ will be close to zero, and so the mean distribution of $[Q]$ will be determined by a balance between the mean and eddy transports of Q and the source or sink terms, $[S]$. It must be recognized, though, that the mean and eddy transports are not necessarily independent and that, in some circumstances, they may nearly cancel out. We will return to this theme in Section 4.3.

The discussion in this section has envisaged the average value of Q at some observing site which is stationary with respect to the Earth's surface. Such an averaging procedure is called 'Eulerian averaging'. For many purposes, it would be preferable to average a quantity following the motion of an individual element of the atmosphere. This is called 'Lagrangian averaging'. Unfortunately, in most circumstances, Lagrangian averaging is attended by formidable practical difficulties. These are associated with the turbulent nature of the atmospheric flow, which means that an initially compact element of fluid rapidly becomes shredded and pulled out until filaments of the original element are inextricably intermingled with the rest of the atmosphere. So Lagrangian averaging is generally a hypothetical concept of little practical utility. However, it is possible to devise some practical averaging procedures which are approximately Lagrangian. For instance, taking the average of fields on isentropic surfaces (surfaces of constant potential temperature), rather than on constant pressure or height surfaces, would give Lagrangian averages for adiabatic motion, and would give roughly Lagrangian averages for typical values of large scale heating in the troposphere.

2.2 The global observing network

Meteorologists are uniquely blessed among scientists in that many thousands of high quality observations are made for them routinely across the globe, from the surface to the stratosphere. These observations are taken by trained observers in order to provide operational meteorological information, and constitute the input to numerical forecast models. They are also used to advise aviators, mariners and many others about the current state of the atmosphere. As a spin off, these observations, archived, quality controlled and analysed, form an important data base for the scientific study of individual weather systems and of their organization into the global circulation. Compare the meteorologist with his or her oceanographer colleague: the oceanographer may spend many years planning a campaign of observations of currents, temperature and salinity in a tiny area of the ocean, many weeks of discomfort on a ship taking the observations and several years analysing them back at the laboratory. All this work is done for the research meteorologist, several times per day on a global basis, who merely has to read the numbers from an archive and construct whatever diagnostic quantity is required. We owe an enormous debt to the dedicated, and all too easily forgotten, efforts of professional observers across the world.

In this section, some of the principal sources of quantitative data about the global circulation will be described. Most of what will be described would apply to the activities of any one of the several meteorological institutes around the world where routine global analyses and forecasts are carried out. A particular emphasis will be placed on the system used at the European Centre for Medium Range Weather Forecasts (ECMWF). Archives of global circulation data have been built up from the ECMWF archives, and form a major source of data used throughout this book to illustrate various aspects of the Earth's global circulation.

The primary observing system is the radiosonde network. Around the world, some 1000 stations launch weather balloons bearing expendable instrument packages. The instruments record temperature, pressure and humidity as the sonde rises. Tracking of the balloon by radar provides data on the horizontal wind components at different levels. Launches are made at least twice daily on the so-called 'synoptic hours', at 0.00 hours GMT (Greenwich Mean Time) and 12.00 hours GMT; some stations also take observations at intermediate times. A 'pilot balloon' is a weather balloon which is merely tracked, and carries no instruments. A pilot ascent therefore returns information only about the horizontal wind vector as a function of height.

The radiosonde system is one of the most accurate atmospheric observing systems. Temperatures are recorded to within ± 1 K, relative humidities to $\pm 10\%$, and winds to ± 3–5 m s^{-1}. Errors become larger at higher levels, where the low density of the air means that the response time of the instruments is longer and shielding them from thermal radiation becomes more difficult. Radiosondes generally sample the entire troposphere, and may reach well into the lower stratosphere. Currently, around 50% of launches reach 10 kPa or higher. A major international effort, led by the World Meteorological Organization, is designed to ensure that the radiosondes used by different meteorological services, and the launching and data recovery routines, are consistent with one another, and all meet the same standards of accuracy. Even so, it is not unknown to see significant discontinuities in reported meteorological variables at national frontiers, particularly at higher levels, reflecting the different instrument systems used by different services.

Radiosonde measurements are thus very accurate, and have a very high resolution in the vertical. Indeed, so much information is returned from a single ascent that it must be smoothed and summarized in some way before distribution. Stations summarize the ascent in terms of the values of the meteorological elements at 'standard' levels, together with information at 'significant' levels, where a major change of a parameter or its gradient is observed. The 'standard levels' are 100 kPa, 85 kPa, 70 kPa, 50 kPa, 40 kPa, 30 kPa, 25 kPa, 20 kPa, 15 kPa, 10 kPa, 5 kPa and 3 kPa. Essentially, though, the radiosonde is sampling a point volume of the atmosphere at each level in its ascent. The accuracy of the observation may be misleading in terms of how representative it is of a substantial volume of atmosphere. For example, a radiosonde which passes through a cloud layer would record a significantly different profile from a launch a few kilometres away or a few minutes later which takes place in the clear air between clouds. These considerations make the humidity measurement especially difficult to interpret, but they also apply to the temperature and wind fields.

Maintaining an upper air station is extremely expensive, which explains the highly inhomogeneous distribution of radiosonde stations revealed by Fig. 2.1. The average distance between adjacent stations is about 700 km, a figure which should be compared with the typical dimension of a midlatitude cyclone which is often taken as 1000 km. But some 800 of the total 1000 stations are located in the northern hemisphere. This means that the average spacing is around 1100 km in the southern hemisphere. But of course, this

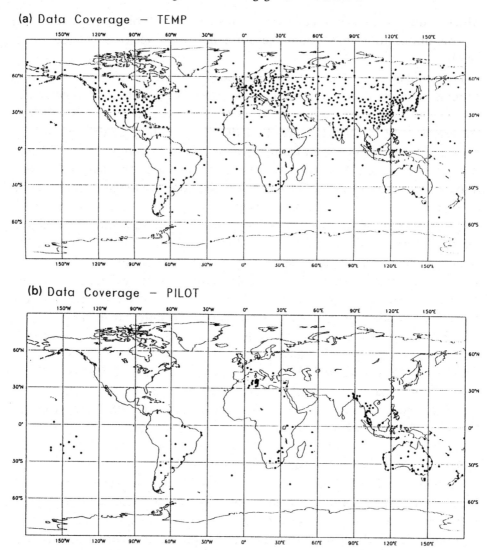

Fig. 2.1. Global distribution of (a) radiosonde (total number of observations = 611 (• 601 land, ∝ 10 ship)); and (b) pilot balloon ascents which were included in the routine analysis for 12.00 GMT on 29 October 1991 by the European Centre for Medium Range Forecasts (total number of observations = 161 (all land)). (Courtesy ECMWF.)

is only one aspect of the data inhomogeneity. The stations, not unnaturally, are heavily biased to land areas, and then to the more wealthy regions of the midlatitudes. Data voids are apparent over the oceans and over sparsely populated areas such as the deserts of North Africa and Arabia. Some midocean islands provide observations, and a handful of permanently manned weather ships are maintained in the North Atlantic and North Pacific Oceans, but nevertheless some areas are very poorly covered. The most notable gap is the midlatitude Pacific in the southern hemisphere; not a single station is to be found between New Zealand and the coast of Chile. It is also an inevitable fact that operational difficulties mean that ascents sometimes give poor results, or communications problems prevent data from reaching the analysis centre. Such problems again tend to have regional biases, and emphasize that the global radiosonde station network is by no means ideal. Nevertheless, that such a network operates, and that data is rapidly exchanged throughout the world, across all kinds of physical and political barriers, is no small achievement.

Many more meteorological stations are operating across the world which report only surface data. Because they are much less expensive to maintain, they are more readily provided, and many such stations can be seen reporting in Fig. 2.2. In addition to measurements of temperature, pressure, humidity and wind, surface stations also report a large number of more qualitative data, such as weather conditions and cloud types. At present, it is difficult to integrate this kind of data with the data analysed by computer for numerical weather prediction purposes. This is a pity since it contains much useful information which is readily put to work by a human analyst. As well as observations at fixed land stations, similar surface synoptic observations are also taken by certain merchant ships wherever they happen to be located at the observation time (so-called 'ships of opportunity'). In recent years, surface observations have been made automatically by unmanned stations. This raises the possibility of extending the data network relatively cheaply, especially into remote areas or hostile environments. Indeed, the deployment of automatic weather stations on drifting buoys in the southern oceans during 1978–9 was an important contribution to the 'First GARP Global Experiment' (FGGE), an ambitious attempt to obtain high quality data for the entire globe for a year. Such systems are now becoming a routine part of the data network. Data from both moored and drifting buoys included in a recent analysis is shown in Fig. 2.3.

Although more dense than the network of upper air stations, the surface network suffers similar shortcomings. The stations are most closely spaced in northern hemisphere land areas. Although the use of ships of opportunity helps to fill in some of the data voids in the oceans, their observations are biased towards the major shipping routes. Furthermore, there is liable to be an understandable systematic bias in that merchant ships tend to avoid severe weather if possible.

Data Coverage — SYNOP/SHIP

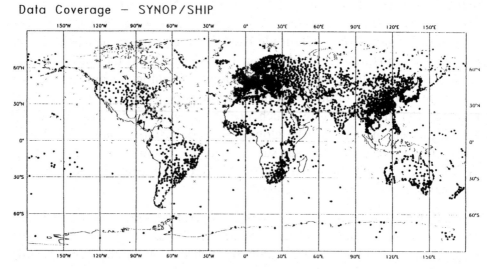

Fig. 2.2. The global distribution of surface observations, both on land and from ships, at 1200 GMT, 29 October 1991, used in the ECMWF analysis (total number of observations = 7983 (• 6993 SYNOP, × 990 ship)). (Courtesy ECMWF.)

Satellite information has become an important data source in recent years, although it is of poorer accuracy (especially in the troposphere) than are conventional measurements. The most widely used data are in the form of temperature soundings retrieved from infrared radiance measurements taken by polar orbiting satellites. The satellite monitors the state of the atmosphere beneath its track with high horizontal resolution. The typical orbital period is about 90 minutes, so that it takes several hours before the entire globe is covered by any one satellite. Thus the data gathered is not taken at the main synoptic hour. The ECMWF analysis system ascribes its data over a six-hour period to the nearest analysis time. The example given in Fig. 2.4(a) shows information from two satellites, between them covering most of the Earth's surface.

It is in the vertical resolution that the most serious shortcomings of satellite soundings are found. The typical vertical resolution is several kilometres. This is satisfactory for studying the upper stratosphere and mesosphere, but is a very coarse resolution for the troposphere. The soundings cease at the cloud tops, so that no data is taken within vigorous weather systems. New microwave sounders will enable information to be retrieved from below the cloud tops, but will do little to improve vertical resolution. Because the measurement is only of temperature (or thickness), the satellite sounding has to be calibrated with the aid of conventional measurements (of surface

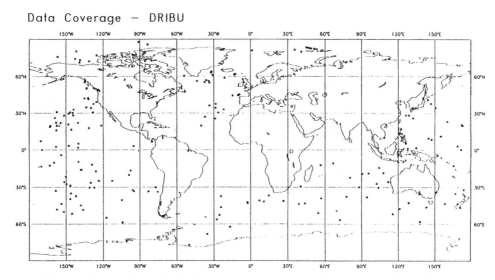

Fig. 2.3. Surface observations collected automatically by moored and drifting buoys which were included in the 1200 GMT ECMWF analysis of 29 October 1991 (total number of observations = 369 (• 335 drifting, ∝ 34 moored)). (Courtesy ECMWF).

pressure, for example) to obtain pressures. Then, some balance condition must be imposed in order to deduce winds from the surface pressure and upper level temperatures. Despite all these shortcomings, satellite temperature soundings are often the most important data source over oceans.

Images from geostationary satellites are used to derive winds by following the motion of identifiable cloud features. An example of the data coverage achieved is shown in Fig. 2.4(b). Once certain sources of error are removed, such as lee wave clouds, which propagate relative to the flow while remaining stationary with respect to the ground, a fairly high resolution map of winds in partly cloudy areas results. But the satellite winds can only be located in the vertical very crudely, and so they are a very inaccurate source of data. Nevertheless, satellite winds do provide important improvements to the analyses over the southern oceans.

There are other rather limited data sources. 'Aireps' are reports of temperature and pressure taken automatically by selected civil airliners. The data are relayed by satellite into the global data network. Measurements are only taken at the flight level of the aircraft, and they are mostly confined to the main air routes. In the example given in Fig. 2.5, most of the aireps are seen in the air lanes of the North Atlantic and the Pacific Oceans.

(a) Data Coverage — SATEM (500km)

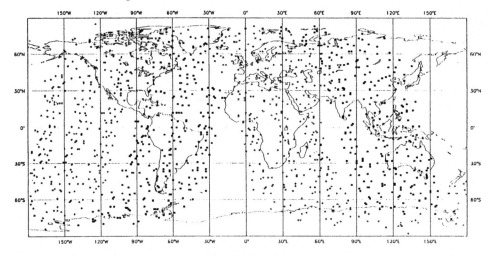

(b) Data Coverage — SATOB

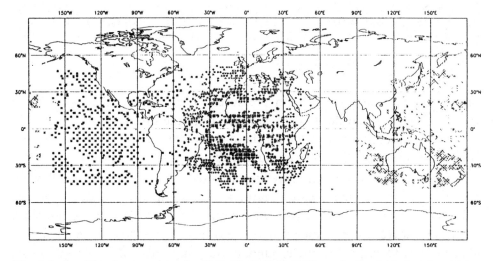

Fig. 2.4. Satellite data used in the 1200 GMT ECMWF analysis of 29 October 1991. (a) Temperature retrievals. Crosses indicate the NOAA 11 satellite and solid squares the NOAA 12 satellite (total number of observations = 1239 (× 674 NOAA11, ■ 565 NOAA12)). (b) Satellite winds from geostationary satellites (total number of observations = 2414 (■ 1472 METEOSTAT, × 522 HIMAWARI)). (Courtesy ECMWF.)

Data Coverage — AIREP

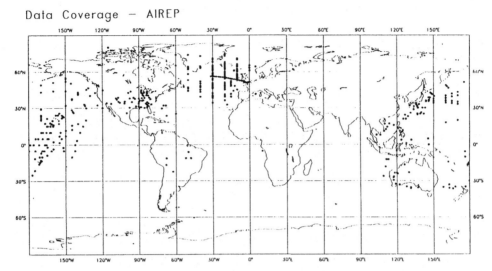

Fig. 2.5. Aircraft reports (AIREPs) included in the 1200 GMT ECMWF analysis of 29 October 1991 (total number of observations = 854 (all AIREP)). (Courtesy ECMWF.)

2.3 Numerical weather prediction models

In order to understand the way in which the various data described above are treated, it is necessary first to examine the principles of numerical weather prediction models. Given initial observed values of atmospheric variables, the underlying concept is to use the governing physical laws, embodied to sufficient accuracy in an equation set such as the primitive equations, Eqs. (1.33a)–(1.36), to predict the values of meteorological variables at some later time. The complication is that the governing equations are far too complicated to be solved exactly. Instead, numerical solutions have to be generated using a large computer. The equations are 'discretized' so that they are written in terms of meteorological variables at a large but finite number of discrete points. In this way, the continuous equations are replaced by an approximately equivalent set of algebraic equations, which can then be used to predict values a small but finite time later. This 'marching procedure' can be repeated as many times as required to generate a forecast at some arbitrary future time.

These principles are well illustrated in terms of the 'linear advection equation' in one dimension:

$$\frac{\partial Q}{\partial t} + u\frac{\partial Q}{\partial x} = 0, \tag{2.14}$$

where $Q = Q(x, t)$ may be any variable, and u is an advecting velocity, considered constant for the present. Denoting the initial distribution of Q as $Q_0(x)$, the solution to this equation is

$$Q(x, t) = Q_0(x - ut). \tag{2.15}$$

That is, any arbitrary initial distribution of Q simply moves at speed u along the x-axis without change of shape. Equation (2.14) is a prototype of several important terms which appear in the primitive equations. The equation may be discretized by defining values of Q on a grid of points in the (x, t) plane. We write:

$$Q(x_0 + n\Delta x, t_0 + m\Delta t) = Q_n^m, \tag{2.16}$$

where n and m are integers and Δx and Δt are grid lengths in the x and t directions respectively. Using Taylor series expansions about any given point on the grid, approximate expressions for the derivatives can be obtained:

$$\frac{\partial Q}{\partial t} = \frac{Q_n^{m+1} - Q_n^{m-1}}{2\Delta t} + O(\Delta t^2), \tag{2.17a}$$

$$\frac{\partial Q}{\partial x} = \frac{Q_{n+1}^m - Q_{n-1}^m}{2\Delta x} + O(\Delta x^2). \tag{2.17b}$$

The last term in each of these expressions is the truncation error in a finite difference approximation to the derivative. Substituting these expressions into the advection equation and re-arranging gives a means of predicting Q at the $(m + 1)$th time level, given its values at the mth and $(m - 1)$th times:

$$Q_n^{m+1} = Q_n^{m-1} - \frac{u\Delta t}{\Delta x}(Q_{n+1}^m - Q_{n-1}^m) + O(\Delta t^2, \Delta x^2). \tag{2.18}$$

A successful integration requires that the truncation error be small. Initially, this is achieved if Δx and Δt are small. Later, we must ensure that the truncation error cannot grow. To explore this requirement, note that the truncation error, denoted ϵ_n^m, must satisfy the same linear Eq. (2.18), as Q itself. Suppose that the error has the form:

$$\epsilon_n^m = Ae^{qt}e^{ikx}. \tag{2.19}$$

If $|e^{q\Delta t}| > 1$, then the truncation error will amplify each timestep, and will quickly dominate the solution. Such a situation is said to be 'computationally unstable'. On the other hand, if $|e^{q\Delta t}| \leq 1$, the errors remain bounded and the numerical scheme is 'stable'. Substituting into Eq. (2.18) and solving the resulting quadratic equation gives:

$$e^{q\Delta t} = \pm\sqrt{1 - \alpha^2 s^2} - i\alpha s_N, \tag{2.20}$$

where $\alpha = u\Delta t/\Delta x$ and $s_N = \sin(2\pi/N)$, $N\Delta x = 2\pi/k$ being the wavelength of the error. Stability is ensured if $\alpha \le 1$. If $\alpha > 1$, errors with some wavelengths can grow exponentially. This condition for stability, which is usually thought of as a restriction on the timestep, is called the 'Courant–Friedrich–Lewy' (CFL) condition:

$$\Delta t \le \Delta x/u \text{ for stability.} \tag{2.21}$$

Other finite difference approximations to the advection equation could have been devised, and these would have had different stability criteria, although most simple schemes involve some stability condition similar to Eq. (2.21). This relationship shows that the maximum timestep becomes small as the grid spacing Δx is reduced. Consequently, any reduction of the truncation by decreasing Δx means that the timestep has to be reduced to ensure stability. All these factors combine to mean that the computational effort required for a numerical solution of the meteorological equations generally rises as a large power (at least the cube) of the number of grid points.

The CFL condition can be generalized to other, more complicated, equation sets. The general principle is that 'information' must not travel more than one grid length in a timestep. Thus, in an equation which describes wave propagation with phase speed c, a condition for stability is $\Delta t \le \Delta x/c$. The largest phase speed of waves is often rather large compared to the flow speeds in the atmosphere, and so this provides a more stringent restriction on the maximum timestep. For instance, the large flow speeds in the troposphere are generally less than $100 \, \mathrm{m\,s^{-1}}$. But the external gravity wave or 'Lamb wave' has a phase speed of around $300 \, \mathrm{m\,s^{-1}}$, and other internal gravity waves have phase speeds in excess of $100 \, \mathrm{m\,s^{-1}}$. In the early days of numerical weather prediction, various 'filtered' equation sets were employed to remove the very fast waves from the system. For instance, equation sets based on the quasi-geostrophic vorticity equation implicitly include a thermal wind balance condition which links the wind and temperature fields, thereby removing gravity waves from the solution. The approach generally favoured today is to use the primitive equations, which admit gravity wave solutions, in conjunction with a 'semi-implicit' integration scheme. An implicit integration scheme is one in which data from the unknown $(m+1)$th time level is used in the formulation of the left hand side of the equation. For example, the linear advection equation may be written:

$$Q_n^{m+1} - Q_n^{m-1} = -\alpha \left(\frac{Q_{n+1}^{m+1} + Q_{n+1}^{m-1} - Q_{n-1}^{m+1} - Q_{n-1}^{m-1}}{2} \right). \tag{2.22}$$

The spatial derivatives are effectively smoothed in time, and it is easy to show that such a scheme is computationally stable for all Δt, though its numerical solutions become increasingly unlike the analytical solutions for $\alpha > 1$. Furthermore, obtaining the Q_n^{m+1} from this finite difference relation requires the solution of a set of coupled simultaneous equations, one for each gridpoint. The computational effort for such an implicit scheme is much greater than the corresponding 'explicit' scheme. In modern semi-implicit schemes, such a formulation is applied just to those terms in the primitive equations which generate fast gravity waves; the remainder are handled explicitly. The limitation on the timestep is then due to the flow speed, rather than the phase speed of the fastest gravity waves.

Even when Δt is sufficiently small to guarantee stability, and for the finite difference approximations to the derivative to be accurate, the numerical solution to the advection equation can be unrealistic. This is readily seen if the phase speed of wavelike disturbances, wavelength $N\Delta x$, is calculated. The analytic solution reveals that the phase speed should, of course, be u for all wavelengths. But for the finite difference approximation to the equation, it is easy to show that

$$\frac{c}{u} = \left(\frac{N}{2\pi\alpha}\right) \sin^{-1}\left\{\alpha\sin\left(\frac{2\pi}{N}\right)\right\}, \tag{2.23}$$

where c is the phase speed of disturbances in the discretized analogue to the advection equation. If N becomes large, c/u tends to 1. But for smaller N, c is always smaller than u. For the smallest wavelength defined by the grid, $2\Delta x$, c is zero. Thus, small scale features are advected appreciably more slowly in the numerical solution than they would be in the true, analytic, solution. An example is shown in Fig. 2.6. What is worse, an arbitrarily shaped disturbance, which may be decomposed into a Fourier series of waves of different wavelengths, will disperse. The smaller wavelength components will move more slowly than the longer wavelength components.

Our discussion so far has been focussed upon a Cartesian coordinate system in which the grid points are spaced uniformly in x and y. Such a grid might be adequate for a local numerical weather prediction model, but will not suffice in a global context. A grid which is regular in (say) latitude ϕ and longitude λ will become singular at the poles, with the grid length becoming very small at high latitudes. This in turn means that the CFL condition will demand exceedingly small timesteps in order to handle the small region near the poles without computational instability. The 'polar problem' is indeed a serious difficulty in global atmospheric modelling. One solution is to design a grid whose spacing in λ varies with latitude. This still requires special treatment of the polar cap itself. Another is to use a regular grid in ϕ

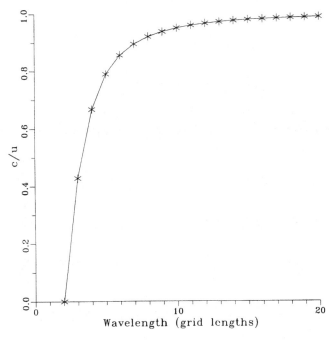

Fig. 2.6. Variation of phase speed with wavelength in a numerical solution of the linear advection equation. In this example, $u\Delta t/\Delta x$ was chosen to be 0.5.

and λ, but to apply a numerical filter at each timestep which removes those disturbances whose wavelengths are such that the CFL condition is violated. None of these solutions is entirely satisfactory, in that they widen the gap between the continous problem and its discretized analogue.

This discussion reveals some of the formidable obstacles to carrying out an accurate numerical simulation of the equations of atmospheric flow. We must be sceptical of features which are only a few grid lengths across in any numerical model, and if possible ensure that all the important weather systems and meteorological features are well resolved. But for a system such as the atmosphere, with its three spatial dimensions, computer resources are at a premium, and a, generally unsatisfactory, compromise between resolution and economy of computing time has to be made.

In recent years, so-called 'spectral methods' have become popular. Each variable is expanded as a series in some convenient orthogonal basis functions. For example, any quantity Q in a periodic channel with length X and width Y might be represented by a Fourier series:

$$Q(x, y) = \sum_{n=0}^{N} \sum_{m=0}^{M} Q_{m,n} \sin(\pi n y / Y) e^{i2\pi m x / X}. \tag{2.24}$$

N and M define the wavenumbers at which the series are arbitrarily truncated. Their choice is equivalent to the choice of Δx and Δy in a finite difference model. Substitution into a linear equation, such as the linear advection equation, reduces the partial differential equation to a set of ordinary differential equations. These can be solved accurately by a variety of techniques. Effectively, the method uses all the available information about Q when computing derivatives, rather than the limited local information used by a finite difference method. The great advantage of the method is that the phase speed of wavelike disturbances is represented accurately for all wavenumbers. Difficulties arise when the governing equations contain non-linear terms. Terms such as $u \partial Q / \partial x$ involve multiplying together two Fourier series, and, unless the truncation is very severe, the computational labour involved is very much larger than for the equivalent finite difference scheme. Spectral methods became competitive with finite difference methods with the development of the 'spectral transform technique'. In this, each timestep is divided into two stages:

(i)   A gridpoint stage, in which all those products of variables which appear in the equations are formed.

(ii) A spectral stage. All variables and their required products are represented as spectral series, and their spatial derivatives are calculated by simple multiplication. A timestep can then be taken.

The geometry demands rather more complicated basis functions on the sphere, but the principle remains the same. The appropriate basis functions are the 'spherical harmonics', which are products of sinusoidal functions representing variations in the zonal direction with Legendre functions representing variations in the meridional direction. Technical details are given in some references in the bibliography. One very significant advantage of the spectral transform method on a spherical domain is that it avoids the polar problem entirely. The spectral representation is isotropic, and resolves features near the pole in exactly the same detail as features at lower latitudes.

The successful implementation of the method requires a fast transform between the spectral and gridpoint representations of the variables. The fast Fourier transform is such a transform suited to periodic domains. Spectral methods are now at least as popular as finite difference methods for weather prediction and global circulation models. At reasonably high resolution, the methods are comparable in computational requirements and are very similar in accuracy.

This section has concentrated on the 'advective' parts of the solutions of equations such as the primitive equations. These terms dominate the solutions for up to the first day or so of a numerical forecast. Beyond this, the effects of unresolved, subgridscale motions, and of the various heating and friction processes, become increasingly important. We will concern ourselves with these topics in Section 2.5.

2.4 The analysis–forecast cycle

Numerical weather prediction requires initial data defined on some mathematically specified grid of points covering the Earth's surface. However, observing stations are not located at mathematically prescribed positions. Rather, their locations are determined by a succession of historical, geographical and economical (not to mention military) accidents. What is worse, the observing stations are not in particularly close proximity over much of the Earth's surface, and barely serve to resolve the most important weather systems. The problem of preparing suitable initial data from the heterogeneous, irregular and possibly inadequate set of available observations is known as 'meteorological data analysis' or simply as 'analysis'.

Figure 2.7 shows a schematic view of the problem. The lines represent the regular grid of data points demanded by a finite difference or spectral model. The stars represent the actual observations. For simplicity, we assume that all observations are taken simultaneously, and that we are only dealing with one level in the atmosphere. If there were many more stars than gridpoints, our problem would be relatively straightforward. We could simply average the nearby observations in some way to provide a representative value at the desired gridpoint. But in practice, the situation is reversed. There are generally many fewer observations than gridpoints, and in some regions, there may be virtually no observations.

The usual technique is to start with some reasonable initial guess Q_G for the actual field Q on the gridpoints. A refined estimate is provided by modifying this initial guess, using information from any observations within a certain distance of each gridpoint:

$$Q_R = \left(1 - \sum_i w_i\right) Q_G + \sum_i w_i Q_i. \tag{2.25}$$

The heart of the method is the appropriate choice of the weights w_i. In general, these will depend upon the distance of the observation from the

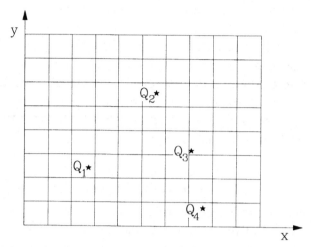

Fig. 2.7. Schematic illustration of the meteorological analysis problem.

gridpoint, and upon its likely accuracy. The initial guess field is usually based upon an earlier forecast field. Thus, in data-sparse regions, the initial guess field will be almost unchanged by the procedure, although the successive analyses of the field in this region should evolve in a physically consistent fashion, within the limitations of the forecast model employed. In data-dense regions, the initial guess field will be completely replaced by a fresh field based upon the observations. In regions where there are only a few observations of dubious accuracy or relevance, a compromise between the observations and the background field is achieved.

An important quality control stage is usually included at this point. Data are checked to ensure that they have no obvious fault. For instance, an observation might be flagged or rejected if it departed by more than a prescribed amount from the background field. This is a dangerous procedure if the criteria are too stringent, since sudden developments might be missed; equally if data are accepted too uncritically, serious errors and inconsistencies will appear in the final analysis. The intervention of a human forecaster is often beneficial in doubtful cases; the forecaster is able to use sources of data (such as satellite images) which the analysis programme is unable to consider when verifying or rejecting suspect observations.

The method just outlined applies to *spatial analysis*, where it is assumed that all observations are taken simultaneously. An increasing amount of data, especially satellite data, is taken at times other than the synoptic observing times. This is most simply handled by assigning it to the nearest synoptic hour, and adjusting the weights to reflect the uncertainty thereby introduced. More sophisticated techniques involve inserting such asynoptic data into the run of a numerical forecast model, modifying the evolving fields wherever and whenever data becomes available.

Most analysis schemes tend to treat the various meteorological variables as independent. Thus the wind and temperature analyses are performed separately. Of course, such fields are not entirely independent. Balance conditions mean that such variables are related to one another. Thermal wind balance, linking the wind and temperature fields is an obvious example, as is gradient wind balance, which relates wind and mass distribution. Similarly, observations have established that on the larger scale in the extratropics, the vertical component of vorticity is at least an order of magnitude larger than the horizontal divergence. This means that the horizontal components of wind are not entirely independent. In the tropics, other balances are important, such as a near balance between heating or cooling and ascent or descent. One goal of more sophisticated schemes is to include such balance conditions wherever possible.

Nevertheless, at the end of the analysis procedure, the fields are not in a very good state of dynamical balance. Were the forecast model to be initialized with these fields, its integration would be dominated by large amplitude gravity waves of high frequency. Such disturbances are rarely observed and so an analysed field which gives rise to them in a numerical integration must be regarded as unphysical. They are removed by including a process of 'initialization' or balancing between the analysis and forecast stage. This process involves making rather small adjustments to the fields, generally considerably smaller than the typical errors of observation. Simple initialization merely ensures that the horizontal divergence and its rate of change is set to zero. More sophisticated initialization schemes generate a pattern of vertical velocity (or divergence) which is consistent with the observed winds and temperature fields and with the heating rates. In the midlatitudes, such consistency is related to the quasi-geostrophic approximation; see Section 1.7. For our purposes in studying the global circulation, these estimates of the unobservable vertical velocity are extremely useful, and enable a truly three-dimensional view of atmospheric circulations to be taken.

A number of the diagnostics to be discussed in this book have been based on the archived initialized analyses carried out at the European Centre for Medium Range Weather Forecasts (ECMWF). The ECMWF performs a global analysis of the atmospheric circulation in the troposphere and lower stratosphere every six hours. Other centres carry out a comparable analysis every 12 hours. Once each day, the analysis is integrated forward

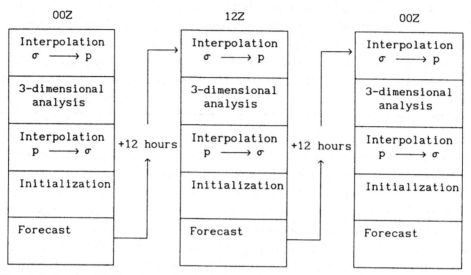

Fig. 2.8. A simplified schematic depiction of the analysis–forecast cycle used at ECMWF.

in time to provide an extended range (up to ten days) forecast. Figure 2.8 gives a schematic description of the forecast–analysis cycle which operates at ECMWF. The archives extend back to late 1979, and so by creating ensemble seasonal averages, climatological fields based on ten years or more of data can be constructed. Over this period, the analysis schemes and the forecast models have been improved, and so the earlier data may not be as reliable as more recent analyses. This is particularly true of the divergent part of the wind field in the tropics. One must also be aware that despite the global appearance of the fields, there is really very little high quality data available in areas such as the southern oceans. The fields there are almost undiluted background field, and have more the status of a numerical simulation of the atmospheric flow than of independent observations.

The original 'ten-year climatology' prepared at Reading University from ECMWF data was based on the period from March 1979 to February 1989. Various improvements were made to the analysis and initialization techniques during this period. The most important improvements concern the divergent part of the tropical wind field. The original initialization scheme attenuated this part of the wind field rather badly, and so any diagnostics which involve the vertical velocity or the divergence at low latitudes are better calculated from a more recent run of data. A 'six-year climatology' using analyses from the period March 1983 to February 1989 has been used to provide many of the diagnostics used in this book.

2.5 Global circulation models

When a numerical model of the atmospheric flow includes sufficient thermal forcing and friction terms to enable it to be run for long periods, and thereby to simulate a mean climate, it is called a global circulation model or 'GCM'. Such models are now vital tools in studying the global circulation. Given sufficient computer resources, a global circulation model provides the opportunity to experiment with the global circulation, to investigate the separate and combined effects of different physical and dynamical processes. GCMs are also used extensively in attempts to predict natural and anthropogenic climate change, and to explore past climates.

In Section 2.3, a numerical weather prediction model was presented as a frictionless, adiabatic simulation, which allowed the observed state of the atmosphere to evolve for a short time. As longer forecasts have been demanded, various friction and radiative transfer schemes have had to be added, making such models really very similar to global circulation models. A similar development has taken place with global circulation models. The early models combined relatively sophisticated simulation of the relevant physical processes with fairly coarse, low resolution simulation of the dynamical processes. It became apparent that higher resolution of the principal weather systems was essential. So as computers became larger and more powerful, global circulation models became more detailed. Today, the differences between numerical weather prediction and global circulation models are small, and in some cases, nonexistent.

We will now consider some of the additions needed to make an adiabatic, frictionless weather prediction model into a global circulation model. There are three important elements, namely, the fluxes of both short wave solar radiation and long wave terrestrial radiation through the atmosphere and at the Earth's surface; the turbulent exchanges of heat, momentum and moisture between the Earth's surface and the atmosphere; and finally, the effects of subgridscale motions on such transports (especially in the vertical). The latter introduces the concept of parametrization.

Electromagnetic radiation in the atmosphere may be partitioned into the short wave flux of solar radiation, and the long wave flux of infrared radiation emitted both by the Earth's surface and the atmosphere itself. The object is to calculate the upward and downward fluxes of both kinds of radiation at each level in the atmosphere; the divergence of the net radiative flux then gives the heating rate due to radiative processes. The concept is simple enough; the implementation can be complex and very demanding of computer resources.

The flux of solar radiation incident on the top of the atmosphere is a straightforward function of latitude, time of day and time of year. During its passage through the clear atmosphere, some is absorbed, some is scattered, but most reaches the surface. In cloudy conditions, the situation is more complex; reflection of sunlight from cloud tops and absorption of sunlight by clouds both attenuate the solar beam. Both effects depend upon the nature of the cloud particles and can generally only be represented very crudely in current models. Other complications include the multiple reflection of sunlight between layers of cloud, or between clouds and a high albedo surface such as snow or ice.

The net flux of long wave radiation is a more complex calculation. The emittance and transmitivity of the atmosphere at these wavelengths is a function of temperature and wavelength. Figure 2.9 illustrates the dependence of absorption upon wavelength in clear conditions, showing the very fine detail in the absorption curves. The absorption is mainly due to the molecular absorption bands of trace constituents, especially water vapour, carbon dioxide and, to a lesser extent, ozone. Other 'radiatively active gases' such as methane are now recognized as playing an important role in the infrared radiative transfer even though their concentrations are very small. To calculate the net flux of infrared radiation in clear sky conditions strictly requires a numerical integration over wavelength for the particular temperature profile at the location being considered, taking account of each of the thousands of molecular absorption lines. The various molecular transitions which generate these lines are known with sufficient accuracy to enable such 'line-by-line' calculations to be carried out with great accuracy. But line-by-line calculations are far too time consuming to form part of a global circulation model. Various approximations, involving the grouping together of large numbers of lines into bands, are used to simplify the integration. The line-by-line calculations are used to test and refine these more approximate schemes. Further saving of computer resources are effected by updating the radiative fluxes every few hours, rather than at every timestep. Even so, the calculations can dominate the computer time of a typical model.

As with short wave radiation, the presence of clouds introduces great uncertainty into long wave flux estimates. Clouds increase the absorption of infrared radiation as well as scattering and reflecting incoming solar radiation. The net effect is quite uncertain, even as to its sign, though it is generally agreed that high cirrus clouds in the tropics have a net warming effect on the atmosphere, while low stratus clouds at higher latitudes have a cooling effect by reflecting sunlight back to space. Possibly the greatest uncertainty in modern global circulation models is the prediction

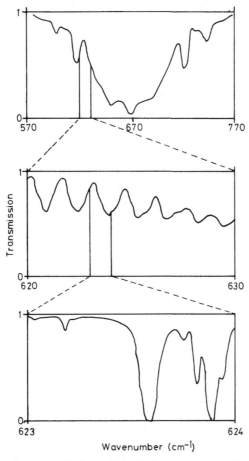

Fig. 2.9. The transmission of infrared radiation by atmospheric gases in clear sky conditions, as a function of wavelength. The diagram shows the important $15\,\mu m$ band of carbon dioxide. The top frame shows the entire band, while the lower frames magnify restricted parts of it at increasing spectral resolutions.

of cloudiness and its radiative impact; it is this uncertainty which makes the prediction of anthropogenic climate change resulting from atmospheric pollution so difficult.

A considerable fraction of the incoming solar radiation reaches the Earth's surface. What happens to it then depends upon the nature of that surface. An ocean surface has a very large thermal capacity compared to the atmosphere. Many global circulation models simply hold the temperature of the ocean surface fixed at the climatological values for the time of year being simulated. This provides an adequate lower boundary condition which can be used to calculate fluxes of heat, moisture and momentum into the atmosphere.

But this approach is probably misleading when undertaking, for example, calculations of climate change. The changing wind patterns will disturb the ocean circulation, in turn leading to modified sea surface temperature fields. Representing these feedbacks requires circulation models both for the atmosphere and for the ocean, and a number of research groups are attempting to develop such coupled models. But for most of the discussion in this book we will restrict ourselves to the simpler case of prescribed sea surface conditions.

A land surface has a much smaller thermal capacity, and its temperature can change considerably as a result of the daily fluxes of energy in or out of it. Some kind of simple soil model is required, which represents the various inputs of energy to the upper level of ground and thereby computes a surface temperature. A typical soil model is illustrated in Fig. 2.10. Essentially, it consists of solving a diffusion equation:

$$\frac{\partial T}{\partial t} = K \frac{\partial^2 T}{\partial z^2}, \tag{2.26}$$

subject to the boundary conditions:

$$T = T_d \text{ at } z = d, \ K \frac{\partial T}{\partial z} = \mathscr{F} \text{ at } z = 0. \tag{2.27}$$

Here T is the soil temperature, z is the depth, and K is a coefficient of thermal conductivity; T_d is called the 'deep soil temperature' and is prescribed from climatological data. The base of the soil model is at depth d where d is a few metres. The net flux of heat out of the ground, $\mathscr{F}$, includes the net radiative flux, the fluxes of sensible and latent heat carried by boundary layer turbulence, and latent heat required to melt any lying snow. Clearly, there is an important feedback between the soil model and the atmospheric model. The surface temperature determines the heat and moisture fluxes out of the surface into the atmosphere. At the same time, the simulated meteorological conditions determine $\mathscr{F}$ and hence the soil temperature. The soil model must also keep an inventory of any water which enters the surface in the form of rain or melting snow. Some will 'run off' and be lost to the grid box. The rest will accumulate and be available for subsequent evaporation. Various physical properties of the surface, such as its albedo and water capacity, have to be prescribed and are based on observed surface and pedological information.

Once heat and moisture have entered the base of the atmosphere, they are transported upwards through the lowest few hundred metres and into the main troposphere. This transport is dominated by turbulent eddy processes.

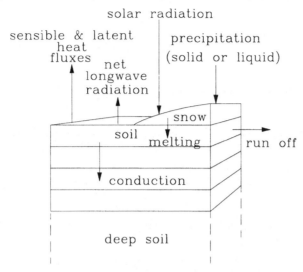

Fig. 2.10. Illustrating the soil model used in a modern global circulation model.

A global circulation model will attempt to characterize these turbulent up-ward fluxes of heat, moisture and momentum in terms of the mean vertical shear of the wind, the stratification of the atmosphere and the gradients of moisture. This procedure is called a 'parametrization' of the transport effects of turbulent eddies. The underlying hypothesis is that there will be a relation-ship between the large scale flow structure and the transports by small scale motions, so that the details of individual turbulent eddies are unimportant. The equations describing turbulent boundary layer flow are highly nonlinear and no general analytical solutions to them are known. Consequently, the basis of such parametrizations are largely empirical. In the following para-graphs, we will consider two important types of parametrization: the first is the turbulent flux of heat, momentum and moisture between the surface and the lowest model layer, and between the model layers themselves. The second is the rapid vertical transport of heat and moisture which takes place in cumulus convection.

The parametrization of boundary layer fluxes is more fully described in texts on the atmospheric boundary layer, to which the interested reader is referred. The basic concept is that, integrated over time and space, eddies act to diffuse heat, momentum or trace constituents, reducing the gradients of these quantities. The momentum equation is written:

$$\frac{D\mathbf{v}}{Dt} + f\mathbf{k} \times \mathbf{v} = -\nabla\Phi - g\frac{\partial \boldsymbol{\tau}_s}{\partial p}, \tag{2.28}$$

where τ_s is the stress due to small scale, turbulent eddies. A typical simple formulation for the surface stress τ_s uses the 'bulk aerodynamic formulae', in terms of which the stress is written:

$$\tau_s = \rho_s c_D |\mathbf{v}_s| \mathbf{v}_s. \tag{2.29}$$

Here, the subscript 's' denotes surface values. The coefficient c_D is a dimensionless drag coefficient whose value generally depends upon the nature of the underlying surface, and also on the vertical wind shear and static stability of the air in the lowest layers, these quantities determining how readily the air can overturn in turbulent eddies. Typical values of c_D are 10^{-3} over oceans and 3×10^{-3} over land. More sophisticated schemes estimate the appropriate value of c_D at each gridpoint and timestep in terms of a 'roughness length', which depends upon the surface properties, and the static stability of the boundary layer. Similar formulae are used to estimate the vertical fluxes of moisture and momentum through the turbulent boundary layer.

Deep cumulus convection dominates the weather systems of the tropics, and the latent heat released as water condenses in convective clouds is the major part of the atmospheric heating at low latitudes. However, cumulus clouds are small features in the global circulation. Their typical horizontal dimension is between 1 and 10 km, and even extremely large clouds do not exceed a few tens of kilometres across. They cannot be resolved explicitly in global circulation models, and their effects must somehow be 'parametrized'. Furthermore, if no attempt is made to neutralize the convective instability, the model will generate vigorous grid scale convection. Such small scales are unlikely to be handled realistically by the model, with the result that grid scale 'noise' rapidly obscures the large scale fields and may eventually cause the integration to become numerically unstable. Representing cumulus convection is important both for computational reasons and in order to simulate an important component of tropical atmospheric circulation.

The conditions for cumulus convection to take place involve both the vertical profiles of temperature and of humidity at a particular gridpoint. Suppose a parcel of air rises adiabatically through the atmosphere a short distance dz, its pressure decreasing as it ascends. If the parcel of air is dry, its decrease of temperature is simply calculated from the first law of thermodynamics, Eq. (1.7):

$$c_p \mathrm{d}T = -g\mathrm{d}z. \tag{2.30}$$

This expression can be integrated to calculate the 'adiabatic lapse rate', which is about $10\,\mathrm{K\,km^{-1}}$. If the parcel of air is saturated, then water vapour condenses as it rises, and its latent heat of condensation contributes to the energy budget of the ascending parcel:

$$c_p \mathrm{d}T = -g\mathrm{d}z + L\mathrm{d}r_s, \tag{2.31}$$

where $\mathrm{d}r_s$, the change in the saturated parcel humidity, is the amount of water vapour condensing as the parcel ascends through the height $\mathrm{d}z$; $\mathrm{d}r_s$ depends upon the temperature change, so this relationship is a rather complicated nonlinear equation for the increment of temperature, $\mathrm{d}T$. It can be solved explicitly making use of the Clausius–Clapeyron equation (Eq. (7.30)) to relate r_s to temperature. The main point of Eq. (2.31) is to show that the temperature of the ascending saturated parcel of air will be larger than that of the ascending dry parcel. Such a profile is sometimes called a 'saturated adiabat'. The first condition for convective instability is that the lapse rate observed in the clear air should be less than the saturated adiabatic lapse rate, for in such conditions a rising parcel of saturated air will always be more buoyant than its surroundings. Most of the tropical atmosphere is in such a state of 'conditional instability'. The second condition for convective instability to break out is that some process must saturate air parcels. This will be achieved if there is net convergence of moisture into the air parcel for sufficiently long periods. The second condition can also be met if some mechanical forcing causes the air parcel to rise adiabatically for a sufficient height. For instance, this might occur if air blows over a suitable mountain. Figure 2.11 illustrates conditional instability. If both conditions are satisfied, then parcels of saturated, cloudy air will rise through the atmosphere, with compensating descent of unsaturated clear air in neighbouring regions.

The problem of incorporating such moist convection into GCMs is that the scale of regions of ascent is very small. This scale is determined by turbulent mixing between the ascending air and the surrounding clear air which moderates the buoyancy excess within the cloud. These processes do not operate on the grid scale or on any scale approaching it. Various schemes, of differing degrees of complexity, have been offered as a means of parametrizing cumulus convection. It is fair to say that all have a heavily heuristic element within them, and the various schemes all lead to rather different tropical climatologies when incorporated into global circulation models. It seems that an ideal way of representing convection has yet to be found.

By way of illustration, a brief outline of the 'Kuo scheme', which is widely used and less elaborate than some schemes will be given. The terminology is explained in the schematic diagram, Fig. 2.12. In this scheme, convection is triggered within a conditionally unstable layer between pressures p_1 and

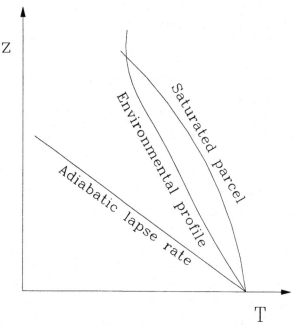

Fig. 2.11. The necessary conditions for instability. 'Conditional instability' is possible when the environmental profile lies between the dry and saturated adiabatic profiles.

p_2 if there is net convergence of moisture into the grid box, that is, if

$$I = -\int_{p_1}^{p_2} \nabla \cdot (\mathbf{v}r)\frac{dp}{g} > 0. \tag{2.32}$$

It is assumed that this moisture flux is carried up by the cumulus clouds. A fraction b is given up to the surrounding unsaturated air by evaporation and turbulent mixing from the sides of the cumulus towers. The remaining moisture flux, $(1 - b)I$, is rained out from the clouds. The fractional area of the grid box occupied by the ascending cumulus towers is assumed to be small. Then the net rates of change of temperature and moisture due to cumulus convection are given by:

$$\frac{\partial T}{\partial t} = a_T(T_C - T_E), \quad \frac{\partial r}{\partial t} = a_r(r_C - r_E), \tag{2.33a}$$

where

$$a_T = \frac{L(1 - b)I}{c_p \int_{p_1}^{p_2} (T_C - T_E)dp/g}, \quad a_r = \frac{bI}{\int_{p_1}^{p_2} (r_C - r_E)dp/g}. \tag{2.33b}$$

The 'detrainment parameter' b is crucial in determining the rainfall and

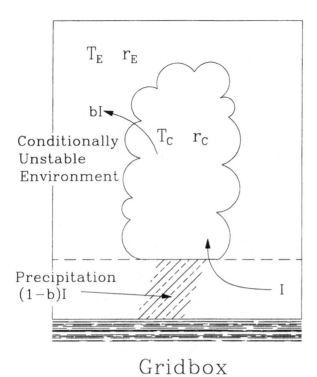

Gridbox

Fig. 2.12. The Kuo convection scheme. The parametrization assumes that there are many small cumulus elements in the grid box. The temperature and humidities of the environment are $T_E(p)$ and $r_E(p)$, respectively, while $T_c(p)$ and $r_c(p)$ are the corresponding cloud values. The moisture flux carried by the clouds is I, of which an undetermined fraction b is detrained into the environment.

modification of the environment air by cumulus convection. But there are no compelling theoretical grounds for its determination. It is generally regarded as a 'tunable parameter', adjusted in test integrations to obtain the best fit to observed fields. Such a procedure is not satisfactory, and begs many questions of how universal a parameter b might be, and how reliable the GCM will be in conditions far removed from the situations for which it was calibrated.

Before concluding this discussion of global circulation models, a simplified global circulation model which has considerable pedagogical value will be introduced; it might be called a 'simplified global circulation model' or 'SGCM'. Results from the SGCM will be introduced at a number of points in later chapters of this book to illustrate primary processes in global circulation. The model is based on a fairly sophisticated numerical scheme for solving the primitive equations, using the spectral transform method described in Section 2.3. It differs from a true GCM in its representation of heating and friction, which are replaced by simple linear terms. The momentum equation may be written:

$$\frac{\partial \mathbf{v}}{\partial t} + \mathscr{L}_M + \mathscr{N}_M = -\frac{\mathbf{v}}{\tau_D} + K\nabla^{2p}\mathbf{v}, \tag{2.34}$$

where $\mathscr{L}_M$ and $\mathscr{N}_M$ represent the various linear and nonlinear terms respectively. The thermodynamic equation is similar:

$$\frac{\partial \theta}{\partial t} + \mathscr{L}_T + \mathscr{N}_T = \frac{\theta_E - \theta}{\tau_E} + K\nabla^{2p}\theta. \tag{2.35}$$

The first term on the right hand side of Eq. (2.34) represents friction. In the absence of any other terms, it would represent an exponential decay of the velocity on the drag timescale τ_D. Such a term is called Rayleigh friction, and is the very simplest parametrization of boundary layer drag that one can imagine. The drag timescale τ_D is of order a day or so near the lower boundary, but is very long at higher levels in the atmosphere. The thermodynamic equation, Eq. (2.35), contains a similar 'Newtonian cooling' term on its right hand side; the potential temperature is relaxed towards a 'convective–radiative equilibrium' value θ_E on a radiative timescale τ_E. The field of θ_E is chosen so that the atmosphere is stably stratified (removing the need for any parametrizations of convection) and zonally symmetric with suitable horizontal temperature gradients. Although the model is dry, and simulates no moist processes explicitly, moisture may be regarded as being implicitly present, since the convective–radiative equilibrium static stability is characteristic of a saturated adiabat in the tropics. Both equations contain a hyperharmonic diffusion term to represent subgridscale motions. Such terms are very simply incorporated when the spectral formulation is used. The diffusion coefficient is chosen so that wavelengths near the limit of the model resolution are dissipated in a few hours. With a large value of p, the larger scales of motion will be virtually unaffected by this term. A frequent choice for low or intermediate resolution models is $p = 4$.

Figure 2.13 compares the time and zonal mean zonal wind, $[\bar{u}]$, for the JJA season, as observed, as simulated in a modern GCM and as represented by the SGCM. The main features, which will be discussed in more detail in Chapter 4, include the weak easterlies in the tropics, the strong jet in the upper troposphere near 30 °S and the weaker jet in the northern hemisphere. The observations also show a second, deeper jet at middle tropospheric levels, with its core near 55 °S. Both models show all these features. They differ from each other and from the observations in the lower stratosphere.

The SGCM makes no attempt to simulate the stratosphere. In the GCM, the temperature field, and hence, by thermal wind balance, the zonal wind field in the lower stratosphere, are sensitive to the details of the radiation scheme and the deep convection scheme. But it is clear from the diagram that the SGCM captures many of the basic processes which determine the gross features of the zonal wind field. Later, we shall see that it also simulates various eddy fluxes of heat and momentum surprisingly realistically.

2.6 Problems

2.1 Show that

$$\overline{Q'R'} = \overline{QR} - \overline{Q}\ \overline{R}.$$

2.2 If Q is some conserved quantity such that $DQ/Dt = 0$, and Q is independent of height, show that the root mean square meridional displacement of air parcels is

$$\overline{\eta'^2}^{1/2} = \frac{\overline{Q'^2}^{1/2}}{\overline{Q}_y}.$$

2.3 Show that a finite difference analogue of the second derivative can be written:

$$\frac{\partial^2 Q}{\partial x^2} = \frac{Q^n_{m+1} - 2Q^n_m + Q^n_{m-1}}{\Delta x^2} + O(\Delta x^2).$$

Using this formula and centred time differencing, show that the finite difference analogue of the linear diffusion equation:

$$\frac{\partial Q}{\partial t} = K\frac{\partial^2 Q}{\partial x^2}$$

is numerically unstable for all Δt. Suggest a modification of this finite difference representation which would be stable.

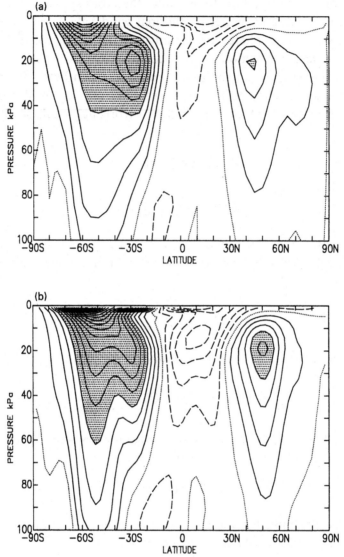

Fig. 2.13. The patterns of time and zonal mean zonal wind $[\bar{u}]$ for the JJA season: (a) from the ECMWF analyses; (b) from the UK Universities' global circulation model.

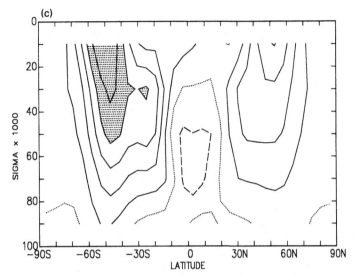

Fig. 2.13 (*cont.*). (c) From a simplified global circulation model described in the text. Contour interval $5\,\mathrm{m\,s^{-1}}$, with negative contours dashed. Shading indicates values in excess of $20\,\mathrm{m\,s^{-1}}$.

3

The atmospheric heat engine

3.1 Global energy balance

In this section, we will discuss some basic principles which will help to describe the thermal forcing of the global circulation. For the present, we consider the global mean energy balance of the atmosphere; in the next section we will consider the geographical variations of this balance. In these sections, we wish to introduce the concept of radiative equilibrium, and to indicate how a timescale for establishing such an equilibrium might be estimated.

The basic physical principle to be used is Stefan's law. This states that the radiant energy emitted per unit area of a perfectly black body is proportional to the fourth power of its temperature:

$$S = \sigma T^4, \tag{3.1}$$

where σ is the Stefan–Boltzman constant and has the value $5.67 \times 10^{-8} \, \mathrm{W \, m^{-2} \, K^{-4}}$. For our purposes, it will be adequate to regard a slab of gas, the surface of the Earth or the surface of the sun as black bodies which obey Stefan's law.

A black body emits radiation with a range of frequencies, but with a maximum at frequency ν_{max}. Wien's displacement law relates ν_{max} to the temperature of the black body:

$$\nu_{max} = W T, \tag{3.2}$$

where W is a constant which has the value $1.035 \times 10^{11} \, \mathrm{K^{-1} \, s^{-1}}$. Figure 3.1 shows graphs of energy versus frequency for black bodies at different temperatures. The Sun, with a surface temperature of around 5750 K, radiates mainly at visible and near infrared wavelengths, with a maximum in the visible part of the spectrum. The clear atmosphere is nearly transparent to

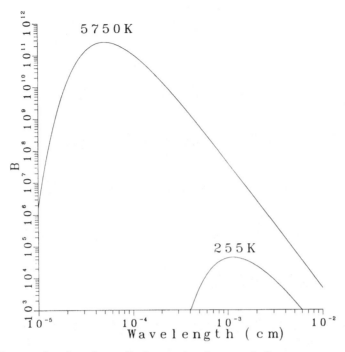

Fig. 3.1. Curves showing the radiation emitted per unit frequency as a function of frequency for black bodies with temperatures of 255 K and 5750 K.

these wavelengths, so most sunlight will reach the ground, or at least the tropospheric levels where the cloud tops are situated. In fact, most gases which form planetary atmospheres transmit sunlight with little absorption. The atmosphere itself, with typical temperatures of 200–300 K, radiates at much longer wavelengths, in the infrared part of the spectrum. The atmosphere is rather opaque at these wavelengths. Trace constituents such as water vapour, carbon dioxide and (at stratospheric levels) ozone, provide most of this absorption. Thus the surface is unable to radiate directly to space, and receives additional long wave radiation from the overlying layers of the atmosphere. Its temperature is raised above the equilibrium expected for an airless body. The atmosphere behaves rather like a blanket, trapping heat near the surface and raising its temperature, an effect popularly mis-named the 'greenhouse' effect.

These concepts can be made quantitative. Let the solar radiation flux incident upon the Earth be S; S is sometimes called the 'solar constant', though in fact it varies through the year as the Earth–Sun distance fluctuates. The mean value of S is 1370 W m^{-2}. The Earth presents an area πa^2 normal to the solar beam but has a total surface area of $4\pi a^2$; the mean flux of solar

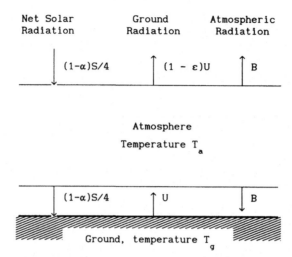

Fig. 3.2. Single slab model of the Earth–atmosphere system.

radiation per unit area of the Earth's surface is therefore $S/4$. A fraction α of the incident sunlight is simply reflected straight back to space; α is the 'albedo' and is 0.29 for the Earth, though there are strong and important local variations. The albedo of cloud or fresh snow is large, as much as 0.90, while that of forest or ocean surface is less than 0.07. A total flux $I = (1 - \alpha)S/4$ is absorbed by the Earth; in the steady state this must balance the infrared radiation emitted by the Earth. Using Eq. (3.1), the mean temperature of the Earth–atmosphere system is therefore:

$$T_B = \left\{ \frac{(1 - \alpha)S}{4\sigma} \right\}^{1/4}. \qquad (3.3)$$

The temperature defined in this way is sometimes called the 'bolometric' or 'brightness' temperature; it is the temperature which a black body would have to have in order to radiate the same flux of infrared radiation. Substituting values for the Earth leads to $T_B = 255\,\text{K}$. This is considerably colder than the mean temperature of the Earth's surface (which is close to 288 K). The bolometric temperature T_B may be thought of as a temperature typical of the higher layers of the troposphere which can radiate infrared radiation directly to space. The surface and the lower layers of the atmosphere can be much warmer, provided the atmosphere is opaque to infrared radiation.

This may be demonstrated if the atmosphere is regarded as a single uniform slab of gas lying above the Earth's surface, as shown in Fig. 3.2. For simplicity, assume that the atmosphere is completely transparent to visible radiation, but that it absorbs a fraction ϵ of the infrared radiation which impinges upon it. The ground emits radiant energy at a rate U, of which a fraction $(1 - \epsilon)$ escapes to space, while the warm atmosphere emits B both into the ground and to space. The temperature of the ground T_g and of the air T_a are obtained from Eq. (3.1):

$$T_g = (U/\sigma)^{1/4}, T_a = (B/\sigma)^{1/4}. \tag{3.4}$$

In the steady state, the fluxes into and out of the ground must balance:

$$(1 - \alpha)\frac{S}{4} + B = U, \tag{3.5}$$

as must the fluxes into and out of the atmosphere:

$$2B = \epsilon U. \tag{3.6}$$

Eliminating B between these equations and using Eq. (3.3), we find

$$T_g = \left\{ \frac{2}{2 - \epsilon} \right\}^{1/4} T_B \tag{3.7}$$

and similarly

$$T_a = \left\{ \frac{\epsilon}{2 - \epsilon} \right\}^{1/4} T_B. \tag{3.8}$$

Thus, in the limit of an atmosphere totally opaque to infrared radiation, i.e., $\epsilon \to 1$, we have that $T_g = 2^{1/4} T_B$ and $T_a = T_B$; the surface is warmer than the overlying atmosphere. Taking T_B as 255 K gives $T_g = 303$ K, somewhat warmer than observed. Using a more modest value of $\epsilon = 0.771$ gives the observed T_g of 288 K. In this case, $T_a = 227$ K, suggesting that a more transparent atmosphere would have an even stronger lapse rate in radiative equilibrium. A more sophisticated calculation, using a multilevel approximation to the vertical structure of the atmosphere would yield a higher surface temperature; such a model is needed to account for the extremely high surface temperatures of Venus. It could include additional processes such as scattering of radiation by aerosols. The action of radiation alone would eventually establish an equilibrium temperature structure, with temperature decreasing with height and depending upon the intensity of the incoming solar radiation. This equilibrium is called 'radiative equilibrium'.

Now consider the time taken to establish such a radiative equilibrium temperature. This can be estimated as follows. Suppose that by some unspecified agency the atmospheric temperature is disturbed away from radiative equilibrium. For simplicity, suppose that the surface has negligible thermal capacity; in that case, Eq. (3.5) still holds. But Eq. (3.6) must be rewritten

$$\frac{c_p p_s}{g}\frac{dT_a}{dt} = \epsilon U - 2B, \qquad (3.9)$$

where p_s is the surface pressure; $c_p p_s/g$ is the thermal capacity of an atmospheric column of unit cross sectional area. Write $T_a = T_{a0} + \Delta T$, where T_{a0} is the equilibrium value calculated from Eq. (3.7), and $|\Delta T| \ll T_{a0}$. Then Eq. (3.9) can be linearized about T_{a0}:

$$\frac{d}{dt}(\Delta T) = -\frac{4(2-\epsilon)\sigma T_{a0}^3 g}{p_s c_p}\Delta T. \qquad (3.10)$$

This equation has a simple and well-known solution: ΔT decays exponentially on a timescale τ_E where

$$\tau_E = \frac{p_s c_p}{4(2-\epsilon)\sigma T_{a0}^3 g}, \qquad (3.11)$$

which is called the 'radiative equilibrium timescale'. If the radiative timescale is long compared to the typical timescale for motions in the atmosphere, then fluid motions can maintain the atmospheric temperature far from radiative equilibrium; on the other hand, if τ_E is short, meteorology will do little to disturb radiative equilibrium. For the Earth, we suppose that ϵ is 1 so that T_B is 255 K; then τ_E from Eq. (3.11) is around 30 days. This is much longer than the typical timescales of 1–5 days associated with advection by large scale tropospheric weather systems, and so we anticipate that the circulation of the atmosphere will greatly modify its thermal structure.

The kind of model which has been described in this section is sometimes called an 'energy balance' climate model. Such models use the global mean incoming and outgoing fluxes of energy to estimate the global mean temperatures. More sophisticated versions of the genre include complicated feedback effects, such as ice–albedo feedbacks. Of course, they give no guidance as to global circulation, which is driven by variations of the temperature fields from place to place in the atmosphere. Indeed, because of the complicated nonlinearities within the climate system, such simple energy balance models do not even give a very reliable guide to the global mean climate, though they have been used to yield interesting possibilities about the sensitivity of climate to various kinds of perturbation.

3.2 Local radiative balance

The arguments of the preceding section were global. But the large scale circulation of the atmosphere is driven by variations of temperature from place to place. So in this section, the geographical variations of the radiation field will be described. These geographical variations will also change with time, thus inducing diurnal and seasonal changes into the atmospheric circulation.

Both the incoming solar radiation and the albedo of the Earth vary strongly with latitude. Figure 3.3(a) shows the annual mean incoming solar radiant energy flux plotted against latitude. We denote this incoming flux of radiation I. The global mean value of I is of course $S/4$, as shown in the last section. Geometrical factors mean that I is large in the tropics but considerably smaller at the poles. Were it not for the axial tilt of the Earth, the polar value would of course be zero. As it is, there is substantial insolation at the poles in the summer. In fact, the local value of I is a maximum at the pole, for a time around the summer solstice. The equator to pole decline of the annual mean I is accentuated by the variations of albedo. The albedo of high latitudes, where the snow and ice cover is large and permanent, is much higher than at lower latitudes. Again, the midlatitudes are more cloudy than the subtropics. As a result, a nearly constant short wave flux of about $100\,\mathrm{W\,m^{-2}}$ is reflected directly back to space at all latitudes. Thus the radiation absorbed by the Earth and its atmosphere varies from around $100\,\mathrm{W\,m^{-2}}$ at the poles to near $400\,\mathrm{W\,m^{-2}}$ at the equator.

The local radiative equilibrium temperature can be calculated in the same way as can the global radiative equilibrium temperature, using Stefan's law. The local bolometric temperature is given by:

$$T_B = (S_L/\sigma)^{1/4}, \tag{3.12}$$

where S_L denotes the local absorbed solar radiation. At the pole, the annual mean value of T_B would be around $205\,\mathrm{K}$ from this formula, while at the equator we find $290\,\mathrm{K}$. Using Eq. (3.7) to estimate a surface temperature, assuming $\epsilon = 0.771$, the equator-pole temperature difference would be about $96\,\mathrm{K}$.

In fact, this temperature difference is much larger than observed. The annual average temperature difference is around $35\,\mathrm{K}$, with the poles considerably warmer than would be predicted from local radiative balance, and the tropics rather cooler. The third curve in Fig. 3.3(a) shows the outgoing long wave radiation plotted against latitude. It is a considerably flatter curve than that for S_L, reflecting that the temperature contrast is smaller than it would be in radiative equilibrium. Of course, the total outgoing radiation, integrated over the globe balances the total incoming radiation. But locally, the imbalances are substantial. These local imbalances are a manifestation

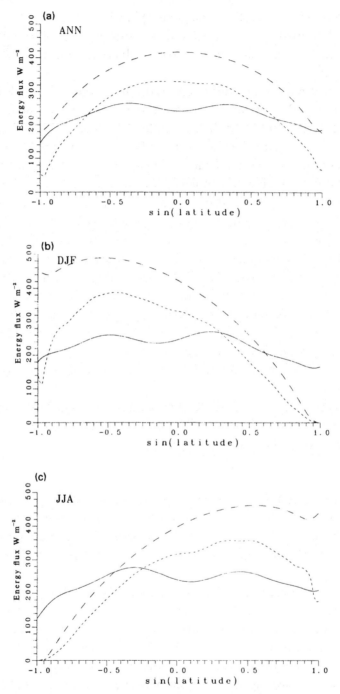

Fig. 3.3. Showing the Earth's radiation budget as a function of latitude. Dashed curves: incident solar radiant energy flux; fine dashed curves: absorbed incident solar radiant energy flux; solid curves: emitted long wave radiant energy flux. (a) Annual mean. (b) DJF. (c) JJA.

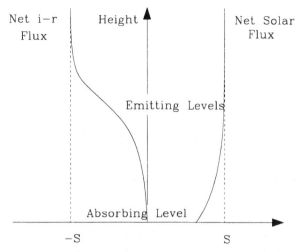

Fig. 3.4. Schematic illustration of the vertical variation of the downward flux of solar radiation and the net upward flux of infrared radiation in a cloud–free atmosphere. The difference between incoming and outgoing radiant energy fluxes in the lower atmosphere is balanced by an upward transport of heat by atmospheric motions.

of the heat–transporting motions within the atmosphere and oceans. The difference between the incoming and outgoing radiation is balanced by the divergence of the heat fluxes. The partitioning of the heat fluxes between the atmosphere and the oceans is still uncertain, though it is generally agreed that not more than half the heat flux is carried by the ocean circulation.

Less generally appreciated is the fact that the vertical transport of heat is also a crucial aspect of the atmospheric circulation. Since the atmosphere is relatively opaque to infrared radiation, radiation escapes to space principally from upper tropospheric levels. But sunlight by and large reaches the lower troposphere. At intermediate levels, there is an imbalance between the net upward flux of infrared radiation and the downward flux of solar radiation. This imbalance is compensated by an upward dynamical flux of heat by motions within the atmosphere. Fig. 3.4 gives a schematic sketch of the imbalance of vertical heat fluxes.

3.3 Thermodynamics of fluid motion

The thermodynamic state of a parcel of dry air is determined by fixing the value of any two of the thermodynamic state variables. These include such quantities as temperature, pressure, density or specific entropy. The remaining variables can than be calculated from the equation of state. For our purposes, it will be most convenient to describe the state of an air parcel by specifying its temperature T and potential temperature θ. The specific entropy s is closely related to θ by:

$$s = c_p \ln \theta, \tag{3.13}$$

while the pressure is obtained from a form of the equation of state for an ideal gas, Eq. (1.9):

$$T = \theta \left(\frac{p}{p_R} \right)^\kappa, \tag{3.14}$$

where p_R is some arbitrary reference pressure, generally taken as $100\,\text{kPa}$ for the Earth's atmosphere.

In order to change the thermodynamic state of the air parcel, heat must in general be supplied or removed. The only exception is the special class of 'adiabatic processes'. Many meteorological processes are approximately adiabatic, because the time scales associated with fluid motion are often fast compared to the timescales associated with radiative and diffusive transport of heat. But of course in discussing the global circulation, departures from adiabatic conditions are crucial in supplying the energy needed to maintain the circulation against friction. The heat taken up by a unit mass of air during a thermodynamic process from state A to state B is

$$Q = \int_A^B T\,\mathrm{d}s. \tag{3.15}$$

It is helpful to represent such a process on a 'thermodynamic diagram' in which s is plotted as abscissa and T as ordinate, shown in Fig. 3.5. The heat supplied to execute the process is simply the area under the curve joining successive states of the air parcel on the thermodynamic diagram. Note that if the specific entropy of state B exceeds that of state A, then Q is positive, that is, heat must be supplied to the air parcel. Conversely, if $s(B) < s(A)$, heat must be extracted.

Immediately apparent from this diagram is that the heat supplied in changing from state A to state B cannot be stated uniquely. There are an infinite number of different paths between the two states, and the area under each will usually be different. In general, if the entropy is increased at higher temperature, then more heat will be required. Now suppose that the air parcel is processed from A to B, and then back to A. As illustrated, more heat is supplied during the A to B process than is extracted in the return process. The air parcel has been returned to its initial state, and yet net heat has been supplied to it. What has happened to this excess heat energy? The answer is that it has been converted into some other form of

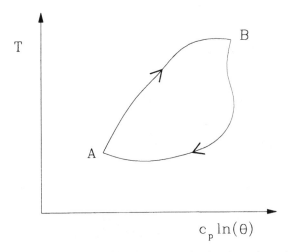

Fig. 3.5. Thermodynamic diagram, with $s = c_p\ln(\theta)$ plotted against T, for a parcel of air whose state changes from state A to state B.

energy, typically into kinetic energy of motion of the air parcel. Any cyclic process which takes the air around a clockwise path on the thermodynamic diagram will generate kinetic energy. A parcel of air executing such a path is an example of a thermodynamic 'heat engine'. Conversely, an anticlockwise path will extract more heat than is supplied; the imbalance is made up by reducing the kinetic energy of the parcel. The cycle is then an example of a 'refrigeration' cycle.

Such thermodynamic cycles are familiar to the physicist or engineer and are of value in discussing the performance of practical heat engines. Obviously, only a fraction of the heat supplied to the air parcel can be converted to mechanical energy; the ratio of mechanical energy obtained to heat supplied is called the 'thermodynamic efficiency' of the heat engine:

$$e = \frac{\oint T \, \mathrm{d}s}{\int_A^B T \, \mathrm{d}s}. \tag{3.16}$$

The thermodynamic efficiency of a heat engine operating between states A and B is a maximum when the entropy is increased at constant temperature T_B and reduced at a lower but constant temperature T_A, the cycle being closed by adiabatic processes. The path on the thermodynamic diagram is therefore a rectangle. Such an ideal cycle is called a 'Carnot cycle', and is illustrated in Fig. 3.6. Real heat engines are always less efficient than an ideal Carnot engine, and atmospheric motions are very much less efficient.

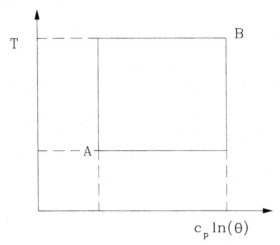

$$c_p \ln(\theta)$$

Fig. 3.6. A Carnot cycle, operating between state A and state B.

In the preceding section, it was suggested that tropospheric motions must transport heat upwards and polewards. The simplest model of such heat transporting motions would consist of rising motion in the tropics, poleward motion at upper levels, sinking at high latitudes and return flow to the equator at lower levels. Cross sections of the observed zonal mean temperature structure are shown in Section 4.1. Projecting such an idealized circulation on to this section, the successive thermodynamic states of an air parcel can be read off and plotted on the thermodynamic diagram. The result is a small loop on the thermodynamic diagram, indicating that such motions would generate kinetic energy; hence they can be maintained against friction by the continual conversion of heat energy into mechanical energy. Circulations which are clockwise on the thermodynamic diagram involve ascent of warmer air and descent of colder air; they are referred to as 'thermodynamically direct' circulations. Circulations in the reverse sense, which must be forced mechanically, are termed 'thermodynamically indirect'.

The kinetic energy generated by each circuit of an air parcel of unit mass is

$$K = \oint T \, \mathrm{d}s. \tag{3.17}$$

If τ_c is the time required for an air parcel to execute a complete circulation in the meridional plane, then the rate at which kinetic energy per unit mass is generated is

$$\frac{\mathrm{d}K}{\mathrm{d}t} \simeq \frac{1}{\tau_c} \oint T \, \mathrm{d}s. \tag{3.18}$$

The kinetic energy per unit mass is related to the typical wind speed, U, by $K = U^2/2$. The generation of kinetic energy must be balanced by frictional dissipation in the long term. Using the 'Rayleigh friction' introduced in Section 2.4 as a schematic way of representing the complicated friction processes, the rate of destruction of kinetic energy by friction is

$$\frac{\mathrm{d}K}{\mathrm{d}t} = -\frac{2K}{\tau_D}, \tag{3.19}$$

τ_D being the frictional timescale with a typical value of around five days for the troposphere. Balancing the creation and destruction of kinetic energy, it follows that in the long term mean:

$$\frac{1}{\tau_c} \oint T \, \mathrm{d}s = \frac{U^2}{\tau_D} \text{ or } U = \left\{ \frac{\tau_D}{\tau_c} \oint T \, \mathrm{d}s \right\}^{1/2}. \tag{3.20}$$

Purely thermodynamic arguments can take us no further. Given τ_D and τ_c, a typical value of U may be estimated. Observations of the mean meridional wind suggest that τ_c is typically around 140 days, leading to U of around 20 m s^{-1}. The global average wind is around 14 m s^{-1}, which is in reasonable agreement with this estimate. But to predict U on purely theoretical grounds requires consideration of the dynamics of the flow as well as its thermodynamics.

3.4 Observed atmospheric heating

The distribution of heating and cooling within the atmosphere which drives the atmospheric circulation will be considered in this section. We must distinguish between adiabatic changes of temperature, which may arise from vertical motions during which no heat enters or leaves the air, and changes which result from heat entering or leaving the air. The latter is sometimes referred to as 'diabatic warming' or, more simply, as 'heating'. Adiabatic temperature changes result in no transport of heat. On the other hand, heating provides the sources and sinks of heat which drive the global circulation. Heating arises from a large number of processes. Ultimately, absorption of short wave radiation is the source of heating, and emission of long wave radiation provides cooling. But exchanges of heat between different components of the climate system, which includes the solid Earth, the oceans and the ice and snow (or 'cryosphere'), as well as the atmosphere, are important for driving the global circulation.

Among the various mechanisms which are important are the following:

(i)   Exchanges of heat with the underlying surface, which may either be the ground or the ocean surface. Much incoming sunlight reaches the surface directly, and heats it up. This heat may find its way back into the atmosphere as a result of turbulent transports through the atmospheric boundary layer. Solid ground has a rather small heat capacity, and so the sunlight reaching a solid surface finds its way into the atmosphere rather quickly. The oceans, on the other hand, have a huge heat capacity, and they therefore represent an important reservoir of heat in the climate system. It is generally assumed that much of the memory of the climate system on timescales longer than a month or so is due to heat stored in the oceans.

(ii) Most of the solar energy reaching the Earth's surface goes to evaporate water, rather than to raise temperatures. Because of the very large latent heat of evaporation of water, and the fact that most of the Earth's surface is moist, very large amounts of heat can be taken up in this way. Measurements reveal that as much as 90% of the incident sunlight at the Earth's surface evaporates water. Even in arid regions, evaporation takes up some 10% of the absorbed incident radiation.

(iii) Water vapour in the atmosphere acts as a means of storing heat which can be released later. As the air circulates, it may rise and cool; if it becomes saturated, water vapour condenses and may rain out of the air. This condensation releases large amounts of latent heat. Indeed, in the tropical atmosphere, the heating is dominated by the release of latent heat in rain clouds.

(iv) Most of the cooling of the atmosphere is due to long wave radiation, though some heat can be extracted locally by contact with a colder underlying surface. The transmission of long wave radiation through the atmosphere is highly variable, being affected by the humidity of the air and its cloudiness.

Measuring all these different processes is a formidable task, and certainly cannot be done routinely over the entire volume of the atmosphere. The parametrizations included in a sophisticated numerical forecast model represent an attempt to estimate the heating directly in terms of the observed scale fields of temperature, wind, moisture, and so on. But, as we have seen in the last chapter, such parametrizations are less reliable than one would wish, and are sometimes subject to considerable errors. It must be concluded that the thermal forcing of the global circulation cannot be observed or calculated directly with any great accuracy. What can be done is to infer the *net* heating from the large scale temperature and wind fields, and their changes over long periods. As we have seen, these fields can be monitored reasonably accurately by the operational observing network.

Take the time average of the thermodynamic equation, Eq (1.12); the result is:

$$\Delta \theta_\tau + \bar{\mathbf{v}} \cdot \nabla \bar{\theta} + \bar{\omega} \frac{\partial \bar{\theta}}{\partial p} + \nabla \cdot \overline{\mathbf{v}'\theta'} + \frac{\partial}{\partial p} \overline{\omega'\theta'} = \bar{\mathcal{Q}}. \tag{3.21}$$

Here, $\Delta\theta_\tau$ represents the change of θ over the averaging period τ. Provided τ is reasonably large, this term will be small, especially for the DJF and JJA periods. The second and third terms represent, respectively, the horizontal and vertical advection of potential temperature by the time average motions. The fourth and fifth terms represent the divergence of the transient eddy fluxes of θ. All these terms must balance the heating. Accordingly, Eq. (3.21) can be rearranged as a diagnostic expression for $\bar{\mathcal{Q}}$. Calculating $\bar{\mathcal{Q}}$ in this way frequently involves much cancellation between the various transport terms, which tend to be of comparable magnitude but varying sign. Such a calculation of $\bar{\mathcal{Q}}$ is called a 'residual method'. Note that it cannot distinguish between the various physical processes which produce heating, but only determines the net heating. Its reliability depends crucially upon accurate estimates of the vertical velocity field. These may be suspect, especially in the tropics and in the stratosphere.

Figure 3.7 shows the zonal mean heating, $[\overline{Q}] = (p/p_R)^\kappa [\overline{\mathcal{Q}}]$, calculated using this residual method. It is based upon just six years of ECMWF analyses (since there is evidence that difficulties with initialization led to underestimates of the tropical vertical velocity, and hence of the tropical heating, during the first few years of operational analysis and forecasting at the centre). The general pattern throughout the troposphere generally conforms to the qualitative account given earlier in this chapter. Away from the deep tropics, there is heating near the ground. This heating is dominated by the turbulent transport of heat out of the ground, which is heated by direct insolation. Regions of cooling, dominated by long wave radiation to space, occupy much of the middle and upper troposphere. The heating is strong and fills the entire depth of the troposphere in the deep tropics. This is largely a signature of deep cumulus convection, releasing latent heat throughout the tropical troposphere. Separate bands of relatively deep heating are found in the midlatitudes. These are where active midlatitude depression systems lead to enhanced precipitation and release of latent heat.

Comparison of the DJF and JJA cross sections reveals the seasonal cycle of heating. The maximum tropical heating rates are in the summer hemisphere. The midlatitude heating is in the winter hemisphere. The large maximum at $70\,°\text{S}$ in JJA is probably spurious, and is a result of extrapolation of the wind and temperature fields beneath the high Antarctic ice sheet. There are substantial differences between the midlatitude heating rates in the two hemispheres. We will return to a discussion of some of the differences between the storm depression latitudes in each hemisphere in Chapter 5.

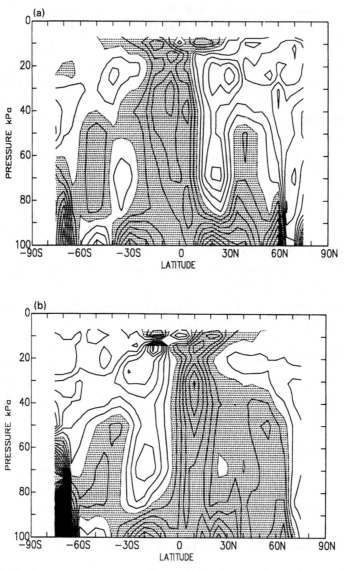

Fig. 3.7. Latitude–pressure cross sections of the time and zonal mean heating, $[\bar{Q}]$, based on six years of ECMWF data. Contour interval $0.2\,\mathrm{K\,day^{-1}}$, positive values shaded. (a) DJF. (b) JJA.

Regional variations of the heating are illustrated if $\bar{Q}$ is integrated with respect to height. Figure 3.8 shows the quantity

$$\{\bar{Q}\} = \frac{1}{g} \int_0^{p_s} \bar{Q}\mathrm{d}p. \qquad (3.22)$$

Since p_s/g is the mass per unit area of a column of the atmosphere reaching from the surface to infinity, $\{\bar{Q}\}$ has the dimensions of $W\,m^{-2}$ and is therefore directly comparable with the insolation and the long wave radiative fluxes discussed earlier in the chapter. In the DJF season, the largest heating is concentrated in the Indonesian region. There are strong maxima over the equatorial continents of Africa and South America. Two midlatitude 'storm track' regions show up as heating maxima over the Pacific and Atlantic Oceans. As might be expected, the largest cooling is close to the winter pole. The maximum heating or cooling rates are generally no more than $100\,W\,m^{-2}$, although the Indonesian maximum reaches $225\,W\,m^{-2}$ over a small area. This is quite small compared with the global average absorbed heat flux of around $240\,W\,m^{-2}$, showing that a considerable fraction of the incoming solar flux is simply re-radiated by the surface, without heating the atmosphere. In JJA, the largest heating rates are found north of the equator, and the maxima associated with the midlatitude storm tracks are considerably weaker. The most striking feature is the large maximum close to the Tibetan plateau. This represents huge latent heat releases as the moist winds of the Asian summer monsoon impinge on the Himalayan mountain ranges.

3.5 Problems

3.1 Use the following data to estimate the radiative equilibrium timescale for Venus, Earth and Mars. In each case, compare this timescale to the planet's solar day and its year.

	Venus	Earth	Mars
p_s	9 MPa	100 kPa	700 Pa
S $W\,m^{-2}$	2550	1370	583
α	0.76	0.29	0.15
g $m\,s^{-2}$	8.9	9.8	3.7
c_p $J\,K^{-1}\,kg^{-1}$	567	1004	567

3.2 The radiative timescale given in Eq. (3.11) was derived for a surface with negligible thermal capacity, whose temperature adjusts instantly to changes of the radiation incident upon it. If the surface has a finite thermal capacity, C_g say, write down equations for the evolution of perturbations ΔT_g and ΔT_a to the temperature of the ground and the atmosphere, respectively, and hence deduce how such perturbations will change in time. If C_g is very

(a)

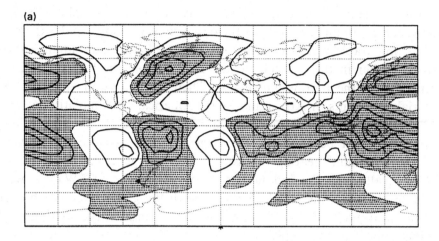

(b)

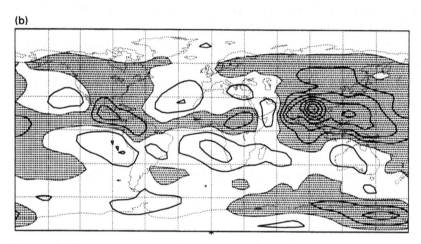

Fig. 3.8. Vertically integrated net heating based on six years of ECMWF data. Contour interval $50\,\mathrm{W\,m^{-2}}$, with positive values shaded. (a) DJF. (b) JJA.

much larger than $c_p p_s/g \equiv C_a$, show that two timescales, corresponding to the equilibrium timescales for the surface and the atmosphere, characterize the evolution of the temperature.

3.3 Suppose that the radiative equilibrium temperature of an atmosphere T_E varies sinusoidally with period τ_s and amplitude ΔT_E. Determine the time between the maximum T_E and the maximum atmospheric temperature, and the amplitude of the atmospheric temperature fluctuation.

3.4 A typical air parcel circulates from low levels in the tropics to high levels in the polar regions and back on a timescale of 120 days. Assume a friction timescale of five days, and a typical windspeed of $15\,\mathrm{m\,s^{-1}}$. Assuming, further, that the upper level flow is close to 30 kPa and that the ascent and descent processes are adiabatic, use the first law of thermodynamics to estimate the pole–equator temperature difference. Look up a cross section of $[\overline{\theta}]$ (given in Fig.4.2) to check and comment upon your result.

4

The zonal mean meridional circulation

4.1 Observational basis

The large scale structure of the atmospheric flow varies most rapidly in the vertical direction, and least rapidly in the zonal direction. Zonal averaging therefore makes the important vertical and meridional variations plain, and has been employed for many years as a compact way of studying the global circulation. Indeed, for many writers, the global circulation is simply the pattern of flow projected on to the meridional plane. In this book, we will take a broader view by attempting to summarize our current understanding of the full, evolving three-dimensional pattern of winds and temperature in the atmosphere. But the traditional zonal mean view is a useful starting point which we will explore in this chapter.

The zonal mean wind and vectors of the mean meridional wind are illustrated in Fig. 4.1, based on ECMWF analyses. Rising motion is seen in the tropics, with the maximum vertical velocity in the summer hemisphere. Strongest descent is at latitudes of around $25-30°$ in the winter hemisphere, with flow towards the equator near the surface and away from the tropics in the upper troposphere, as is required by continuity. Such an axisymmetric circulation is the most obvious response of the atmospheric flow to the net heating excess in the tropics and the deficit at high latitudes discussed in the preceding chapter. Halley, in 1689, and Hadley, in 1735, both suggested the existence of such a circulation in order to account for the trade winds blowing towards the equator at the surface. Their work is of great historical importance, representing some of the first attempts to account for the global circulation in terms of simple physically based models.

The pattern of zonal wind $[\bar{u}]$ is closely related to the distribution of potential temperature $[\bar{\theta}]$. This is illustrated by Fig. 4.2, which shows the same contours of $[\bar{u}]$ as Fig. 4.1, but with contours of potential temperature, $[\bar{\theta}]$. The DJF and JJA cases are shown for completeness. Thermal wind balance, Eq. (1.53), holds to a good approximation for the zonal mean state. Hence, strong horizontal temperature gradients are related to strong vertical wind shears throughout middle and high latitudes. In the tropics, the horizontal temperature gradients are small; since f tends to zero also, the wind is not determined by the thermal wind relationship in the deep tropics.

Comparison of Figs. 4.1 and 4.2 shows that the low latitude Hadley circulation has ascent where the temperature is greatest, and descent where it is less. By the arguments of Chapter 3, such a circulation will generate kinetic energy; it is termed 'thermally direct'. Thermally direct circulation is also seen at high latitudes, especially in the winter southern hemisphere. In the midlatitudes, the mean meridional circulation is 'thermally indirect'. This axisymmetric thermally direct overturning is absent in the midlatitudes; there it is replaced by a thermally indirect circulation which is called the 'Ferrel cell'. Thermodynamically, this cell, characterized by descent in warm regions and ascent in colder regions, represents a sink of kinetic energy. It must be forced by some form of mechanical stirring. Its study led to a considerable emphasis being placed on the role of the midlatitude eddies in driving the circulation, especially by Victor Starr and his coworkers at MIT in the 1940s and 1950s.

The confinement of the Hadley circulation to the tropics is related to the rotation of the Earth. Nearly inviscid axisymmetric motion would tend to conserve angular momentum with the result that unrealistically strong zonal winds would develop in the upper troposphere poleward of $20°$ or so. Defining the specific angular momentum of a ring of air at latitude ϕ as $(\Omega a \cos \phi + [u]) a \cos \phi$, conservation of angular momentum for a ring whose zonal velocity is zero at the equator implies that the zonal wind at other latitudes is given by:

$$[u] = \Omega a \sin^2 \phi / \cos \phi. \qquad (4.1)$$

It can quickly be verified that this formula suggests winds of $56 \, \mathrm{m \, s^{-1}}$ at $20°$ and $127 \, \mathrm{m \, s^{-1}}$ at $30°$ of latitude. Nevertheless, a consideration of the diagram suggests that while angular momentum conservation is clearly a hopeless description of the zonal wind throughout middle and higher latitudes, the upper tropospheric wind in the tropics and subtropics does indeed vary in a way which is reminiscent of Eq. (4.1). It increases in a roughly quadratic fashion away from the equator, with a maximum of 35–$40 \, \mathrm{m \, s^{-1}}$ in the so-called subtropical jet. The fact that this subtropical jet maximum coincides with the poleward limit of the upper branch of

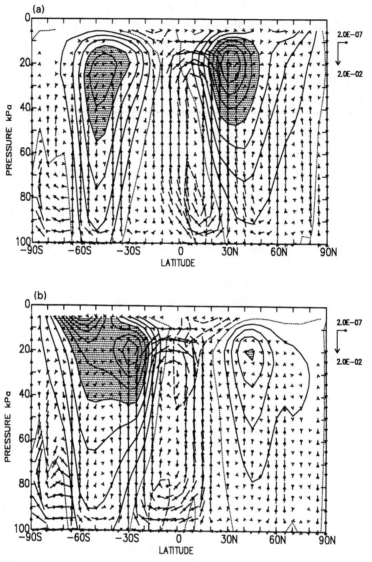

Fig. 4.1. The zonal mean wind $[\bar{u}]$ and vectors of the meridional wind for (a) December–January–February (DJF); (b) June–July–August (JJA).

the Hadley circulation implies that the Hadley cell can indeed be modelled crudely as an angular momentum conserving axisymmetric overturning.

Before describing a simple model of the Hadley circulation based upon these principles, it is worth remarking that the simple Hadley/Halley model of the circulation is now seen to be more realistic than diagrams such as Fig. 4.1 suggest. The data used for that diagram have been averaged in an

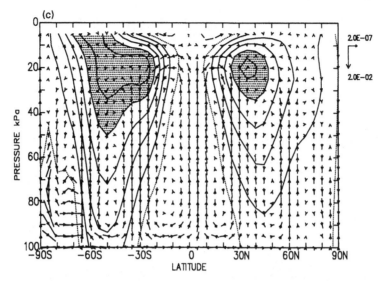

Fig. 4.1 (*cont.*). (c) The annual mean. Based on six years of ECMWF data. Contour interval $5\,\mathrm{m\,s^{-1}}$, values in excess of $20\,\mathrm{m\,s^{-1}}$ shaded. The horizontal sample arrow indicates a meridional wind of $3\,\mathrm{m\,s^{-1}}$, and the vertical sample arrow a vertical velocity of $0.03\,\mathrm{Pa\,s^{-1}}$.

Eulerian fashion, that is, a time series of the winds at a particular latitude and level have been summed to provide a seasonal mean wind. But a very different picture emerges if the averaging can be done in a Lagrangian way, by averaging the velocity of individual fluid elements as they circulate in the atmosphere. Such averaging is very difficult to carry out, and the current database is inadequate for the task. Nevertheless it is possible to carry out averaging which is more nearly Lagrangian. For example, Johnson has advocated the analysis of wind data on to surfaces of constant potential temperature before carrying out time and zonal averaging operations. Since potential temperature is generally conserved by fluid elements on timescales of less than five days or so, such isentropic averaging will follow fluid elements for short periods of time. In particular, it is able to track fluid elements through the typical depression systems of the midlatitudes. Figure 4.3 contrasts an analysis of the meridional streamfunction for a particular period during the FGGE, using traditional Eulerian averaging, with a similar analysis on θ surfaces. Allowing for the scaling of the vertical coordinate, the tropical pattern is fairly similar in both cases. But in the midlatitudes, where the transient depression systems are concentrated, they differ completely. The Ferrel cell is absent from the isentropic picture, which reveals instead

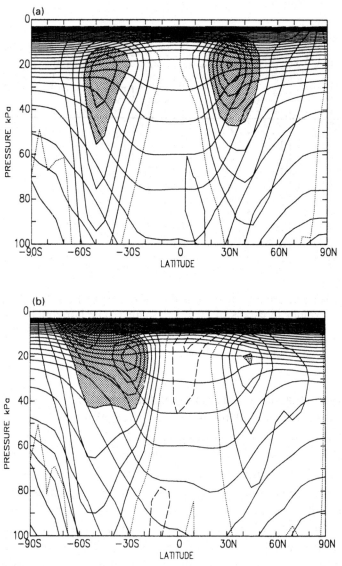

Fig. 4.2. Contours of $[\bar{u}]$ and $[\bar{\theta}]$ for (a) DJF; and (b) JJA, based on six years of ECMWF data. Contour interval for $[\bar{u}]$ as in Fig. 4.1. Contour interval for θ is 10 K.

that fluid elements do indeed circulate from tropical to polar regions in a thermodynamically direct fashion.

Angular momentum conservation does not hold, even approximately, for this circulation at higher latitudes. The reason is that local zonal pressure gradients associated with the midlatitude transients exert strong zonal torques on the atmosphere. So it is impossible to account for the

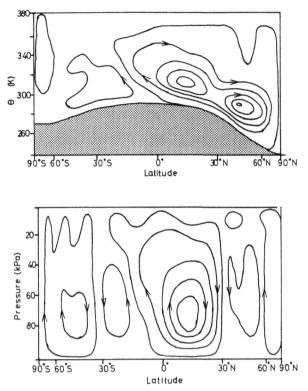

Fig. 4.3. Contrasting the meridional mass streamfunction for the January 1979 period using (a) quasi-Lagrangian averaging on potential temperature surfaces; and (b) traditional Eulerian averaging on pressure surfaces. (Redrawn from Townsend and Johnson 1985.)

zonal wind field associated with this 'diabatic circulation' without taking the transient disturbances fully into account. The effects of the eddies on the zonal flow will be discussed in Section 4.4. The origin of the eddies will be a major theme of Chapters 5 and 6.

4.2 The Held–Hou model of the Hadley circulation

Perhaps the simplest and most physically illuminating quantitative model of the Hadley cell was published by Held and Hou (1980). It uses the principles of angular momentum balance and thermal wind balance for circulating air parcels to predict both the latitudinal extent of the Hadley cell and its strength. Figure 4.4 shows the system envisaged by the model. It is essentially a two-level model of the tropical troposphere. Flow towards the equator takes place near the surface, where friction near the ground is

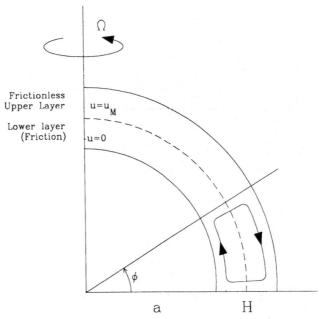

Fig. 4.4. Schematic illustration of the Held–Hou model.

supposed to reduce any zonal wind which develops as a result of angular momentum conservation to negligible values. The return poleward flow occurs at a height H above the ground. The thermal structure is described by a potential temperature θ at the middle level $H/2$.

The flow is driven by Newtonian cooling towards some radiative equilibrium temperature distribution $\theta_E(\phi)$ on a timescale τ_E. That is, the thermodynamic equation is written

$$\frac{\mathrm{D}\theta}{\mathrm{D}t} = \frac{(\theta_E - \theta)}{\tau_E} \tag{4.2}$$

and θ_E is taken to be

$$\theta_E(\phi) = \theta_0 - \frac{2}{3}\Delta\theta P_2(\sin\phi). \tag{4.3}$$

Here, $P_2(\sin\phi) = (3\sin^2\phi - 1)/2$ is the second Legendre polynomial, θ_0 is the global mean radiative equilibrium potential temperature and $\Delta\theta$ is the equilibrium pole–equator temperature difference. Such a distribution is smooth and continuous and preserves the important symmetry property that $\partial\theta/\partial\phi = 0$ at the poles and at the equator. In fact, the spherical geometry turns out merely to be a complicating factor in this analysis and introduces no new physical principle into the model; accordingly, we will assume that $\phi = y/a$ is so small that $\sin \phi$ can be replaced by y/a. Equation (4.3) is therefore conveniently rewritten

$$\theta_E(\phi) = \theta_{E0} - \frac{\Delta\theta}{a^2}y^2. \tag{4.4}$$

The actual temperature distribution will differ from this radiative equilibrium distribution, with advection by the air motion balancing the diabatic tendencies implied by Eq. (4.2). The essence of the Held–Hou model is in predicting the actual temperature from angular momentum balance considerations. Assume the wind at the upper level is given by Eq. (4.1) which in the small latitude limit becomes

$$U_M = \frac{\Omega y^2}{a}. \tag{4.5}$$

The subscript M reminds us that this is a zonal wind derived on the basis of angular momentum conservation. The low level zonal wind is taken to be zero as a result of friction. Then

$$\frac{\partial u}{\partial z} = \frac{U_M}{H} = \frac{\Omega y^2}{aH}. \tag{4.6}$$

But the vertical wind shear is related to the horizontal temperature gradient by the thermal wind relationship. Under the assumption of steady axisymmetric flow and hydrostatic equilibrium, thermal wind balance must hold even at low latitudes (see Eq. (1.53)). In the present notation, and using height as the vertical coordinate, it can be written

$$2\Omega \sin \phi \frac{\partial u}{\partial z} = -\frac{g}{\theta_0} \frac{\partial \theta}{\partial y}, \tag{4.7}$$

or substituting for $\partial u/\partial z$ from Eq. (4.6):

$$\frac{\partial \theta}{\partial y} = -\frac{2\Omega^2 \theta_0}{a^2 gH}y^3. \tag{4.8}$$

By integrating this relation, the actual potential temperature distribution required by this pattern of zonal wind is

$$\theta_M = \theta_{M0} - \frac{\Omega^2 \theta_0}{2a^2 gH}y^4. \tag{4.9}$$

Once more, the subscript M emphasizes that this potential temperature distribution was derived using angular momentum conservation arguments. The constant θ_{M0} is a constant of integration which has yet to be determined;

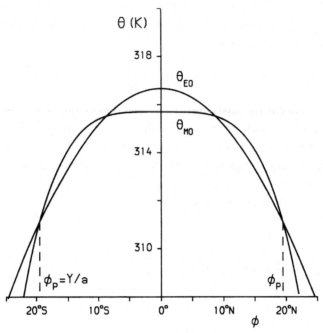

Fig. 4.5. Showing θ_E and θ_M as a function of poleward distance for the Held and Hou model. θ_{M0} must be chosen so that the areas between the two curves are equal, i.e., so that there is no net heating of air parcels.

it represents the equatorial temperature and we anticipate that the heat transport by the Hadley cell means that it will be less than θ_{E0}.

The distributions of θ_E and θ_M are compared in Fig. 4.5. The temperature profile is much flatter than the radiative equilibrium profile close to the equator. At higher latitudes, it falls off more rapidly. A suitable choice of θ_{M0} means that there is heating between the equator and the first crossing point of the curves and cooling between the first and second crossing points. At higher latitudes, heating is implied; clearly this is thermodynamically impossible and so we conclude that poleward motion ceases at this second crossing point, which must represent the poleward limit of the Hadley circulation; its latitude is denoted Y. At higher latitudes, $\theta = \theta_E$ in this strictly axisymmetric model. The latitude of the poleward edge of the cell, Y, is determined by choosing θ_{M0} so that there is no net heating of circulating air parcels, so that (from Eq. (4.2)):

$$\int_0^Y \theta_M \mathrm{d}y = \int_0^Y \theta_E \mathrm{d}y$$

or

$$\theta_{M0} - \frac{\Omega^2 \theta_0}{10a^2 g H} Y^4 = \theta_{E0} - \frac{\Delta\theta}{3a^2} Y^2, \tag{4.10}$$

assuming that τ_E does not vary with latitude. Furthermore, $\theta_M = \theta_E$ at $y = Y$, giving a second equation in Y and θ_{M0}:

$$\theta_{M0} - \frac{\Omega^2 \theta_0}{2a^2 g H} Y^4 = \theta_{E0} - \frac{\Delta\theta}{a^2} Y^2. \tag{4.11}$$

These two equations contain two unknown quantities Y and θ_{M0}. It is straightforward to solve for either of them; the resulting formulae are:

$$Y = \left(\frac{5\Delta\theta g H}{3\Omega^2 \theta_0} \right)^{1/2} \tag{4.12}$$

and

$$\theta_{E0} - \theta_{M0} = \frac{5\Delta\theta^2 g H}{18a^2 \Omega^2 \theta_0}. \tag{4.13}$$

Let $\theta_0 = 255\,\mathrm{K}$ and $\Delta\theta = 40\,\mathrm{K}$ (a typical observed value). Then we find $Y = 2200\,\mathrm{km}$ and $\theta_{E0} - \theta_{M0} = 0.8\,\mathrm{K}$. Comparison with Fig. 4.1 shows that the estimate of the Hadley cell width is at least roughly in agreement with observations, albeit slightly on the small side.

The model also leads to a picture of the upper tropospheric zonal winds associated with the Hadley circulation. For $y \leq Y$, the upper level zonal wind is simply equal to U_M, Eq. (4.5). For $y > Y$, the temperature is equal to the radiative equilibrium temperature, and the zonal wind can be calculated by applying thermal wind balance to θ_E; this wind may be denoted u_E and, when the small latitude approximation is applied, it is easily shown to be a constant, equal to $gH\Delta\theta/a\Omega\theta$. There is a discontinuity of the zonal wind at $y = Y$. This prediction has both realistic and unrealistic elements. It suggests the existence of a subtropical jet at the poleward edge of the Hadley cell. Indeed, Fig. 4.1 shows just such a jet. But the discontinuity of wind speed beyond the edge of the Hadley cell is not observed, and indeed would be violently unstable.

The difference between θ_{M0}, the actual equatorial temperature, and θ_{E0}, the radiative equilibrium equatorial temperature, can be used to estimate the meridional flow speeds associated with the Hadley circulation. On the equator, by symmetry, there must be a balance between vertical advection and heating, so that

$$w\frac{\partial\theta}{\partial z} = \frac{\theta_{E0} - \theta_{M0}}{\tau_E}$$

or

$$w = \frac{g}{\theta_0 N^2} \frac{\theta_{E0} - \theta_{M0}}{\tau_E}. \tag{4.14}$$

Here, N is the Brunt–Väisälä frequency of the atmosphere. Taking τ_E as 15 days, and assuming $N = 10^{-2}\,\mathrm{s}^{-1}$, we find that $w \simeq 0.24\,\mathrm{mm\,s}^{-1}$. From continuity, a typical horizontal velocity in the subtropical part of the Hadley cell will be

$$v \sim \frac{Y}{H} w. \tag{4.15}$$

With the present parameters, this works out to be $0.5\,\mathrm{cm\,s}^{-1}$. A comparison with Fig. 4.1 shows that the observed meridional winds in the Hadley cell are closer to $1\,\mathrm{m\,s}^{-1}$. So we may say that our simple theory has provided a reasonable estimate of the geometry of the Hadley cell, but a very poor estimate of its strength.

However, the Held–Hou theory is perhaps better than these simple estimates suggest. We have been considering the annual mean Hadley cell, which is symmetric about the equator. But during the solstices, the distribution of solar heating is asymmetric about the equator, with a maximum of heating in the summer hemisphere. The theory can fairly easily be generalized to this situation. The radiative equilibrium temperature distribution can be written:

$$\theta_E(\phi) = \theta_0 + \frac{\Delta\theta_{NS}}{2} \sin\phi + \Delta\theta_{EP}(3\sin^2\phi - 1), \tag{4.16}$$

where $\Delta\theta_{EP}$ is the mean pole–equator temperature difference and $\Delta\theta_{NS}$ is the temperature difference between the summer and winter poles. The ascending motion will now not be at the equator. Consequently, angular momentum conservation will lead to upper level easterlies over the equator and (by thermal wind balance) to maximum θ_M away from the equator at the latitude of maximum ascent. Figure 4.6 illustrates the forms of θ_E and θ_M in this asymmetric case. The lack of symmetry means that the maximum ascent need no longer coincide with the latitude of maximum radiative equilibrium temperature, nor with the streamline dividing the northern and southern hemisphere cells. Consequently, we have a more complicated system with four unknown parameters, determined from four matching conditions in the manner of Eqs. (4.12) and (4.13).

The results of these asymmetric calculations are summarized in Fig. 4.7. The cell rapidly becomes highly asymmetric as the radiative equilibrium maximum is moved off the equator, with a small weak cell in the summer hemisphere and a much stronger cell with ascent in the summer hemisphere

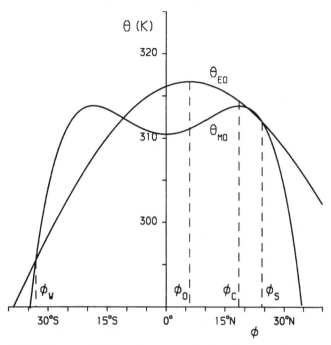

Fig. 4.6. The Held–Hou model for the case of a heating maximum away from the equator. The latitudes ϕ_s, ϕ_N and ϕ_D as well as the equatorial temperature θ_{M0} are to be determined.

and descent in the winter hemisphere. The mass flux carried by the two cells (proportional to the area between the θ_E and θ_M curves in Fig. 4.5) differs by a very large factor for even modest asymmetry of the heating, and we would expect the annual mean Hadley circulation to be dominated by the two winter cells. If this is so, our estimate of Y will be increased somewhat, but our estimates of w and v will be increased by an order of magnitude, bringing them more in line with the observations.

An important principle is illustrated by these calculations. The strength and character of the cells react in a highly nonlinear fashion to changing the latitude of the heating maximum. Consequently, the annual mean Hadley circulation is very different from the response to the annual mean forcing which we would predict. In various forms, this problem of 'nonlinear averaging' is a central difficulty in parametrizing many diabatic forcing processes in the atmosphere. It means that a complex general circulation model has to be run even if one is only interested in the annual and zonal mean circulation.

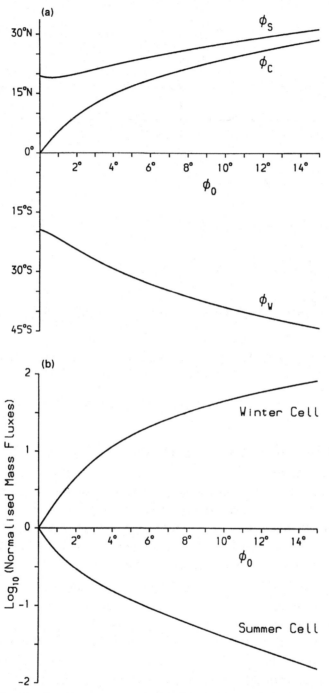

Fig. 4.7. Results for the asymmetric Hadley cell, assuming the same parameters as previously: (a) variation of ϕ_N, ϕ_S and ϕ_D as a function of ϕ_0, the latitude of maximum radiative equilibrium temperature; (b) variation of the mass flux carried by the winter and summer cells as a function of ϕ_0.

4.3 More realistic models of the Hadley circulation

There were two glaring omissions in the discussions in the preceding sections. The first was that no effects of friction in the upper layer of the atmosphere were included. Since the overturning timescale of the Hadley cell predicted by the model was several radiative timescales, even very weak dissipation could modify the flow significantly. The second was that the effects of latent heat release by condensation of water vapour were ignored. In fact, latent heat release dominates heating in the tropics. This is not to say that using such models is folly. Indeed, the aim of any scientific modelling is to separate crucial from incidental mechanisms. Comprehensive complexity is no virtue in modelling, but, rather, an admission of failure. The Held–Hou model is remarkable, not for the effects it omits, but because even with such minimal assumptions it reproduces so many observed features of the meridional circulation and the zonal wind field. But in this section, it is necessary to examine how much of an effect these various complicating factors might have.

We will examine the effects of friction using a variant of the 'simple global circulation model' introduced in Section 2.4. This variant is axisymmetric, with all variations in the longitudinal direction suppressed. A slightly more complicated form of friction is introduced, with the momentum equation written:

$$\frac{\partial u}{\partial t} + \mathcal{N}_M + \mathcal{L}_M = K \frac{\partial^2 u}{\partial z'^2}. \tag{4.17}$$

Here, $\mathcal{L}_M$ and $\mathcal{N}_M$ represent the linear and nonlinear dynamical terms, respectively, and K is a constant vertical diffusion coefficient. The diffusion term on the right hand side of this equation is a crude parametrization of the turbulent transports of momentum in the planetary boundary layer, and K is sometimes called an 'eddy viscosity' coefficient. The effect of such a diffusion term is to introduce a classical Ekman boundary layer into the flow, in which the depth of the boundary layer is given by

$$D = \left(\frac{2K}{f}\right)^{1/2}. \tag{4.18}$$

Noting that the midlatitude planetary boundary layer is around $1\,\text{km}$ in depth, it follows that a reasonable value for K is around $10\,\text{m}^2\,\text{s}^{-1}$. Such an Ekman boundary layer induces secondary circulations which dissipate vorticity in the fluid above the boundary layer by stretching or shrinking of vortex tubes. Figure 4.8 illustrates this. The typical timescale for this 'spinup'

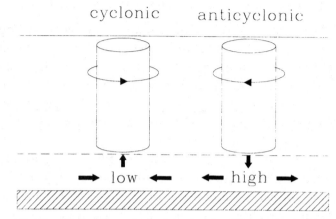

Fig. 4.8. The dissipation of vorticity by the Ekman boundary layer.

process is

$$\tau_D = \left(\frac{2H^2}{fK}\right)^{1/2}, \tag{4.19}$$

which is a few days for typical midlatitude conditions. In this way, if the overturning timescale of the Hadley cell is long compared to the spinup timescale, the strong meridional shears which develop in the vicinity of the subtropical jet in the Held–Hou model will be moderated by the presence of a midlatitude boundary layer.

The results of two integrations of the simple global circulation model with this friction term are shown in Fig. 4.9. The first has θ_E symmetric about the equator, as in the simplest form of the Held–Hou model. There is a weak Hadley circulation extending into the subtropics, with an associated subtropical jet. The zonal wind speeds are much less than those predicted by the frictionless model, and there is no discontinuity of wind associated with the edge of the Hadley cell. Indeed, the Hadley circulation does not have a sharp poleward boundary, although it does become very weak at high latitudes. The dimensions of the Hadley cell are very much in line with the predictions of the Held–Hou model. But the zonal winds generated by the circulation are considerably weaker. The second integration is for solstitial conditions, with the midlatitude temperature gradients around twice as large in the winter hemisphere as in the summer hemisphere. The maximum of θ_E is at $6\,°N$. The asymmetry between the summer and winter Hadley cells is marked, though not as extreme as in Fig. 4.7. The summer cell has only a weak subtropical jet at its poleward edge, consistent with the dominant effect of friction on such a slowly overturning flow. The winter cell has a much more vigorous circulation. The jet is correspondingly much stronger, though not as strong as it would be if the flow conserved angular momentum exactly, and it shows signs of very sharp shears on its poleward flank. As we will see in later chapters, this flow would be violently unstable if the axisymmetric assumption were relaxed. The resulting eddy fluxes of heat and momentum would quickly modify the jet profile into a more stable form.

Now consider the effects of moisture. Satellite images show that the Hadley cells in each hemisphere are separated by a band of cumulonimbus clouds. This band forms at the line of convergence where the low level trade winds originating in each hemisphere meet; it is sometimes called the intertropical convergence zone (or 'ITCZ'). Figure 4.10 shows a schematic diagram of the formation of the ITCZ and its relationship with the Hadley circulation. Descending air at the poleward edge of the Hadley cell reaches the boundary layer with an extremely low humidity. Air returns towards the equator in the low level flow, picking up heat and moisture from the underlying surface as it goes. When it meets the air from the opposite hemisphere, it is forced to rise. As it rises, it quickly becomes saturated, so that condensation and latent heat release take place. The release of latent heat is balanced by ascent, and so the ITCZ is characterized by deep convection extending through the depth of the troposphere. The principal effect of latent heat release is to concentrate most of the ascending motion of the Hadley circulation into the narrow ITCZ.

Calculating the nature of the Hadley circulation associated with this moist model would appear to be extremely difficult, since it involves various boundary layer exchanges of heat and moisture, as well as deep cumulus convection in the ITCZ. As we saw in Section 2.4, parametrizing such processes is a crucial but poorly resolved problem in global circulation modelling. However, placing some general bounds on the width and strength of the Hadley circulation is in fact more straightforward than might initially be expected.

Let us continue to work with the Held–Hou two-layer formulation, with the lower layer representing the moistening boundary layer, and the upper layer representing the frictionless poleward flow in the middle and upper troposphere. For simplicity, we return to the case of a Hadley circulation which is symmetric about the equator. The zonal wind in the upper layer will, as before, be determined by angular momentum conservation, and hence the variation of potential temperature with latitude will be given by θ_M, defined by Eq. (4.9). The cooling due to the radiation of long wave radiation to space can be related to θ_M/τ_E. Equally, in the absence of fluid motion, the

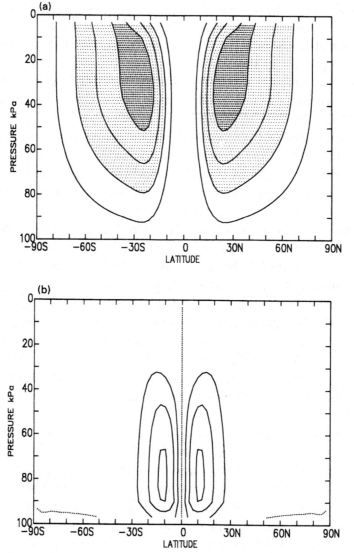

Fig. 4.9. The Hadley circulations and zonal winds set up by a simple axisymmetric global circulation model. (a) and (b) show a case for which θ_E is symmetric about the equator, and (c) and (d) for a case where the maximum of θ_E is located at about 10°N.

temperature would be θ_E, the radiative equilibrium potential temperature distribution, given by Eq. (4.4). The distribution of short wave radiative heating is simply θ_E/τ_E; the total heating is given by

$$H = \int_0^Y \frac{\theta_E}{\tau_E} dy. \tag{4.20}$$

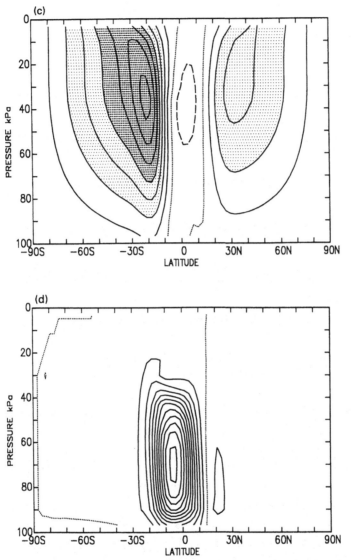

Fig. 4.9 (*cont.*). (b) and (d) show the mass streamfunctions with a contour interval of $5 \times 10^9 \, \mathrm{kg \, s^{-1}}$, while (a) and (c) show zonal winds, contour interval $5 \, \mathrm{m \, s^{-1}}$, heavy shading denoting values in excess of $20 \, \mathrm{m \, s^{-1}}$.

Now let us make the extreme assumption that all incoming solar radiation is used to evaporate water into the boundary layer. This heat is realized as latent heat of condensation in the ITCZ, where the total heat released will be H, Eq. (4.20). As in Section 4.2, the width of the Hadley cell can be determined by the requirement that the heating and cooling integrated over

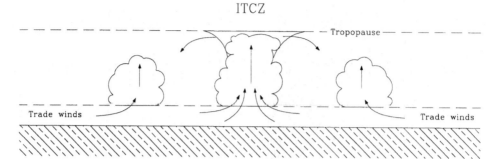

Fig. 4.10. Schematic illustration of the ITCZ and the Hadley circulation.

the width of the Hadley cell must balance, i.e., that

$$\int_0^Y \frac{\theta_M}{\tau_E} \mathrm{d}y = H = \int_0^Y \frac{\theta_E}{\tau_E} \mathrm{d}y. \qquad (4.21)$$

But we have met this equation before, as Eq. (4.10). It is identical to the zero net heating condition for the dry Held–Hou model. The other condition, namely $\theta_M = \theta_E$ at $y = Y$, is also identical in the two models. In this case, the inclusion of moisture leaves the width and mass flux of the Hadley cell unchanged; the difference from the Held–Hou model is simply that the ascent and heating is concentrated into a narrow region at the ITCZ while there is descent and cooling throughout the rest of the cell. If we made a less extreme assumption, that some of the incoming sunlight went to evaporate water, while the remainder led to sensible heating, we would still obtain the same result for the width and strength of the cell, but with some ascent over the tropical part of the cell and a stronger concentration of the ascent at the ITCZ. Figure 4.11 illustrates these possibilities.

To summarize, the effect of moisture is to introduce an asymmetry between large scale ascent and descent in the Hadley circulation. In the dry case, roughly equal areas of the cells are ascending and descending. As condensation and latent heat release become more important, the vertical motions become stronger and more concentrated, while the bulk of the cells are characterized by descent. The total circulation of the cells remains unchanged, however, because there is no direct change in the ultimate thermal forcing, which is provided by the incident flux of solar radiation. The final width of the ITCZ depends upon the structure of the individual cumulus towers which make it up; turbulent mixing between the strongly ascending towers and the surrounding clear air determines the size of the cumulus elements and hence the width of the ITCZ. Observations show that the meridional

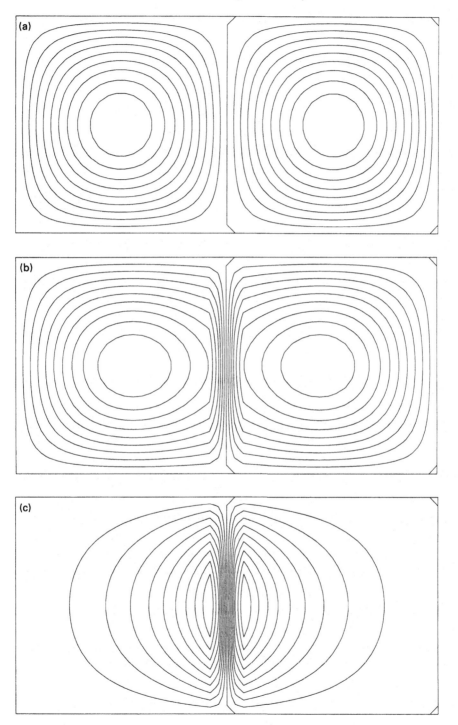

Fig. 4.11. Schematic illustration of the effects of including moisture in the Held–Hou Hadley cell model: (a) dry case; (b) some incoming energy evaporates water; (c) all incoming radiation evaporates water.

scale of the ITCZ is no more than 100 km or so; mesoscale processes, which lie beyond the scope of this book, are involved in determining this scale.

This reasoning suggests that the basic mechanism in the Held–Hou model, namely, angular momentum mixing and thermal wind balance, gives a general form of Hadley circulation which is insensitive to the details of how the heat enters the system. The actual location of the ITCZ and its strength will depend upon details of the tropical boundary layer, and the fluxes of moisture out of the surface; these processes need to be represented in GCMs and weather prediction models. Much of the flux of moisture into the boundary layer is controlled by the sea surface temperature. Because of the large thermal capacity of the ocean, this has a smaller seasonal cycle than the land, and so the asymmetry of the circulation about the equator is somewhat reduced by the moist processes. Figure 4.1 shows that it is nevertheless quite substantial. We will return for a further discussion of convection and heating processes on the large scale tropical circulation in Chapter 7. Meanwhile, we will move to higher latitudes, and consider the meridional circulation beyond the confines of the Hadley cell.

4.4 Zonal mean circulation in midlatitudes

Our study of the Hadley circulation of subtropical latitudes was simplified by the observation that eddy fluxes are small equatorward of 30° of latitude. So the circulation can be thought of as approximately axisymmetric at low latitudes. But in the midlatitudes, the eddies are large and the fluxes of heat and momentum that they carry are a major part of the global circulation. In describing the zonal mean circulation at these latitudes, we must recognize the role of the eddies explicitly. We then face a conceptual difficulty. We mentally partition the flow into eddies and a zonal mean part which we try to examine separately. But this distinction is artificial. The eddies induce changes in the mean flow which in turn affect the distribution and vigour of the eddies. In recent global circulation studies, a good deal of effort has been devoted to isolating the effects of eddies from the effects of other, genuinely axisymmetric, processes. In this section, we will discuss a traditional approach to the effect of eddies on the mean flow. We will suppose that the fluxes of heat and momentum associated with the eddies are prescribed (for example, by reference to observations) so that we can consider the response of the zonal mean flow to these forcings.

Our starting point is the quasi-geostrophic equation set, described in Section 1.7. The quasi-geostrophic vorticity equation, using the pressure as vertical coordinate, may be written:

$$\frac{\partial \xi_g}{\partial t} + \frac{\partial}{\partial x}(u\xi_g) + \frac{\partial}{\partial y}(v\xi_g) + \beta v = f\frac{\partial \omega}{\partial p} + F, \qquad (4.22)$$

where ξ_g is the relative geostrophic vorticity $v_{gx} - u_{gy}$, and $F = \mathcal{F}_{1y} - \mathcal{F}_{2x}$ is a schematic form of the friction term, with $\mathcal{F}_1$ and $\mathcal{F}_2$ the x- and y-components of acceleration due to friction respectively. Note that we will ignore the effects of the Earth's curvature, save through the βv term, at this stage. If we average with respect to the zonal direction, we may write:

$$\frac{\partial}{\partial t}[\xi_g] + \frac{\partial}{\partial y}[v\xi_g] + \beta[v] - f\frac{\partial}{\partial p}[\omega] = -\frac{\partial}{\partial y}[\mathcal{F}_1] \qquad (4.23)$$

or, from continuity,

$$\frac{\partial}{\partial t}[\xi_g] + \frac{\partial}{\partial y}[v\xi_g] + \frac{\partial}{\partial y}\{f[v]\} = \frac{\partial}{\partial y}[\mathcal{F}_1]. \qquad (4.24)$$

But since $[\xi_g] = -[u]_y$, we may write:

$$\frac{\partial}{\partial y}\left\{-\frac{\partial}{\partial t}[u] + [v\xi_g] + f[v] + [\mathcal{F}_1]\right\} = 0. \qquad (4.25)$$

Integrating with respect to y, and determining the constant of integration from the condition that $\partial[u]/\partial t = 0$ in the absence of friction, Coriolis acceleration or dynamical fluxes, we have a quasi-geostrophic version of the zonal mean momentum equation:

$$\frac{\partial}{\partial t}[u] + \frac{\partial}{\partial y}[vu] - f[v] = [\mathcal{F}_1]. \qquad (4.26)$$

The momentum flux $[uv]$ may be written as $[u][v] + [u^*v^*]$; scaling arguments such as those of Section 1.7 show that $u^* \sim [u]$, whereas $v^* \gg [v]$, so that the mean momentum flux may be neglected compared to the eddy flux. The final form of the zonal momentum equation that we will use is therefore:

$$[u]_t = f[v] - [u^*v^*]_y + [\mathcal{F}_1]. \qquad (4.27)$$

where the subscripts t and y indicate differentiation. Alternatively, this equation could have been derived by taking the zonal mean of Eq. (1.65a) and using quasi-geostrophic scaling arguments. The first term on the right hand side represents changing $[u]$ due to angular momentum conservation by axisymmetric meridional motion; the second represents accelerations due to eddy fluxes; and the last, of course, represents friction.

The thermodynamic equation is given as Eq. (1.71); on zonal averaging, it reduces to

$$[\theta]_t + [v^*\theta^*]_y = \frac{s^2}{h}[\omega] + [\mathcal{Q}]. \tag{4.28}$$

Once again, $[v]$ has been neglected compared to v^*, so that the poleward eddy potential temperature flux dominates over the vertical eddy potential flux or mean poleward flux. The vertical advection of the basic stratification dominates the mean advection.

Equations (4.28) and (4.27) are linked through the zonal mean form of the thermal wind equation. This is conveniently written in the form:

$$f[u]_p = h[\theta]_y. \tag{4.29}$$

The meridional circulation which is required to maintain a state of thermal wind balance despite the unbalancing tendencies of eddy temperature and momentum fluxes, friction and heating can be calculated from the set of Eqs. (4.27), (4.28) and (4.29). It is convenient to define a meridional streamfunction ψ such that

$$[v] = \frac{\partial\psi}{\partial p}, \quad [\omega] = -\frac{\partial\psi}{\partial y}, \tag{4.30}$$

where the sign convention has been chosen so that a maximum of ψ corresponds to a direct circulation, with ascent at low latitudes (small y) and descent (large y) at higher latitudes in the northern hemisphere. Then multiply the momentum equation by f and differentiate with respect to p, and multiply the thermodynamic equation by h before differentiating with respect to y; subtract the two equations to eliminate the derivatives by forming the balance condition:

$$\psi_{yy} + \frac{f^2}{s^2}\psi_{pp} = \frac{f}{s^2}[u^*v^*]_{yp} - \frac{h}{s^2}[v^*\theta^*]_{yy} - \frac{f}{s^2}[\mathcal{F}_1]_p + \frac{h}{s^2}[\mathcal{Q}]_y. \tag{4.31}$$

This elliptic equation relates the mean meridional circulation to a number of source terms. It can be solved if suitable boundary conditions on ψ are specified. We will apply $[v] = 0$ at bounding values of y (which may be taken to be the equator and pole) and $[\omega] = 0$ at $p = 0$ and $p = p_s$, that is, $\psi = $ constant (which may without loss of generality be taken to be zero) along the meridional and vertical boundaries. The elliptic operator means that a maximum in the source terms on the right of Eq. (4.31) is associated with a minimum in the meridional streamfunction, that is, with an indirect circulation in the northern hemisphere. Figure 4.12 illustrates this. Furthermore, the meridional streamfunction will reflect the larger

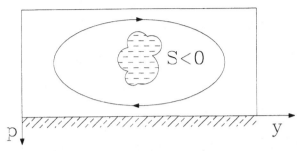

Fig. 4.12. Schematic illustration of the solution of Eq. (4.31).

scale features of the source terms while smoothing out their smaller scale structures.

Equation (4.31) is a diagnostic relationship. It states what meridional circulations must take place in order to maintain thermal wind balance when nonzero source terms are present. It is analogous to the ω-equation discussed in Section 1.7 and, indeed, may be derived from a zonal average of the ω-equation which is then integrated with respect to y.

Consider a simple application of this diagnostic relationship. Suppose that friction and eddy fluxes may be ignored, leaving only the source term associated with heating. The heating is taken to be positive for small y and negative for large y, with a maximum of $-[\mathcal{Q}]_y$ in midlatitudes. Hence, the source term $-(h/s^2)[\mathcal{Q}]_y$ will be negative throughout the midlatitudes. ψ will therefore have an associated maximum value in midlatitudes, implying a thermally direct circulation in which air rises where the heating is large and descends where it is small, as illustrated in Fig. 4.13. The interpretation of this result needs care, however. The buoyancy terms have been scaled out of the quasi-geostrophic set, so it is not a simple matter of warm air rising. Rather, the heating generates horizontal pressure gradients which result in meridional acceleration of the air. Continuity then implies ascent at low latitudes and descent at high latitudes.

This meridional circulation accomplishes two things. First, the ascent at low latitudes results in upward advection of potentially colder air, tending to offset the temperature rises due to heating. A converse argument holds at high latitudes. Thus, the temperature gradient increases by less than we might expect if the atmosphere were unable to circulate. Second, the Coriolis forces acting on the poleward moving air at upper levels impart a westerly acceleration to the flow, while at low levels an easterly acceleration occurs. In this way, the wind shear is increased so that a thermal wind balance with the evolving temperature field is maintained.

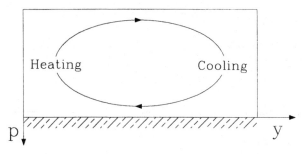

Fig. 4.13. The meridional circulation generated in response to a gradient of heating.

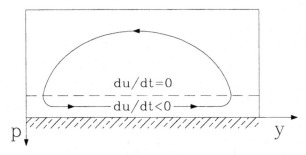

Fig. 4.14. The meridional circulation induced by surface friction.

As a second example, let us explore the effect of friction. In the midlatitudes, surface winds are observed to be westerly. We expect that friction will impart a strong easterly acceleration to the flow at low levels (large p) but will be less effective at higher levels. Hence, the source term will be positive and, this time, an indirect circulation will be forced. Figure 4.14 illustrates the argument. In physical terms, a flow is induced from low to high latitudes at low levels. The Coriolis force acting on this poleward flow produces westerly acceleration, offsetting the easterly acceleration which results directly from the friction. At the same time, descent causes warming in the tropics, while ascent cools the higher latitudes. In this way, the temperature field is brought back into thermal wind balance with the increased vertical wind shear at the top of the boundary layer.

The friction induced circulation illustrates an important principle concerning the role of boundary layer friction on the global circulation. To the south of the jet core, where the vorticity associated with the zonal wind is negative, the friction induces descent. To the north of the jet core, the relative vorticity is positive and friction induces ascent. Such boundary layer pumping is important in leading to the rapid spin up of the flow in the upper troposphere. For simple forms of the friction, analytic relationships between the vertical pumping velocity at the top of the boundary layer and the interior vorticity can be derived. For example, when the friction term is of the form, Ku_{zz}, K being a constant 'eddy viscosity' coefficient, the vertical pumping velocity is

$$w = \xi \left| \frac{K}{2f} \right|^{1/2}. \qquad (4.32)$$

The resulting vortex compression or extension will dissipate relative vorticity on a time scale $(2H^2/fK)^{1/2}$, H being a typical height scale (i.e., the troposphere depth). Such a boundary layer is called an 'Ekman layer'. It occurs readily in laboratory experiments with rotating tanks of fluid in which the boundary layer flow is laminar, but is a rather crude model of the turbulent atmospheric boundary layer. However, even with more realistic parametrizations of the boundary layer friction, a qualitatively similar pumping still takes place. The Ekman boundary layer model provides a rationale for the Rayleigh friction in the simple global circulation models which was introduced rather arbitrarily in Section 2.4.

Now let us consider the role of the eddy flux terms in Eq. (4.31). The poleward eddy temperature fluxes are largest at lower levels in the midlatitudes (see Section 5.2 for some examples). In the region of maximum flux, it follows that $[v^*\theta^*]_{yy}$ must be negative. The temperature flux will therefore contribute a positive source term to the right hand side of the circulation equation, Eq. (4.31), and will force an indirect circulation. The result, shown schematically in Fig. 4.15, is easily appreciated if the reader has followed the argument given above for the effect of a gradient of heating. The eddies tend to shift heat from low latitudes to higher latitudes. There will therefore be a poleward gradient of heating associated with the eddies. An indirect circulation is therefore required to reduce the upper level westerlies relative to the lower level winds and so maintain thermal wind balance. At the same time, the vertical motions will offset the heating by the eddies.

The pattern of eddy momentum fluxes is more complex. The transient momentum fluxes for the northern hemisphere winter season are small in the lower troposphere, but increase with height, with maximum values near the tropopause. At this level, there is a general convergence towards latitudes around 50°N, with poleward flux to the south and equatorward flux at more northerly latitudes. Thus $[u^*v^*]_{yp}$ is generally positive in midlatitudes, leading to a positive source term in the circulation equation and hence to an indirect circulation. Thus, the tendency of the eddy fluxes to accelerate the westerly flow at midlatitudes near the tropopause is opposed by equatorward flow at

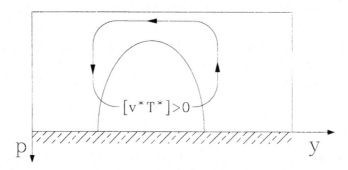

Fig. 4.15. The meridional circulation induced by poleward eddy temperature fluxes.

upper levels, and poleward flow at low levels. The effect is to reduce the vertical shear and make the flow more barotropic. The same effect becomes extremely important in the decay stages of the lifecycle of a midlatitude depression (see Section 5.5) when it leads to a massive build up of barotropic kinetic energy.

From these examples, a general rule of thumb emerges. In the quasi-geostrophic framework, any forcing of the zonal mean flow induces a meridional circulation which offsets the effect of that forcing. This is true whether the forcing is a direct result of friction or heating, or due indirectly to eddy transports. At the same time, the zonal accelerations or temperature changes induced by the circulation are such as to restore thermal wind balance.

For a typical midlatitude location, we conclude that the typical observed patterns of eddy fluxes and friction will all induce indirect circulations. The diabatic heating will induce direct circulation. There is the possibility that there could be a good deal of cancellation between the various source terms. Let us estimate the typical meridional winds induced by the midlatitude eddies. Ignoring all effects save those of the poleward temperature flux, and assuming that ψ_{yy} and $\partial(f^2\psi_p/s^2)/\partial p$ are comparable, then we have, from Eq. (4.31):

$$\psi \simeq \frac{h}{2s^2}[v^*\theta^*]. \tag{4.33}$$

Midlatitude winter values of $[v^*\theta^*]$ have a maximum of around $16\,\mathrm{K\,m\,s^{-1}}$, while $s^2/h \equiv -\partial\theta_R/\partial p$ is typically around $5 \times 10^{-4}\,\mathrm{K\,Pa^{-1}}$. The typical maximum value of ψ is therefore $2 \times 10^4\,\mathrm{Pa\,m\,s^{-1}}$. The poleward component of the zonal mean wind is estimated to be

$$[v] \simeq \psi/\Delta p, \tag{4.34}$$

where Δp is the typical pressure difference between the ground and the midtroposphere. The poleward wind induced by the midlatitude eddy temperature flux is therefore around $0.3 \, \mathrm{m \, s^{-1}}$. Similar reasoning can be applied to the circulation induced by heating; in this case

$$\psi \simeq \frac{L}{2} \frac{h}{s^2} [\mathscr{Q}], \tag{4.35}$$

where L is typical horizontal distance. Typical values of $[\mathscr{Q}]$ in the troposphere are around $1.5 \, \mathrm{K \, day^{-1}}$ and so $[v]$ can again be estimated; a reasonable value is $0.3 \, \mathrm{m \, s^{-1}}$. Thus the thermally direct and indirect components of the midlatitude circulation tend to cancel out. All this proves is that a much more careful calculation is needed if the direction of circulation is to be estimated. Careful numerical solutions of Eq. (4.31) reveal that, indeed, the observed Ferrel circulation is driven by the midlatitude eddy temperature and momentum fluxes.

4.5 A Lagrangian view of the meridional circulation

The discussion in the preceding section was couched in traditional Eulerian terms. That is, fluid properties were measured above some fixed point on the Earth's surface, and the zonal means were calculated around each latitude circle. In this way, we were able to distinguish between the tropical, thermally direct, Hadley circulation, and the indirect Ferrel circulation of midlatitudes. But we should be aware that this analysis may give quite a different result from averaging in a Lagrangian sense, that is, by tracing the motions of individual fluid elements. It turns out that this distinction is very acute in the midlatitudes; indeed, the Ferrel circulation may be regarded as an artifice of Eulerian averaging, and is a misleading description of the mean circulation if (for example) we wish to discuss the advection of tracers around the atmosphere.

Historically, this distinction first showed up in attempts to account for the observed distribution of ozone in the lower stratosphere. The formation of stratospheric ozone is a result of photolysis of ordinary diatomic oxygen molecules by the ultraviolet components of sunlight (see Section 9.3). This process is most effective at heights of around 50 km in the upper troposphere and will proceed most rapidly where there is plenty of sunlight, that is, in the tropics and in the summer hemisphere. Yet the maximum concentrations of ozone are observed in the lower stratosphere, near the pole and in the early spring. Clearly, advection must deposit the ozone in these regions. A thermally direct circulation (now known as the 'Brewer–Dobson' circulation) with descent near the winter pole, was postulated to account for the spring maximum. Yet the winter stratosphere is characterized by a disturbed westerly jet at around 60 °N (the 'polar night jet'). The arguments of the preceding section would suggest a thermally indirect Ferrel type of zonal mean circulation, with ascent near the pole. There is of course no contradiction between these views. The ozone is virtually a passive tracer in the lower stratosphere and its distribution indeed indicates the Lagrangian circulation of fluid parcels. The Ferrel circulation is an artifice based on Eulerian averaging. In Eulerian terms, we must say that the poleward eddy fluxes of ozone more than compensate for the equatorward, mean meridional transports.

To explore the Lagrangian circulation of the atmosphere, let us consider an idealized midlatitude jet such as that shown in Fig. 4.16. In the upper troposphere, fluid moves through the waves from west to east. The waves transport heat polewards, so there must be northward advection of warm air and equatorward advection of colder air. To deduce the Lagrangian trajectory of fluid particles in the meridional plane, we must consider the vertical velocity field associated with the wave. Application of the ω-equation, Eq. (1.76), shows that there will be ascent associated with the ridges of the wave. This upward advection of potentially colder air partially offsets the heating due to the poleward advection of warm air. Similarly, descent is associated with the troughs. Consider a parcel of air initially at point A in the ridge. At this point, it will experience rising motion. As it moves further east, it leaves the ridge, and starts to move equatorwards. At the same time, it moves into a region of weaker ascent. At point B, the ascent is zero. As it moves further into the trough, the parcel begins to descend, with maximum descent at point C. Finally, it moves back through point D to its original latitude and level. The projection of this motion onto the meridional plane is shown in Fig. 4.16(b). The parcel describes a clockwise, elliptical orbit in the y–p plane.

By itself, this motion implies no mass transport. Individual parcels of air describe small elliptical orbits but will return to their initial pressure and latitude at the end of each wave period. To deduce the mass circulation, it is necessary to consider higher order effects than are included in this simple model. The eddies are observed to have maximum amplitude in the midlatitudes in the vicinity of the tropopause; at other levels and latitudes they die away, as indicated in Fig. 4.17. The diagrams in Section 5.1 show some observations of the eddy kinetic energy, a convenient measure of the eddy activity. Now consider a parcel of air to the south of the maximum of the eddy activity. As it orbits in the meridional plane, it moves from a region of slightly lower eddy activity to a region of slightly higher eddy

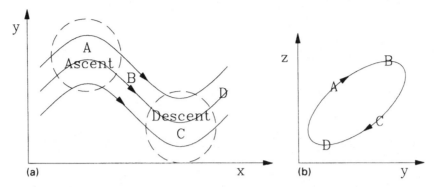

Fig. 4.16. (a) Schematic view of the passage of a fluid parcel through a sinuous jet in the upper troposphere. (b) Projection of the same fluid trajectory on to the y–p plane.

activity. Hence, it will rise slightly more quickly in the northern part of its orbit than it sinks in the southern part. The result is shown schematically in Fig. 4.17; each successive orbit of the parcel will be slightly higher than its predecessor. To the north of the maximum, the opposite effect will occur and the orbits will descend. This slow drift of the fluid parcel over many periods is called 'Stokes drift' and it is well known in the case of surface waves in shallow water. The total mass transport is the sum of the Stokes drift and the indirect Ferrel circulation induced by the eddy temperature and momentum fluxes, and, clearly, there will be a degree of cancellation between the two components.

It is not easy to give simple dynamical arguments concerning which effect will dominate. We will not present a detailed quantitative argument in this book; the interested reader is referred to texts such as that by Andrews *et. al.* (1987) for a more advanced account of these matters. However, we note that the discussion of Chapter 3 led us to conclude on thermodynamic grounds that a net thermally direct circulation is required to maintain the kinetic energy associated with the general circulation against the dissipative effects of friction. So we anticipate that the Stokes drift will prove stronger than the induced indirect circulation. In fact, careful analysis shows that in 'nonacceleration conditions', the Stokes drift balances the Ferrel circulation exactly, so there is no net mass circulation. Nonacceleration conditions require the waves to be frictionless, adiabatic and steady. In the midlatitude troposphere, none of these conditions holds, and it can be deduced that a substantial direct circulation of mass must exist. It is this circulation which helps to account for the spring maximum of ozone in the lower stratosphere. Recognizing this thermally direct mass flux is important in discussing the

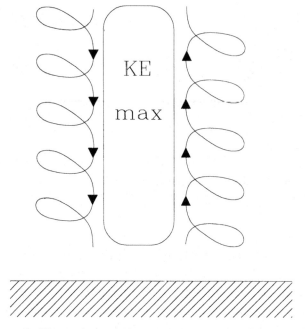

Fig. 4.17. Schematic illustration of the tropospheric distribution of eddy kinetic energy, and the Stokes drift of air parcels to north and south of the storm track latitudes.

transport of any conserved tracer (which may include potential vorticity), and is needed to complete a thermodynamically consistent account of the global circulation.

4.6 Problems

4.1 Using the data in Fig. 4.1, estimate typical values of $[v]$ in the Hadley cell. Hence estimate the time taken for a parcel of air to circulate in the Hadley cell. Compare this time to the seasonal timescale.

4.2 Repeat this calculation for the Ferrel circulation.

4.3 Compare θ at $100\,\text{kPa}$ and $30\,\text{kPa}$ at selected latitudes. Estimate the Brunt–Väisälä frequency N at these latitudes, where

$$N^2 = \frac{g}{\theta}\frac{\partial \theta}{\partial z}.$$

Discuss which processes might determine N.

4.4 By comparing the wind speed at the tropopause with the poleward temperature gradient at selected latitudes, verify that the $[u]$ and $[\theta]$ fields are in thermal wind balance.

4.5 Using the data in the following table, calculate the radiative equilibrium temperatures at the pole, equator and for the globe. Use the results to confirm the estimate of Y which was derived following Eq. (4.12).

	Albedo	Insolation W m^{-2}
Globe	0.29	344
Pole	0.72	182
Equator	0.27	436

4.6 The annual mean precipitation over the Pacific is about 3000 mm per year, concentrated in a strip $7°$ of latitude wide. Estimate the latent heat release implied and compare this with the heating rate at the equator according to the Held–Hou model. Use this estimate to evaluate w. Hence, deduce the typical meridional wind speed in a Hadley cell which includes condensation.

4.7 From Fig. 4.7(a), showing the variation of ϕ_w, ϕ_s, etc., plot a graph showing the strength of the subtropical jet as a function of y_0.

4.8 Estimate typical values of the static stability parameter s^2 in the troposphere. Derive a relationship between s^2 and the Brunt–Väisälä frequency N. State the units of s^2.

4.9 Using the cross sections of $[\overline{v'T'}]$ and $[\overline{u'v'}]$ from Figs. 5.5 and 5.7, estimate the magnitude of terms typical of the forcing of the mean meridional circulation by midlatitude heat and momentum fluxes in the winter midlatitudes. Which term dominates? Estimate a typical poleward velocity $[v]$ induced by midlatitude eddies, and compare this with the meridional velocity associated with the Ferrel cell, as shown in Fig. 4.1.

4.10 It has been suggested that breaking, vertically propagating gravity waves have an effect equivalent to a drag on the zonal flow, with a maximum value just above the midlatitude tropopause. Using the meridional circulation equation, Eq. (4.31), *sketch* the nature of the circulation you expect to be induced by breaking gravity waves. What deceleration due to breaking gravity waves would be required to induce a meridional wind $[v]$ of $1 \, \mathrm{m \, s^{-1}}$ at the breaking level at 50°N?

5

Transient disturbances in the midlatitudes

5.1 Timescales of atmospheric motion

The circulation of the atmosphere is intrinsically unsteady; fluctuations on all timescales are observed. In the last chapter, it was shown that the fluxes of temperature, momentum, and so on, carried by such transients play an important part in determining the time mean circulation of the atmosphere. Our task in this chapter will be to describe the transients on various timescales, and to discuss the mechanisms which can give rise to transient behaviour. Just as the atmosphere contains a wide range of spatial scales, from the molecular to the global, so the atmospheric circulation exhibits a wide range of timescales, ranging from timescales of just a few seconds associated with the overturning of small turbulent eddies, to geological timescales for major climate changes.

Some of the frequencies observed are directly related to the frequencies of periodic forcings. For example, diurnal and semi-diurnal variations of temperature and wind are associated with the diurnal variation of solar heat input. These 'thermal tides' are important at high levels in the atmosphere and can be detected in the lower atmosphere. More importantly for our purposes, the annual cycle of radiative forcing has a profound effect on large scale atmospheric circulation. This seasonal cycle of meteorological quantities affects nearly all parts of the globe.

But in addition to these externally imposed periods, the atmospheric flow itself generates all kinds of timescales internally. Various wave motions have characteristic frequencies, while irregular, aperiodic fluctuations arise from the chaotic, quasi-turbulent character of much atmospheric flow. Why such irregular fluctuations arise in the first place is an intriguing question. The superficial answer is that the time mean flow is unstable to small amplitude disturbances. But how is the mean flow maintained in an unstable state? Why do the developing instabilities not wipe out the shears and temperature gradients which make instability possible, leaving the atmosphere in a quiescent state of near neutral stability? These are questions to which we will return in Section 7.4.

In this chapter, we will examine these 'transient' features of the atmospheric flow. We will discuss how they are generated and how they contribute to the transport of heat and other quantities across the globe. We will attempt to illustrate how individual weather systems contribute to the global circulation of the atmosphere. In the last chapter, we noted that the Hadley circulation reduced the temperature gradients in the tropics, but that it actually increased the temperature gradient and horizontal wind shears in the midlatitudes. The transients of the midlatitudes are responsible for reducing these temperature gradients and transporting heat and momentum out of the subtropical latitudes and into the higher latitudes.

Before turning to a more detailed discussion of the dominant scales and shapes of atmospheric transients, consider the frequency and wavenumber characteristics of the observed transients. We recall the notation introduced in Chapter 2, in which the time mean of any quantity Q is $\overline{Q}$, while the deviation from the time mean is denoted Q'. The kinetic energy associated with the transient eddies is:

$$K = \frac{1}{2}\left(\overline{u'^2 + v'^2}\right). \tag{5.1}$$

The pressure–latitude section in Fig. 5.1 shows the zonal mean distribution of eddy kinetic energy on all timescales. The transients are small in the tropics but become much more important in the midlatitudes, with maximum values near the tropopause and somewhat poleward of the subtropical jet core. The transients become weaker towards the pole. The winter season has stronger transients than the summer; this seasonal cycle is a good deal more marked in the northern hemisphere. Larger values are also seen in the stratosphere, close to the equator and at high latitudes in the winter hemisphere.

The transient eddy kinetic energy varies not only with height and latitude, but also with longitude. In the northern hemisphere winter, maxima of the eddy kinetic energy are located over the western side of the ocean basins, with minima over North America and Asia. The pattern is shown in Fig. 5.2a. In the southern hemisphere winter, a single maximum is observed over the south Atlantic and Indian Oceans, with lower values over the Pacific. These maxima are coincident with the regions where developing and mature midlatitude depressions occur most frequently. On the eastern side of the

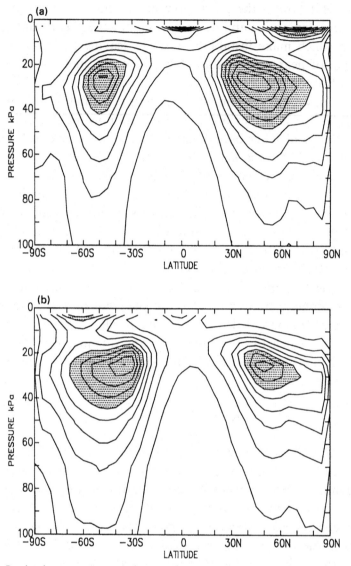

Fig. 5.1. Latitude–pressure sections of transient eddy kinetic energy for (a) December–February and (b) June–August. Contour interval $25\,\mathrm{m^2s^{-2}}$; shading indicates values in excess of $300\,\mathrm{m^2s^{-2}}$. Based on six years of ECMWF data.

ocean basins, systems tend to be occluded and decaying, and so are much less vigorous over the continents.

This pattern is accentuated considerably if the time series of each velocity component is filtered so as to remove the lower frequencies before constructing the variances. In general, numerical filtering of a time series involves replacing the nth member of the time series with a suitably weighted average of the neighbouring members:

$$Q_n^F = \sum_{i=-j}^{j} w_i Q_{n+i}. \tag{5.2}$$

The properties of the filter are defined by the choice of the filter weights w_i. For example, a crude low pass filter, which removes the higher frequencies but leaves the lower frequency variance to survive intact, results if the weights are constant and simply equal to $1/(2j+1)$. In this case, the mean is a simple running mean of the time series. Conversely, by defining Q' simply to be the deviation from some such running mean, a crude high pass filter is established. More sophisticated filters are designed to produce an extremely sharp response, so that only a precisely defined range of frequencies is passed. But since the time spectrum of atmospheric motions is smooth and continuous with no obvious spectral gaps, there is little physical reason to use a particularly elaborate filter for our purposes. The high pass filtered eddy kinetic energy (shown in Fig. 5.2b) was constructed by the use of a running mean filter such as that suggested above. It had the effect of removing much of the variance associated with transients with periods longer than about six days. The filtered transient eddy kinetic energy is a good deal less than the unfiltered equivalent, but the maxima are very clearly defined. They form zonally elongated areas running across the Atlantic from the North American coast, and across the Pacific from the Asian coast. Similar maxima are found if other measures of transient eddy activity, such as geopotential height variance or low level temperature flux, are plotted. The tracks of developing depression centres tend to run along the long axes of these areas, for which reason they are commonly called 'storm tracks'. These storm track regions can be seen in the heating fields in Fig. 3.8, especially in the Northern Hemisphere winter. Chapter 7 contains a fuller discussion of the midlatitude storm tracks.

The higher frequency transients of the midlatitudes, by which we mean transients with periods between one and ten days or so, owe their existence to a hydrodynamic instability of the zonal flow called 'baroclinic instability'. In the first part of this chapter, we will discuss the mean size and structure of the transient eddies. Then we will discuss the energy associated with the transients, and how it can be generated. The theory of baroclinic instability will be introduced in Section 5.4, while the extensions and limitations of this theory will be described in Section 5.5.

The lower frequency transients have a more complex origin, which is

(a)

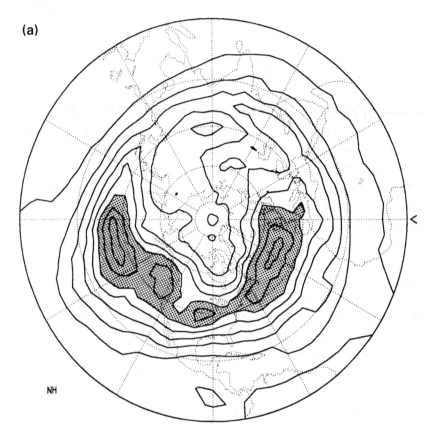

Fig. 5.2. Transient eddy kinetic energy at 25 kPa for the period December–February, northern hemisphere. (a) Total transient eddy kinetic energy, contour interval 50 m^2 s^{-2} with values in excess of 300 m^2 s^{-2} shaded.

still imperfectly understood. Low frequency behaviour results in part from the internal dynamics of the atmospheric flow, particularly the nonlinear interactions between different scales of motion which lead to a cascade of energy towards lower frequencies. But low frequencies are also forced externally, by interactions between the atmospheric circulation and more slowly varying systems, such as the ocean. As the frequency becomes lower, the internal dynamics is generally supposed to become less important, while the role of external forcing becomes dominant. Describing low frequency variability is difficult because the data records are not generally long enough to produce statistically reliable patterns of variances and covariances for the lower frequencies. Gaining greater insight into the underlying mechanisms

(b)

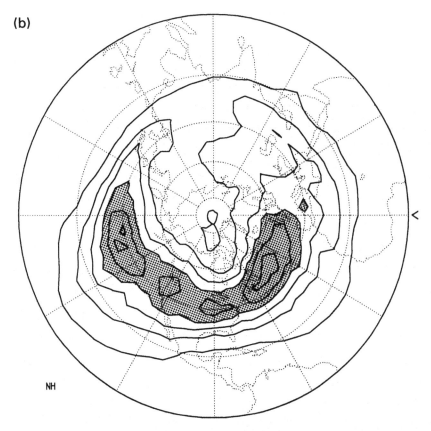

Fig. 5.2 (*cont.*). (b) High frequency transient eddy kinetic energy, contour interval 25 m^2 s^{-2} with values in excess of 150 m^2 s^{-2} shaded. The eddies contributing to (b) have periods less than about 6 days. Based on six years of ECMWF data.

is a goal of much current research. We will look in more detail at low frequency variability in Chapter 8. In the remainder of this chapter, we will focus on the higher frequency transients.

5.2 The structure of transient eddies

Scale analysis shows that large scale motions in the Earth's midlatitudes are nearly nondivergent; the geostrophically balanced rotational part of the wind dominates over the small, mainly divergent ageostrophic wind. The ratio of the geostrophic to ageostrophic wind speeds is of order Rossby number. So it is useful to model the horizontal structure of transient eddies by a geostrophic streamfunction which is supposed to vary sinusoidally in the x- and y-directions:

$$\psi' = \Psi e^{i(kx+ly)}. \tag{5.3}$$

More properly, Eq. (5.3) should be regarded as one component of a Fourier summation over all permitted k and l. In this chapter, we will interpret the equation as describing the typical transient eddy; the values of k and l are regarded as some kind of average for all the observed transients. We may regard the amplitude Ψ as a sinusoidal function of time; in that case, the average over a complete period of the wave will be the same as the average over a complete wavelength. The dimension of the individual vortices is half a wavelength in each direction, that is, π/k in the x-direction and π/l in the y-direction.

The northward and eastward components of the wind are then related to the streamfunction by:

$$u' = -\frac{\partial \psi'}{\partial y} = -il\Psi e^{i(kx+ly)}, \tag{5.4a}$$

$$v' = \frac{\partial \psi'}{\partial x} = ik\Psi e^{i(kx+ly)}. \tag{5.4b}$$

The appearance of i in these expressions indicates that the velocity wave and the streamfunction wave are $\pi/2$ out of phase. The variance of u and v indicate the typical magnitudes of u' and v'. The streamfunction is not measured directly, but it is closely related to the geopotential height Z, which can be observed and analysed. The geostrophic streamfunction is related to the geopotential height by:

$$\psi = \frac{g}{f}Z. \tag{5.5}$$

Hence, the velocity variances and the geopotential height variance are related by:

$$\overline{u'^2} = \frac{l^2 g^2}{f^2}\overline{Z'^2}, \ \overline{v'^2} = \frac{k^2 g^2}{f^2}\overline{Z'^2}, \tag{5.6}$$

assuming that f is constant over the width π/l of the eddy. Thus, a comparison of velocity variance and geopotential height variance will lead to an estimate of the typical wavenumber of the transients, that is, to their spatial scale. A comparison of the u- and v-variance leads to a measure of the typical shape of the eddies, since:

$$\frac{\overline{v'^2}}{\overline{u'^2}} = \frac{k^2}{l^2}. \tag{5.7}$$

If $\overline{u'^2} > \overline{v'^2}$, the eddies will be zonally elongated; if, on the other hand, $\overline{v'^2} > \overline{u'^2}$, they will be meridionally elongated.

Zonal means of $\overline{u'^2}$, $\overline{v'^2}$ and $\overline{Z'^2}$ are shown in Fig. 5.3. Taking typical values for the midlatitude upper troposphere, $\overline{v'^2} = 280\,\mathrm{m^2\,s^{-2}}$, $\overline{u'^2} = 250\,\mathrm{m^2\,s^{-2}}$ and $\overline{Z'^2} = 4 \times 10^4\,\mathrm{m^2}$. Thus the eddies are slightly elongated meridionally. The typical wavenumber is around $1 \times 10^{-6}\,\mathrm{m^{-1}}$; this corresponds to zonal wavenumber 5 at $40°$ of latitude. When filtering is applied, this picture changes. The high frequency eddies are distinctly meridionally elongated and their scale is somewhat smaller.

So far, we have considered eddies with their longer axes oriented either east–west or north–south. There is no reason why they should not be orientated at some intermediate angle. The orientation is related to the poleward flux of momentum carried by the wave. Qualitatively, this can be seen by considering Fig. 5.4, which shows a schematic view of an eddy which tilts from south-west to north-east. Along section AB, both the u and v components of eddy velocity are large and positive; the eddy is carrying westerly momentum northwards. Along section CD, easterly momentum is being carried southwards. The net effect of both these sections is to transport westerly momentum northwards. The shorter sections BC and DA are characterized by weaker winds and by southward transport of westerly momentum. In the limit of extremely elongated tilted eddies, these two sections would make a negligible contribution to the momentum transport by the eddy. Averaging over the entire wavelength, it can be seen that there is net poleward momentum transport, that is, that $\overline{u'v'}$ is positive for such a tilted eddy. By the same arguments, an eddy tilted in the opposite direction would transport westerly momentum southwards. There would be no net transport of momentum by an untilted or exactly circular eddy, since the contributions from the various segments would exactly cancel in those cases.

The poleward momentum flux is closely related to the orientation of the eddies. Figure 5.5 shows latitude–pressure cross sections of the poleward momentum fluxes for the DJF and JJA seasons. The eddy momentum flux is largest in midlatitudes near the tropopause. The momentum flux is poleward at lower latitudes, but tends to be equatorward at higher latitudes, suggesting that the eddies are orientated in opposite directions to north and south of the main cyclone belt. Such a configuration is especially clear in the southern hemisphere, but is also important in the northern hemisphere. This distribution of momentum flux implies, using the arguments of Section 4.4, that an indirect Eulerian mean circulation is associated with the observed eddy momentum fluxes.

The typical angle of tilt of the eddies can be calculated if the momentum

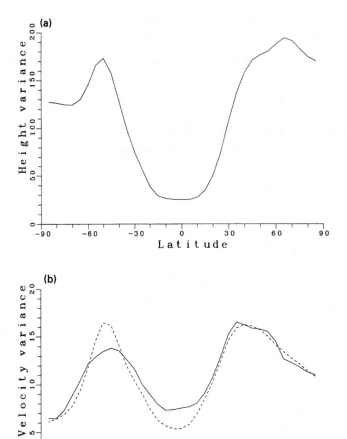

Fig. 5.3. Showing (a) zonal mean height variance $[\overline{Z'^2}^{1/2}]$ and (b) the variance of velocities $[\overline{u'^2}^{1/2}]$ (solid) and $[\overline{v'^2}^{1/2}]$ (dashed).

flux and the velocity variances are known. Let us calculate the velocity components in a frame of reference, denoted by $(\tilde{x}, \tilde{y})$, which is rotated through some angle φ relative to the basic (x, y) frame of reference. Using the usual formulae for rotation of coordinates, the transformed eddy velocities are

$$\tilde{u}' = u' \cos \varphi - v' \sin \varphi, \tag{5.8a}$$

$$\tilde{v}' = u' \sin \varphi + v' \cos \varphi. \tag{5.8b}$$

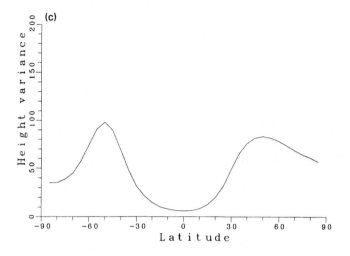

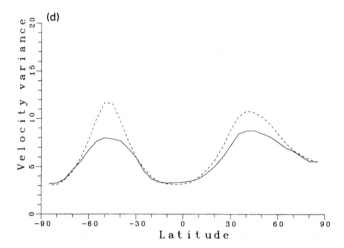

Fig. 5.3 (*cont.*). (c) and (d) as (a) and (b) respectively, but for the high frequency eddies. Based upon six years of ECMWF data at 25 kPa, DJF season.

Then the momentum flux in the rotated frame is:

$$\overline{\tilde{u}'\tilde{v}'} = \frac{1}{2}\left(\overline{u'^2} - \overline{v'^2}\right)\sin 2\varphi + \overline{u'v'}\cos 2\varphi. \tag{5.9}$$

When φ is chosen so that the eddy has no tilt in the rotated frame of reference, the transformed momentum flux $\overline{\tilde{u}'\tilde{v}'} = 0$. Hence the angle of tilt will be given by:

$$\tan 2\varphi = \frac{2\overline{u'v'}}{\overline{v'^2} - \overline{u'^2}}. \tag{5.10}$$

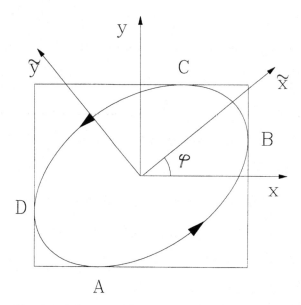

Fig. 5.4. Schematic illustration of a tilted eddy.

This is also satisfied for $\varphi + \pi/2$, indicating that, of course, the eddy has both a major and a minor axis.

The unfiltered transients have $\overline{v'^2} \simeq \overline{u'^2}$, so that $\varphi = 45°$. The high pass filtered eddies have $\overline{u'v'}$ typically of 25 m^2 s^{-2} in the upper troposphere. This implies that they must have a tilt of about 26°. However, the properties of the eddies do vary from place to place. In Fig. 5.6, the anisotropy of the eddies and the orientation of their major axes are shown. The 'anisotropy' will be defined in Section 7.4 (Eq. (7.15) *et seq*); it is simply a dimensionless number ranging from 0 for perfectly circular eddies to 1 for infinitely elongated eddies. The unfiltered eddies are most anisotropic in the tropics, a manifestation of low frequency tropical disturbances, the theory of which will be discussed in Section 7.1. The high frequency eddies are meridionally elongated in the midlatitudes, in the so-called 'storm track' latitudes.

We now turn to an examination of the poleward eddy heat transport due to the transients. This is related to the vertical structure of the eddies. Figure 5.7 shows cross sections of the transient eddy temperature fluxes. The fluxes are poleward in both hemispheres, and are largest in the lower troposphere at midlatitudes. The transient temperature flux is unimportant throughout the tropics, where the heat transport is mainly accomplished by the Hadley cell and by ascent in small scale convective systems. The maximum zonal mean temperature flux in the troposphere is around 15 K m s^{-1}, with somewhat

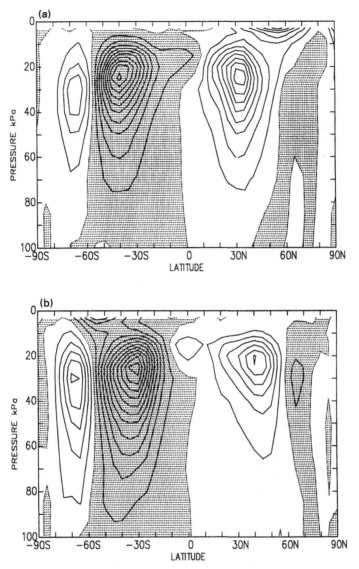

Fig. 5.5. Latitude–pressure cross sections of the poleward transient eddy momentum fluxes observed in (a) the DJF and (b) the JJA seasons. Contour interval 5 m^2 s^{-2}, negative values shaded. Based on six years of ECMWF data.

larger values in the winter stratosphere. The divergence of the temperature flux gives an estimate of the heating or cooling due to transient eddies; this is typically 0.5 K day^{-1}, equivalent to 60 W m^{-2}. The seasonal cycle in the temperature flux is marked in the northern hemisphere, but less so in the southern.

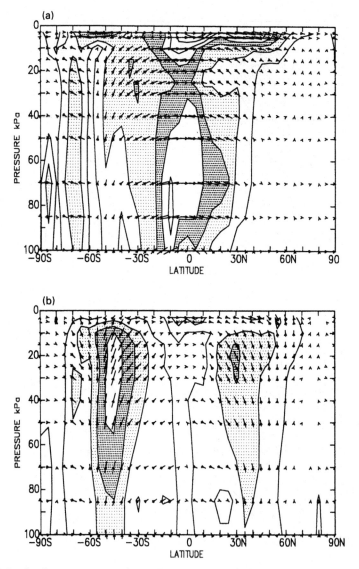

Fig. 5.6. Latitude–pressure sections showing the eddy anisotropy and the orientation of the mean major axes of the eddies. Contours show the eddy anisotropy, contour interval 0.1. Shading indicates values between 0.2 and 0.4. Vectors indicate the orientation of the major axes of the eddies, vertical vectors indicating a north–south orientation. (a) Unfiltered transient eddies, (b) High frequency transient eddies. Based on six years of ECMWF data, DJF season.

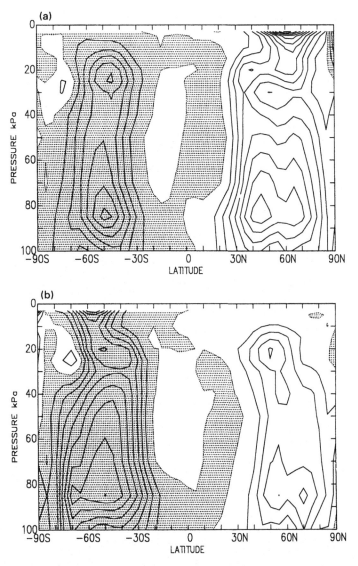

Fig. 5.7. Latitude–pressure sections showing the poleward transient eddy temperature flux, $[\overline{v'T'}]$ for (a) DJF and (b) JJA. Contour interval 2 K m s^{-1}, negative values shaded. Based on six years of ECMWF data.

Now imagine an idealized geostrophically and hydrostatically balanced disturbance. The geometry of the situation envisaged is shown in Fig. 5.8. Three surfaces at pressure $p - \Delta p$, p and $p + \Delta p$ are shown. The geostrophic streamfunction is related to the geopotential height by

$$\psi' = \frac{g}{f} Z' \tag{5.11}$$

so that the wind components are

$$u' = -\frac{g}{f}\frac{\partial Z'}{\partial y}, \quad v' = \frac{g}{f}\frac{\partial Z'}{\partial x}. \tag{5.12}$$

The temperature on the pressure surface is related to the thickness of a layer by the hydrostatic relation; it follows that the temperature on level 0 is:

$$T_0' = \frac{p_0 g}{R}\frac{Z_1' - Z_{-1}'}{2\Delta p}. \tag{5.13}$$

Suppose that the geopotential height varies sinusoidally in x. Furthermore, suppose that the amplitude is constant with height, but that there may be a change of phase with height. We may write

$$Z_i = Z_{Ri} + A\sin(kx + i\delta), \quad i = -1, 0, 1, \tag{5.14}$$

where A is the amplitude of the wave. A sign convention has been chosen so that positive δ means the wave is tilting to the west with height. Then the eddy temperature field at level 0 is

$$T_0' = \frac{p_0 g}{\Delta p R} A\cos(kx)\sin(\delta). \tag{5.15}$$

No fluctuation of temperature with x occurs unless there is a phase tilt. The poleward wind is:

$$v' = \frac{kg}{f} A\cos(kx). \tag{5.16}$$

Note that the temperature and poleward velocity waves are in phase. Hence the poleward temperature flux is:

$$[\overline{v'T'}] = \frac{1}{2}\frac{p_0 g^2 A^2 k}{fR\Delta p}\sin\delta. \tag{5.17}$$

Inspection of Fig. 5.8 makes this result intuitively obvious. The westward vertical phase tilt means that thickness, and hence temperature, is a maximum where the poleward wind is a maximum. Thus there must be a net poleward temperature flux. If the phase tilt were reversed, the wave would transport heat equatorwards. Careful attention to the sign convention for y reveals that these statements are equally true in the southern hemisphere.

An example of the use of Eq. (5.17) is to calculate the vertical phase tilt of a typical transient from the $\overline{Z'^2}$ and $\overline{v'T'}$ fields. A collection of circulation statistics reveals that at 70 kPa and 45 °N, the high frequency height variance $\overline{Z'^2}^{1/2}$ is 35 m, the meridional wind variance is 20 m^2 s^{-2} and the poleward temperature flux is 4 K m s^{-1}. From Eq. (5.6), the zonal wavelength is found to be 1070 km. Substituting values into Eq. (5.17), it is found that the phase

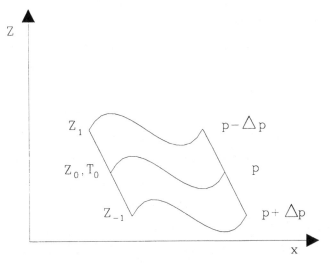

Fig. 5.8. Schematic illustration of a wave-like disturbance with westward phase tilt with height.

difference between the 90 kPa level and the 50 kPa level is about 12° of longitude.

As well as transporting heat polewards, the transient eddies also transport heat vertically. Indeed, the demands of thermal wind balance ensure that there is a close linkage between poleward and vertical temperature fluxes. Figure 5.9 demonstrates the way in which this must happen. If an eddy is characterized by warm advection at one longitude, and cold advection at another, temperature gradients will develop in the zonal direction. This in turn means that to maintain thermal wind balance, the upper level velocity field must change. Such accelerations can be generated by ageostrophic circulations in the height–longitude plane. The resulting flow is correlated with the temperature field in such a way as to lead to a net upward flux of temperature. We will show in the next section that such vertical temperature fluxes are required if transient eddies are to develop spontaneously in a zonal flow.

As discussed in Section 2.4, modern analysis and initialization methods lead to a fairly reliable estimate of the unobservable vertical velocity field. This can be used to calculate the vertical temperature flux associated with the transient eddies. A cross section of the zonal mean transient vertical temperature flux, $[\overline{\omega'T'}]$, for the DJF season, is shown in Fig. 5.10. The most striking feature of this plot is the way in which the distribution of the flux echoes that of the poleward temperature flux, in just the way that the

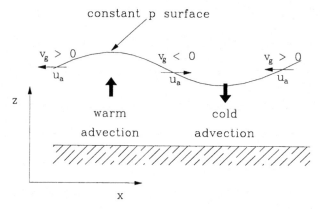

Fig. 5.9. Schematic longitude–height section through a wave which is transporting heat polewards. If thermal wind balance is to be maintained, the warm and cold advections associated with the eddy lead to vertical circulations which transport heat upwards.

arguments of the preceding paragraph would suggest. The flux is almost uniformly upwards, with only small regions of very weak downward flux away from the main storm track regions. Put another way, the temperature flux vector has a rather constant gradient, pointing polewards and upwards. The typical gradient is around 1 in 1000.

5.3 Atmospheric energetics

In order to gain some insight into the processes which generate the transient eddies observed in the midlatitude atmosphere, we will present an approach to large scale atmospheric energetics. We have already discussed in Chapter 3 some general thermodynamic arguments which showed that a net thermally direct circulation is required to generate kinetic energy in the atmosphere. In this section, we will present a discussion based on the quasi-geostrophic dynamics of the large scale flow. This will, of course, confirm the deduction from thermodynamics (which we would expect, since our equations include the basic laws of thermodynamics) but will additionally give insight into the types of dynamical structures needed to release the energy which thermo-dynamics shows is available. The approach is a traditional one, pioneered by Lorenz in the 1950s. The mathematical details will be expounded only briefly since they are available in many other text books.

The kinetic energy of a unit mass of the atmosphere is simply $(u^2 + v^2)/2$ (we will ignore the very tiny increment of kinetic energy associated with vertical motions). Then the total atmospheric kinetic energy $\mathscr{K}$ is obtained

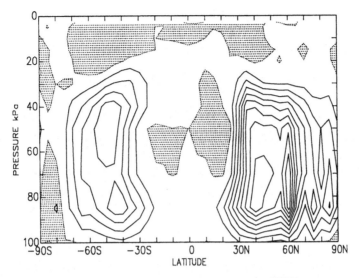

Fig. 5.10. Cross section of the vertical temperature flux, $[\overline{\omega' T'}]$, for the DJF season. Contour interval $0.02 \, \mathrm{K \, Pa \, s^{-1}}$, downward (positive) fluxes shaded. Based on six years of ECMWF analyses.

simply by integrating over the entire atmosphere. Using pressure as a vertical coordinate, we have

$$\mathscr{K} = \frac{1}{g} \int_A \int \int_0^{p_s} \frac{u^2 + v^2}{2} \mathrm{d}p \mathrm{d}x \mathrm{d}y, \qquad (5.18)$$

where $\int \mathrm{d}p/g$ represents an integration with respect to mass; the domain A is taken to be the entire surface of the globe, or least some fraction of that area so extensive that there is no advection of air into or out of it. The result of such an integration is generally awkwardly large, so it is conventional to divide $\mathscr{K}$ by the area of the Earth's surface $4\pi a^2$, giving a mean energy per unit area of the Earth's surface. We will use the notation:

$$\frac{1}{4\pi a^2 g} \int_A \int \int_0^{p_s} Q \mathrm{d}p \, \mathrm{d}x \, \mathrm{d}y = \langle Q \rangle, \qquad (5.19)$$

Q being any meteorological quantity, as a convenient shorthand. Having defined the global kinetic energy, consider how it might be altered. Consistent with the quasi-geostrophic approximation, the momentum equation may be written:

$$\frac{\partial \mathbf{v}}{\partial t} + \mathbf{v} \cdot \nabla \mathbf{v} + \omega \frac{\partial \mathbf{v}}{\partial p} + f \mathbf{k} \times \mathbf{v} = -\nabla \Phi - \frac{\mathbf{v}}{\tau_D} \qquad (5.20)$$

(see Eqs. (1.65a) and (1.65b)). Note that the horizontal advection term can be rewritten as $\nabla(\mathbf{v} \cdot \mathbf{v}/2) = \nabla K$, where K is the local kinetic energy per unit mass of the atmosphere. The last term is an idealization of real frictional processes, in the form of a Rayleigh drag. Take the scalar product of this equation with the horizontal components of the wind:

$$\frac{\partial K}{\partial t} + \mathbf{v} \cdot \nabla K + \omega \frac{\partial K}{\partial p} = -\mathbf{v} \cdot \nabla \Phi - \frac{2K}{\tau_D} \tag{5.21}$$

which is an energy equation for the flow. Note that the Coriolis force can do no work, since it acts perpendicularly to the velocity. The last term states that the kinetic energy is dissipated at twice the rate of the momentum. We are left to deal with the vertical advection of kinetic energy, and the horizontal advection of the geopotential. Consider the latter. After using the continuity equation,

$$\mathbf{v} \cdot \nabla \Phi = \nabla \cdot (\Phi \mathbf{v}) + \frac{\partial}{\partial p}(\omega \Phi) - \omega \frac{\partial \Phi}{\partial p}. \tag{5.22}$$

But $\partial \Phi / \partial p$ is related to the potential temperature through the hydrostatic relation, Eq. (1.68):

$$\frac{\partial \Phi}{\partial p} = -h(p)\theta. \tag{5.23}$$

Hence, writing the advection terms in flux form, and incorporating Eqs. (5.22) and (5.23) the kinetic energy equation becomes:

$$\frac{\partial K}{\partial t} + \nabla \cdot (\mathbf{v}(K + \Phi)) + \frac{\partial}{\partial p}(\omega(K + \Phi)) = -h\omega\theta - \frac{2K}{\tau_D}. \tag{5.24}$$

Now integrate the kinetic energy equation over the globe to give an expression for the rate of change of global kinetic energy. By the divergence theorem, the integral with respect to area of the second term can be written:

$$\int_A \int \nabla \cdot (\mathbf{v}(K + \Phi)) \, \mathrm{d}A = \oint \mathbf{v}(K + \Phi) \cdot \mathrm{d}\mathbf{l} \tag{5.25}$$

where A denotes the horizontal area of the atmosphere and L denotes the circuit around the edge of the domain. But this is zero, since there is supposed to be no inflow or outflow across the boundaries (or, more elegantly for a spherical world, it is zero because $\mathbf{v}$ and $(K + \Phi)$ satisfy periodic boundary conditions). Similarly, the vertical boundary conditions are $\omega = 0$ at $p = p_s$ and $p = 0$, so that the vertical integral of the third term is zero. So after some algebraic labour, the final expression for the rate of change of global kinetic energy is found to be rather straightforward:

$$\frac{\mathrm{d}}{\mathrm{d}t}\langle K \rangle = \langle -h\omega\theta \rangle - \langle 2K/\tau_D \rangle. \tag{5.26}$$

But this is simply the statement made in Chapter 3 on thermodynamic grounds; kinetic energy is generated if there is a net correlation between ascent and positive temperature anomalies, and vice versa. That is, there must be a preponderance of thermally direct circulations. This generation of kinetic energy may be offset by the frictional dissipation of kinetic energy.

In order to generate this kinetic energy, we must have depleted some potential energy. But the potential energy of the atmosphere is not a very useful concept. It would be minimized by compressing all the air into an infinitely thin layer near the ground. But such a compression is physically unrealistic. It would require enormous amounts of work to be done against pressure forces, and this work would go to increase the internal energy, that is, the temperature, of the air. Only if this internal energy could be radiated steadily away to space would the potential energy be reduced, and it would have gone, not into kinetic energy, but into thermal radiation. In other words, most of the potential energy associated with the mass distribution in the atmosphere is unavailable for conversion into kinetic energy. That part of the energy which is available for conversion, the 'available potential energy', is a small fraction of the total. The available potential energy is defined as that part of the potential energy which can be released without any change of the internal energy of the atmosphere. From the first law of thermodynamics, and remembering that the total mass of the atmosphere does not change, it must be released by adiabatic mixing of air parcels.

Formally, the available potential energy is defined as the maximum amount of potential energy which could be destroyed by adiabatic mixing of the atmosphere. Figure 5.11 illustrates the concept. Whether such mixing is in fact possible depends upon the dynamic constraints to which the system is subject. In Fig. 5.11(a), the isentropes are not parallel to the geopotentials. We may imagine adiabatic processes which could exchange the wedge of potentially warmer air marked A for the wedge of potentially colder air marked B. The result would be to replace the mass of air B with somewhat less dense air from A, while the air mass A would be replaced by denser air from B. The mean mass distribution would have been lowered slightly and so a certain amount of potential energy would have been destroyed. Once the isentropes are parallel to the geopotentials, no further adiabatic exchanges of air masses with different potential temperatures are possible. Accordingly, the potential energy released in going from the state shown in Fig. 5.11(a) to the state in Fig. 5.11(b) is indeed the 'available potential energy'.

These arguments are restricted to the usual situation of a statically stable atmosphere, in which the static stability is not supposed to vary on pressure surfaces. This is a consistency requirement for a quasi-geostrophic descrip-

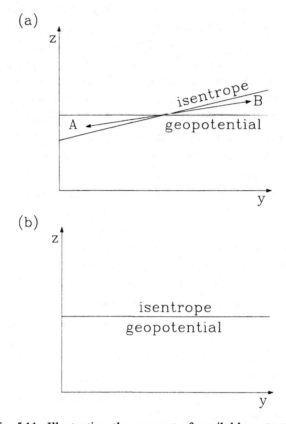

Fig. 5.11. Illustrating the concept of available potential energy.

tion of the dynamics. More elaborate definitions of available potential energy which overcome these restrictions have been proposed, and are mentioned in the bibliography.

We may now proceed, with the benefit of a degree of hindsight, to a more formal mathematical definition of available potential energy in the quasi-geostrophic framework. The thermodynamic equation can be written:

$$\frac{\partial \theta}{\partial t} + \nabla \cdot (\mathbf{v}\theta) = \frac{s^2}{h}\omega + \mathcal{Q} \tag{5.27}$$

in a form which is consistent with the quasi-geostrophic approximation. The potential temperature anomaly θ_A is defined as

$$\theta_A = \theta - \theta_R(p), \tag{5.28}$$

where θ_R is the reference atmosphere, consisting of the global mean θ at each level. Note that since θ_R does not vary in time or in the horizontal, θ

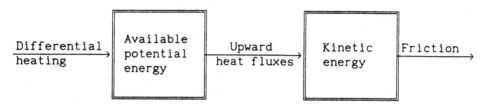

Fig. 5.12. The generation, conversion and dissipation of energy in the global circulation.

can be replaced by θ_A in Eq. (5.27). Take the product of this equation with $h^2\theta_A/s^2$, giving:

$$\frac{\partial}{\partial t}\left(\frac{h^2}{2s^2}\theta_A^2\right) + \nabla \cdot \left(\mathbf{v}\frac{h^2}{2s^2}\theta_A^2\right) = h\omega\theta_A + \frac{h^2}{s^2}\theta_A \mathscr{Q}, \qquad (5.29)$$

and integrate over the globe. As in the discussion of the kinetic energy equation, the global integral of the second term is zero by virtue of the boundary conditions, and so we have:

$$\frac{\mathrm{d}}{\mathrm{d}t}\left\langle\frac{h^2}{2s^2}\theta_A^2\right\rangle = \langle h\omega\theta_A\rangle + \left\langle\frac{h^2}{s^2}\theta_A\mathscr{Q}\right\rangle. \qquad (5.30)$$

The first term on the right hand side is equal in magnitude but opposite in sign to the kinetic energy generation term in Eq. (5.26). Therefore, $\langle h^2\theta_A^2/2s^2\rangle$ is the quantity depleted as kinetic energy is generated; accordingly, it must be an expression for the available potential energy A. It is seen that A depends essentially on the temperature variance on a given geopotential surface, consistent with the schematic picture given in Fig. 5.11. Equation (5.30) states that available potential energy is depleted by net thermally direct circulations, in which upward motion is correlated with warm temperature anomalies. Available potential energy is generated in circumstances in which the heating accentuates warm anomalies and cools cold anomalies.

Putting Eqs. (5.26) and (5.30) together, an idealized picture of the global circulation emerges, shown in Fig. 5.12. Differential heating creates available potential energy, which is converted to kinetic energy by thermally direct circulations. Friction balances the energy budget by dissipating kinetic energy. As it stands, this description adds very little to the results obtained in Chapter 3 on general thermodynamic grounds. A more useful form of energetics is obtained when the kinetic and available potential energies are split into their zonal and eddy parts.

Consider, first, the zonal part of the available potential energy. The zonal mean of the thermodynamic equation is:

$$[\theta]_t + [\bar{v}][\bar{\theta}]_y = \frac{s^2}{h}[\bar{\omega}] - [\overline{v^*\theta^*}]_y + [\bar{\mathcal{Q}}], \qquad (5.31)$$

where the subscript notation is used to denote partial differentiation and the eddy terms have been put on to the right hand side. For brevity, the eddy fluxes have been written in a form which includes both the transient and steady eddy contributions, since, to a good approximation:

$$[\overline{v^*\theta^*}] = [\bar{v}^*\bar{\theta}^*] + [\overline{v'\theta'}]. \qquad (5.32)$$

The zonal available potential energy, denoted AZ, is obtained by multiplying Eq. (5.31) by $h^2[\theta]/s^2$ and integrating over the entire globe. The multiplication yields:

$$\frac{\partial}{\partial t}\left(\frac{h^2[\bar{\theta}]^2}{2s^2}\right) + [\bar{v}]\frac{h^2}{s^2}\frac{\partial}{\partial y}[\bar{\theta}]^2 = h[\bar{\theta}][\bar{\omega}] + \frac{h^2}{s^2}[\bar{\theta}][\overline{v^*\theta^*}]_y + \frac{h^2}{s^2}[\bar{\theta}][\bar{\mathcal{Q}}]. \quad (5.33)$$

The second term on the left hand side integrates to zero, by arguments similar to those used above which make use of the boundary conditions on $[v]$. The global average of the second term on the right can be rewritten in the more convenient form:

$$\left\langle \frac{h^2}{s^2}[\bar{\theta}][\overline{v^*\theta^*}]_y \right\rangle = -\left\langle \frac{h^2}{s^2}[\bar{\theta}]_y[\overline{v^*\theta^*}] \right\rangle. \qquad (5.34)$$

This identity is easily established by integrating by parts with respect to y and using the conditions that $v^* = 0$ at the edges of the domain (or, equally, that the boundary conditions on all variables are cyclic in y). Finally, from Eq. (5.33) we can show that:

$$\frac{d}{dt}(AZ) = \langle h[\bar{\theta}][\bar{\omega}]\rangle - \left\langle \frac{h^2}{s^2}[\bar{\theta}]_y[\overline{v^*\theta^*}] \right\rangle + \left\langle \frac{h^2}{s^2}[\bar{\theta}][\bar{\mathcal{Q}}] \right\rangle. \qquad (5.35)$$

With the exception of the second term, all the terms are very straightforwardly related to the zonal mean version of the various terms in Eq. (5.30). This second term is a new effect; it represents the conversion of zonal into eddy available potential energy by the action of an eddy component of meridional wind in the presence of a basic poleward temperature gradient.

   The difference between Eqs. (5.35) and (5.30) yields an expression for the rate of change of the eddy available potential energy, AE, where

$$AE = \left\langle \frac{h^2}{2s^2}[\overline{\theta^{*2}}] \right\rangle. \qquad (5.36)$$

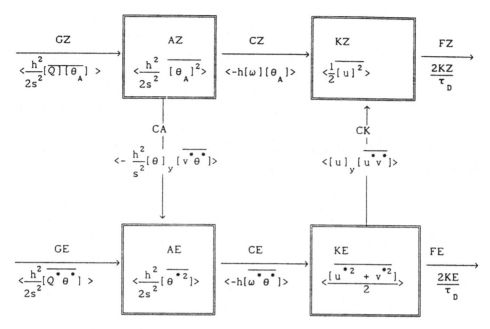

Fig. 5.13. The Lorenz energy cycle, indicating the notation and the form of the various energies and conversion terms.

Similar arguments can be used to split the kinetic energy into its zonal mean and eddy parts. It would be tiresome to give all the algebraic detail (which may be found in other text books); instead the results are summarized in Fig. 5.13. The first derivation of this energetics scheme was due to Lorenz and so it is sometimes called the 'Lorenz energy cycle'. The conversion between zonal and eddy kinetic energy is analogous to the conversion between zonal and eddy available potential energy. It can be written in a form involving the correlation between the horizontal wind shear and the poleward momentum flux.

The directions and relative magnitudes of the energy flow around the Lorenz diagram can be used to summarize the dominant processes giving rise to motions in the atmosphere. The simplest example is provided by Hadley cell circulations, as shown in Fig. 5.14(a). If there were no eddies, this would be a straightforward conversion of energy from zonal available potential energy to zonal kinetic energy. The AZ is generated by differential solar heating, creating temperature gradients between the equator and subtropics. The KZ is associated principally with the zonal winds of the subtropical jet. The conversion term requires a positive correlation between the ascent and temperature anomaly which is provided by the thermally direct circulation of the Hadley cell.

During the 1930s and 1940s, there was considerable debate about the origin of eddy motions in the atmosphere. One possibility was that the strong shears associated with the poleward edge of the subtropical jet would prove hydrodynamically unstable. Such 'barotropic instability' would result in conversion of KZ to KE and would require that the momentum fluxes would be negatively correlated with the wind shears. In terms of the eddy structures, we would see eddies tilting against the background wind shear. Fig. 5.14(b) illustrates the Lorenz energy diagram associated with this view of the global circulation.

The second possibility, which rapidly became the accepted mechanism following the work of Charney and Eady in the latter part of the 1940s, was that 'baroclinic instability' generated the eddies. The energetics of baroclinic instability are illustrated in Fig. 5.14(c); temperature fluxes correlated with a poleward temperature gradient convert AZ to AE. Upward heat fluxes then convert AE to KE. Baroclinic instability theory, which will be outlined in Section 5.4, indicates the conditions under which such a pattern of conversions is dynamically consistent.

Figure 5.15 gives observed values of the various energies and conversions, estimated from observed data. The vertical velocities were inferred from the horizontal wind and mass fields during initialization prior to carrying out a numerical forecast integration (see Section 2.2). It will be seen that the dominant conversions are those associated with baroclinic instability, although only quite a small fraction of AZ is actually converted into other forms of energy. The conversion between AZ and KZ is small and its sign is somewhat uncertain. This arises from the near cancellation between a negative contribution to $\langle h[\bar{\theta}][\bar{\omega}] \rangle$ from the Hadley cell and the positive contribution from the thermally indirect Ferrel circulations of the midlatitudes. There is no trace of barotropic instability, the conversion CK having the wrong sign for this. Rather, the momentum fluxes are associated with the transient eddies returning kinetic energy to the zonal flow and so helping to maintain the tropospheric jets in midlatitudes.

An analysis of the energetics of the global circulation is helpful in revealing the dominant pattern of energy conversions in the atmosphere. If care is taken, these may be regarded as the signatures of basic physical processes occurring, such as baroclinic instability. But the usefulness of the energetics approach is limited in a number of respects.

First, it must be appreciated that the kinetic energies and available potential energies cannot vary independently in the midlatitudes. For instance, in Section 5.2, we found that the poleward thermal advection of a developing baroclinic wave is inevitably accompanied by an upward temperature flux.

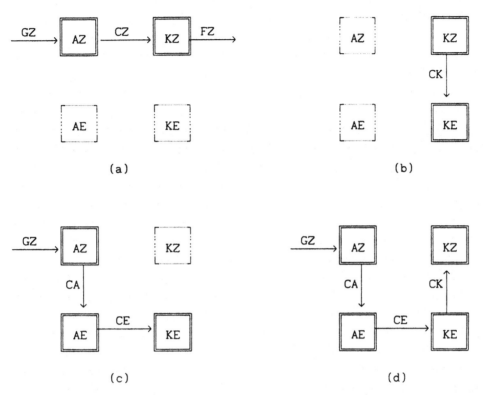

Fig. 5.14. Paradigms of the Lorenz energy cycle. (a) Axisymmetric Hadley circulations (b) barotropic instability, (c) baroclinic instability, (d) observed global circulation.

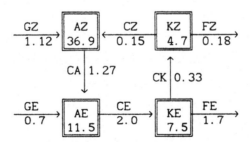

Fig. 5.15. Lorenz energy cycle as observed for the December–February period. Units: $10^5 \, \mathrm{J\,m^{-2}}$ (energies) and $\mathrm{W\,m^{-2}}$ (conversions).

This is necessary to maintain thermal wind balance. Similarly, the generation of *AZ* associated with the forcing of an equator–pole temperature contrast must lead, by thermal wind balance, to the generation of westerly wind shear in midlatitudes, and hence to an associated *KZ*. The *KZ* generated by eddy momentum fluxes, on the other hand, is barotropic in structure and so need

not be accompanied by any change of *AZ*. It is perhaps preferable to reject
the distinction between eddy available potential and kinetic energy, defining
instead a single 'eddy energy':

$$\frac{dE}{dt} = \frac{d}{dt} \left\langle \frac{1}{2} \left[\overline{u^{*2}} + \overline{v^{*2}} + \frac{s^2}{h^2} \overline{\theta^{*2}} \right] \right\rangle = - \left\langle \frac{h^2}{s^2} [\overline{\theta}]_y [\overline{v^*\theta^*}] + [\overline{u}]_y [\overline{u^*v^*}] \right\rangle.$$

(5.37)

Eddy energy is created by the negative correlation between the temperature
flux and the temperature gradient, and is destroyed if there is a positive correlation between momentum flux and horizontal wind shear. But, of course,
this is only a diagnostic relationship. In order to decide whether baroclinically unstable waves can form on any particular zonal flow, it is necessary
to show that energy releasing disturbances, with the necessary phase tilts
and other structures, can be maintained in a dynamically consistent fashion
in that flow. Our discussion in the next section will address these issues.

Second, the analysis is a truly global diagnostic. It is very difficult to
construct an energetics scheme for a limited area. The boundary fluxes
rapidly become the dominant terms. But the global budget is likely to
include a number of processes with quite different signatures in different
regions. An example was given above, where it was shown that the Hadley
and Ferrel cells make large contributions to the *AZ* to *KZ* conversion which
nearly cancel. The net *AZ* to *KZ* conversion is therefore a small residual
between large effects of opposite sign arising from the Hadley cell in the
tropics and the Ferrel cell in the midlatitudes. The magnitude and even the
sign of this conversion are very uncertain.

5.4 Theories of baroclinic instability

The energetics discussed in the last chapter revealed that, because of the
horizontal temperature gradients in a stably stratified atmosphere, potential
energy is available in principle for conversion into eddy energy. The eddy
fluxes required for the necessary conversions of energy also define the sort of
structures that must characterize such growing eddies. But these considerations alone do not guarantee that growing eddies can in fact be sustained. The
demands of thermal wind balance restrict the scale and shape of dynamically
possible eddies. In this section, we will outline a more predictive approach
which will define the conditions under which unstable eddies can be expected,
and which will determine their scale and rate of growth.

The philosophy behind these theories is to imagine small perturbations
growing on a zonally uniform basic state. It is found that certain classes

of disturbance can grow, and it is presumed that the most rapidly growing disturbances will come to dominate the observed flow field. The mathematics is common to a large class of hydrodynamic instability problems, and will not be discussed very deeply. The reader is referred to books such as that by Pedlosky (1987) for a full mathematical discussion. The basic principle is to linearize the governing equations around the zonal mean state by neglecting the products of eddy quantities. A 'normal mode' solution to the linearized equations is then sought, leading to a relationship defining the time behaviour of the various normal modes permitted.

A number of idealized mathematical models of the baroclinic instability problem have been derived. Perhaps the simplest and most elegant is that due to Eady; accordingly we will discuss this extensively. Other calculations give rather similar results, though often at the expense of a considerable increase in algebraic complexity. The basic state envisaged in the Eady model is illustrated in Fig. 5.16. It proves convenient to formulate the problem using 'pseudo-height' (or the logarithm of pressure) as vertical coordinate:

$$z' = -H \ln(p/p_R), \qquad (5.38)$$

where H is the atmospheric pressure scale height and p_R is the reference atmospheric pressure in the system. Boundaries are located at $z' = \pm H/2$. The lower boundary represents the Earth's surface; the upper boundary can be interpreted as a representation of the tropopause, which behaves roughly as a rigid lid on the tropospheric motions. The Eady problem neglects the variation of the Coriolis parameter f with latitude; solutions will be sought which are periodic in both the x- and y-directions. Baroclinic instability is made possible in the system by a uniform temperature gradient in the y-direction which does not vary with height. Equivalently, by thermal wind balance, the zonal wind increases linearly with height; the vertical wind shear is denoted $\Delta U/H$. The description of the system is completed by specifying a stratification, measured by the Brunt–Väisälä frequency N where

$$N^2 = \frac{g}{\theta_R} \frac{\partial \theta_R}{\partial z'}. \qquad (5.39)$$

In principle, N^2 may vary with z', but for simplicity we will restrict our attention to the case where N is a constant.

The evolution of the system can be described using the quasi-geostrophic equations of Sections 1.7 and 1.8, but expressed in log–pressure coordinates. These may be succinctly summarized by conservation of potential vorticity:

$$\frac{\mathrm{D}_g}{\mathrm{D}t} q = 0, \qquad (5.40)$$

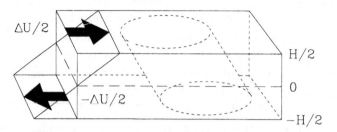

Fig. 5.16. Schematic diagram of the configuration envisaged for the Eady model of baroclinic instability.

provided friction and heating can be ignored. The potential vorticity of the undisturbed zonal flow is given by

$$q = f + \frac{\partial^2 \psi}{\partial x^2} + \frac{\partial^2 \psi}{\partial y^2} + \frac{f^2}{N^2}\frac{\partial^2 \psi}{\partial z'^2}, \tag{5.41}$$

where the streamfunction ψ is defined by

$$\psi = -\Delta U y z'. \tag{5.42}$$

Substituting into Eq. (5.41), it is quickly seen that the potential vorticity of the basic state is constant and equal to f. Since Eq. (5.40) states that this potential vorticity cannot be altered by the subsequent development of the flow, it follows that the perturbation potential vorticity q^* remains zero for all time:

$$q^* = \frac{\partial^2 \psi^*}{\partial x^2} + \frac{\partial^2 \psi^*}{\partial y^2} + \frac{f^2}{N^2}\frac{\partial^2 \psi^*}{\partial z'^2} = 0. \tag{5.43}$$

This relationship serves to determine the structure of any wavelike perturbations. We assume that the initial eddy is periodic in x, y and t (this involves no loss of generality, since any arbitrary disturbance could be resolved into a Fourier series of periodic disturbances), and has the form:

$$\psi^* = \Phi(z')e^{i(kx+ly-\omega t)}, \tag{5.44}$$

where $\Phi(z')$ is a function of height to be determined. Then, since the potential vorticity perturbation is zero for all time, it follows that:

$$\frac{d^2\Phi}{dz'^2} = \frac{N^2K^2}{f^2}\Phi = \frac{1}{H_R^2}\Phi, \tag{5.45}$$

where $K^2 = k^2 + l^2$. The solutions to this ordinary differential equation in z' are most conveniently written as:

$$\Phi = A\cosh(z'/H_R) + B\sinh(z'/H_R), \tag{5.46}$$

where the constants of integration A and B are yet to be determined. The first term in this solution is symmetric about the midlevel $z' = 0$, while the second is antisymmetric.

Equation (5.46) determines the spatial structure of the eddies, but says nothing about their time variations. To determine these, we must apply the boundary conditions at $z' = \pm H/2$. The boundary conditions are simply $w = 0$ at each boundary. However, w does not occur in the potential vorticity Eq. (5.40). To complete the problem, we will use the perturbation thermodynamic equation to relate the vertical velocity to the eddy potential temperature and hence to the eddy streamfunction. The usual linearization, whereby all products of eddy quantities are dropped, will be assumed. Making use of the hydrostatic relation to relate temperature and geostrophic streamfunction, the quasi-geostrophic thermodynamic equation is:

$$\left(\frac{\partial}{\partial t} + \frac{\Delta U z'}{H}\frac{\partial}{\partial x}\right)\left(\frac{\partial \psi^*}{\partial z'}\right) - \frac{\Delta U}{H}\frac{\partial \psi^*}{\partial x} = -w\frac{N^2}{f}. \tag{5.47}$$

Thus setting $w = 0$, we have

$$\left(\frac{\partial}{\partial t} \pm \frac{\Delta U}{2}\frac{\partial}{\partial x}\right)\left(\frac{\partial \psi^*}{\partial z'}\right) - \frac{\Delta U}{H}\frac{\partial \psi^*}{\partial x} = 0 \text{ at } z' = \pm H/2. \tag{5.48}$$

Substituting for ψ^* using Eq. (5.46) gives the relationships between the frequency ω of the disturbance and its spatial structure:

$$\frac{KN}{f}\left(-\omega \pm \frac{k\Delta U}{2}\right)(\pm As + Bc) - \frac{k\Delta U}{H}(Ac \pm Bs) = 0 \text{ at } z' = \pm H/2,$$

where

$$c = \cosh(KNH/2f), \quad s = \sinh(KNH/2f). \tag{5.49}$$

These two relationships can be used either to eliminate the unknown constants A and B, or to eliminate the unknown frequency ω. In the first case, a 'dispersion relation' linking the frequency to the wavenumbers of the eddy results:

$$\omega^2 = k^2 \Delta U^2 \left(\frac{1}{2}\tanh\left\{\frac{K}{2K_R}\right\} - \frac{K_R}{K}\right)\left(\frac{1}{2}\coth\left\{\frac{K}{2K_R}\right\} - \frac{K_R}{K}\right), \tag{5.50}$$

where $K_R = f/(NH)$. In the second case, the ratio B/A is obtained, thereby completing the description of the vertical structure of the modes:

$$\frac{B^2}{A^2} = \frac{\left(\frac{1}{2}\tanh(K/2K_R) - K_R/K\right)}{\left(\frac{1}{2}\coth(K/2K_R) - K_R/K\right)}. \tag{5.51}$$

Consider, first, the frequency given by Eq. (5.50). For small K, the expression for ω^2 is negative, indicating that the frequency is imaginary. That is, the time variation is either exponential growth or exponential decay, depending on which sign of the square root is taken. In a physical system, the exponentially growing modes will quickly dominate. In this case, the fluid is hydrodynamically unstable, and we will show that the mode of instability is indeed the baroclinic instability whose energetics were described in the last section. For $K > 2.399K_R$, ω is real, and we have neutrally stable, propagating disturbances. The most unstable modes are for $l = 0$ and $k = 1.61K_R$, when the growth rate σ is given by

$$\sigma = i\omega = 0.31K_R\Delta U = 0.31\frac{f}{NH}\Delta U. \tag{5.52}$$

Figure 5.17 shows the variation of growth rate with both k and l. Taking typical values of f and N for the midlatitude troposphere, we find that the wavelength of the most unstable mode is around 4000 km, and the growth rate is about 0.5 day^{-1}. This is very similar to the scales for high frequency transients which were deduced in Section 5.2.

Equation (5.51) shows that the ratio $(B/A)^2$ is negative for the unstable waves, that is, (B/A) is imaginary. This implies a vertical phase tilt, and, consequently, a nonzero poleward temperature flux. The stable waves, with $K > 2.399K_R$, have no phase tilt and zero temperature flux. We concentrate on the unstable modes. Denote $b = i(B/A)$. From the perturbation streamfunction, it is straightforward to deduce the eddy poleward velocity and the eddy temperature fields, which are:

$$v^* = kA\left\{\coth\left(\frac{NKz'}{f}\right)\sin kx - b\sinh\left(\frac{NKz'}{f}\right)\cos kx\right\}, \tag{5.53a}$$

$$T^* = \frac{T_0NKA}{g}\left\{\sinh\left(\frac{NKz'}{f}\right)\sin kx - b\cosh\left(\frac{NKz'}{f}\right)\cos kx\right\}. \tag{5.53b}$$

Figure 5.18 shows plots of these fields as a function of x and z'. The poleward velocity shows a characteristic westward phase tilt with height, amounting to 90° for the most unstable wave. A similar phase tilt is found for the streamfunction perturbation. The temperature perturbation, on the other hand, has an eastward phase tilt with height of 48° between the bottom and top boundaries. The temperature and poleward velocity waves are exactly in phase at the midlevel. Towards the boundaries, the phase difference is larger, but the amplitudes also increase. In fact, using the mathematics of Section 5.2, it is easy to show that these effects exactly compensate, so that

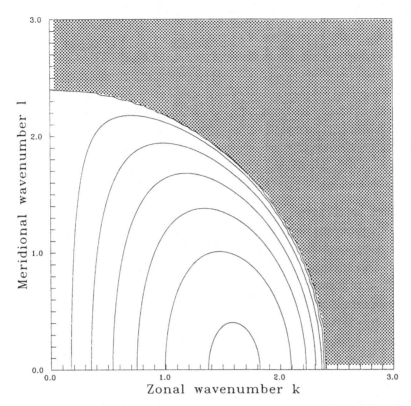

Fig. 5.17. Growth rate of waves with zonal wavenumber k and meridional wavenumber l according to the Eady model of baroclinic instability. Contour interval is $0.05\,K_R\Delta U$.

the poleward temperature flux is constant with height for all unstable waves:

$$[v^*T^*] = \frac{T_0 N}{g} A^2 bKk. \tag{5.54}$$

Within the quasi-geostrophic framework, there must be a small vertical velocity associated with the developing wave. This is easily calculated from the thermodynamic equation, Eq (5.47), suitably rearranged:

$$w^* = -\frac{f}{N^2}\left\{\frac{\partial}{\partial t}\frac{\partial \psi^*}{\partial z'} + \frac{\Delta U z}{H}\frac{\partial}{\partial x}\frac{\partial \psi^*}{\partial z'} - \frac{\Delta U}{H}\frac{\partial \psi^*}{\partial x}\right\}. \tag{5.55}$$

The vertical velocity field is shown in Fig. 5.19. From this expression and from expression (5.53b) for the perturbation temperature field, the vertical temperature flux $[w^*T^*]$ can be calculated; the result is more portentous than enlightening, but is quoted for the sake of completeness:

$$[w^* T^*] = \frac{A^2 f T_0 K}{Ng} \left\{ S_1 \sinh^2 \left(\frac{z'}{H_R} \right) + S_2 \cosh^2 \left(\frac{z'}{H_R} \right) + S_3 \sinh \left(\frac{2z'}{H_R} \right) \right\},$$

(5.56)

where

$$S_1 = \frac{\sigma}{H_R}, \; S_2 = \frac{\sigma b}{H_R},$$

and

$$S_3 = \frac{\Delta U k}{2H} \left\{ b \left(1 - \frac{z'}{H_R} \right) + 1 + \frac{z'}{H_R} \right\}.$$

By comparing $[w^* T^*]$ and $[v^* T^*]$, it can be seen that the growing Eady wave has a temperature flux which is directed at an angle typically midway between the geopotentials and the surfaces of constant potential temperature (or density). This feature is characteristic of the baroclinic instability process, which is why it is sometimes called 'sloping convection'. It is also characteristic of the observed midlatitude transients and, as we showed in Section 3.3, such an upward and poleward temperature flux is required on general thermodynamic grounds. A plot of w^* for the most unstable Eady wave is shown in Fig. 5.19. A comparison of this field with the temperature anomaly shown in Fig. 5.18(b) reveals that upward motion tends to be correlated with warm temperatures, and downward motion with cold temperatures. Consequently, the baroclinically unstable Eady wave transports heat upwards as well as polewards.

In terms of the energetics of the preceding section, the horizontal eddy temperature flux is constant and poleward, while the poleward temperature gradient is constant through the y–z' plane in the Eady model. There is a continual conversion of zonal available potential energy into eddy available potential energy, in just the way suggested for baroclinic instability. The associated vertical temperature fluxes are required to maintain thermal wind balance in the developing wave, and they ensure that there is conversion of eddy available potential energy to eddy kinetic energy. The existence of the same energy conversions in observations of the global circulation strongly suggests that baroclinic instability is indeed the source of the transient eddy activity observed in the midlatitudes. The fact that the Eady theory predicts the wavelength and growth time of unstable disturbances to be close to those observed for the midlatitude transients is further evidence that baroclinic instability is dominating this aspect of the global circulation.

A number of alternative physical interpretations of the nature of the baroclinic instability process can be given. The sloping nature of the temperature

Fig. 5.18. Structure of the meridional velocity perturbations (left) and the temperature perturbations (right) in growing Eady waves. The contour interval is 0.1 v^*_{max} or 0.1 T^*_{max}, as appropriate. Negative values indicated by dashed contours. (a) $K = 0.1\,K_R$. (b) $K = 1.61\,K_R$ (the most unstable wave). (c) $K = 2.3\,K_R$, near the short wave cutoff.

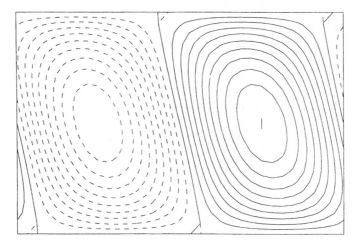

Fig. 5.19. Vertical velocity in a developing unstable Eady wave with $K = 1.61\,K_R$, contour interval $0.1\,w^*_{\max}$. Downward motion indicated by dashed contours.

flux vector suggests one argument, based on energy considerations, which is illustrated in Fig. 5.20. The basic state of the Eady model is characterized by the tilt of the surfaces of constant potential temperature relative to the surfaces of constant z'. If two parcels of air are exchanged along a trajectory which lies in the narrow wedge between the geopotential and the isentrope, then the element with lower θ is moved down, and will be colder than the ambient air, while the element which has been moved up will be warmer than its surroundings. In other words, the potential energy associated with the two elements has been reduced; consequently, kinetic energy must have been generated. The displaced parcels have no tendency to return to their home locations, and their displacement will increase with time. If the motions in an unstable Eady wave are zonally averaged, then parcels of air at some given value of y and z' are dispersed in just this way, along trajectories which lie between the zonal mean isentrope and the zonal mean geopotential. The Eady wave, then, provides a balanced adiabatic motion which can release available potential energy and convert it into kinetic energy of the eddy motions. Plausible as it is, this sloping convection picture of the instability mechanism is incomplete. It does not indicate why there should be a preferred zonal scale for the unstable disturbances; indeed it suggests that instability is always possible if there are horizontal temperature gradients.

An alternative, and in some ways preferable, interpretation exists in terms of interfering trains of waves. To see this, consider the case of the Eady basic state, but with the upper boundary removed, so that the atmosphere

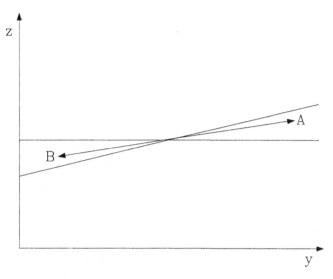

Fig. 5.20. Schematic illustration of the 'sloping convection' explanation of the Eady instability. Parcel A has lower potential temperature than parcel B. Consequently, exchanging them reduces the potential energy of the system.

extends to $z' = +\infty$. We take the lower boundary to be at $z' = 0$. From Eq. (5.45), the form of the disturbances is:

$$\psi^* = A \exp\{-z'/H_R\} \exp\{i(kx + ly - \omega t)\} \tag{5.57}$$

since we require that they die away as $z' \to \infty$. As previously, the thermo-dynamic equation is used to express the lower boundary condition $w = 0$ in terms of ψ and hence to determine the dispersion relationship for wavelike disturbances. From this, it is found that the frequency ω is always real, so that instability is not possible. The phase speed ω/k is:

$$c = \frac{\omega}{k} = \frac{f\Delta U}{NH}\frac{1}{K}, \tag{5.58}$$

indicating that the wave moves at the same speed as the fluid a distance H_R from the boundary. Exactly the same arguments would apply to a rigid boundary at the top of an infinitely deep atmosphere. The structure of such a neutral mode is shown in Fig. 5.21. The neutral modes for the Eady problem beyond the short wave cutoff, when $K \gg K_R$, tend towards just this structure; they become so shallow that we have essentially independent trains of waves trapped on the upper and lower boundaries. For longer wavelengths, the upper and lower waves in the Eady problem become deeper until they have significant amplitude at the opposite boundary. At this point, they will begin

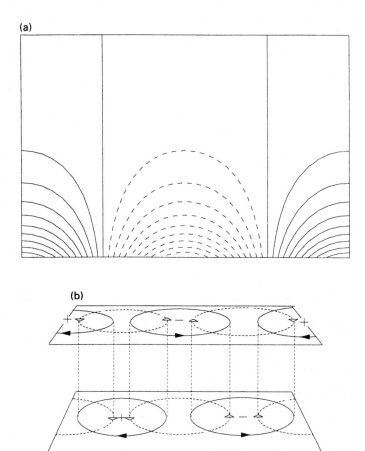

Fig. 5.21. (a) The perturbation streamfunction for a neutral Eady wave trapped on the lower boundary of an infinitely deep atmosphere. (b) Schematic diagram showing how phase locking between two such waves can lead to instability.

to interact with each other. Now suppose that the phase speeds of the upper and lower waves are such that they do not move relative to one another. The circulations induced by the lower wave can induce further meridional displacements in the upper wave, and vice versa, provided the upper trough is to the west of the lower trough. When such phase locking is possible, the waves will be growing, unstable disturbances.

Two other analytical models of baroclinic instability have been used extensively for theoretical studies. They will be outlined much more briefly than the Eady model. The first is the 'two-layer' model. Figure 5.22 illustrates the configuration of this model. The vertical structure of the flow is represented by defining the streamfunction on just two levels in the vertical; the various vertical derivatives which appear in the quasi-geostrophic equations are represented by finite difference formulae. One advantage of this system is that it is straightforward to retain the β-effect in the vorticity equation, and also to add in such effects as friction. Such flexibility has led the two-layer system to be called a 'white mouse' model; that is, it is simple and expendable but can be used for a variety of experiments which would not be possible with a more elaborate system! The disadvantage is that the extreme simplification of the vertical structure can give misleading results, for example by predicting stability when a more elaborate model would show instability of very shallow disturbances. A dispersion relation is derived from the linearized vorticity and thermodynamic equations for the system; the boundary conditions on the vertical velocity are incorporated through the finite difference representation of the stretching terms in the vorticity equation and the vertical advection term in the thermodynamic equation. The growth rate for wavelike disturbances, including β but excluding friction, is:

$$\sigma = k\Delta U \left(\frac{2K_R^2 - K^2}{2K_R^2 + K^2} - \frac{\beta^2 K_R^4}{\Delta U^2 K^4 (2K_R^2 + K^2)^2} \right)^{1/2}. \tag{5.59}$$

Figure 5.23 shows some plots of growth rate versus wavenumber for various values of β. Like the Eady model, this growth rate has a short wave cutoff. When $K \to 0$, there is also a long wave cutoff. The effect of β is to reduce the growth rates. For $\beta = 0$, the maximum growth rate and the short wave cutoff are at similar, but not identical, wavenumbers as in the Eady model; the maximum growth rate is also similar. The dynamics of the two-layer system can be interpreted in terms of two interacting waves, one confined to the upper layer and one to the lower layer. Instability occurs when they can phase lock, with a westward tilt with height.

The final model that will be mentioned is Charney's model. In fact, this was the first model of baroclinic instability to be published. It is based on a configuration which is very similar to that of the Eady model. The two differences are that the upper boundary is removed to infinity, and that the β term is retained in the vorticity equation. These apparently simple modifications lead to considerable complications in the mathematics. The actual results, in terms of the growth rates and structures of the normal modes, are similar to the results from the two-level system or from the Eady model. In the Charney model, unstable modes have a large amplitude and temperature flux, etc., below a certain height which depends upon the wavelength of the mode. Above this level they die away. The phase speed is nonzero, and is equal to the flow speed at a given level, called the 'steering

level
variables
pressure
0
0
1
u,v
2
T,ω
$p_0/2$
3
u,v
4
p_0

Fig. 5.22. The configuration of the two-layer model of baroclinic instability.

Growth rate
0.0
0.2
0.4
0.6
0.0
1.0
2.0
Zonal wavenumber

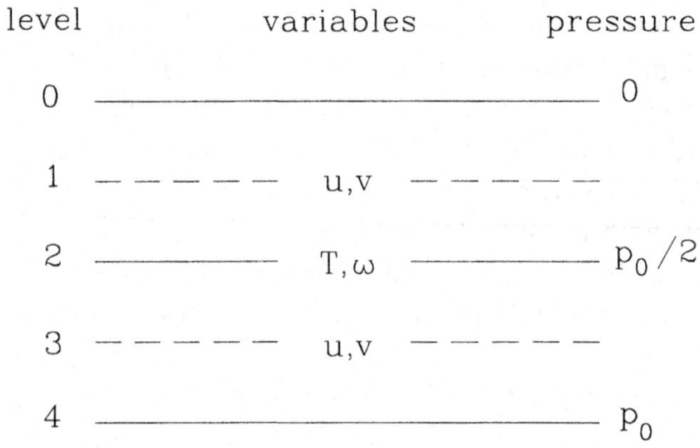

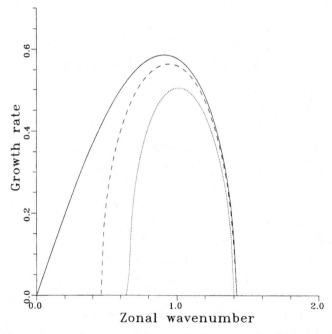

Fig. 5.23. Growth rate versus wavenumber for the two-level model for various values of β, calculated from Eq. (5.59). Growth rates are scaled by $K_R \Delta U$ and wavenumbers by K_R. Basic parameters are for a β-plane centred at 45° with $N^2 = 10^{-4}\,\mathrm{s}^{-2}$ and a vertical shear of $40\,\mathrm{m\,s}^{-1}$, roughly corresponding to northern hemisphere winter. Solid curve: $\beta = 0$. Dashed curve: $\beta = 8.1 \times 10^{-12}\,\mathrm{m}^{-1}\,\mathrm{s}^{-1}$. Dotted curve: $\beta = 1.6 \times 10^{-11}\,\mathrm{m}^{-1}\,\mathrm{s}^{-1}$.

level', in the atmosphere. The steering level is also the level above which the mode dies away. Mathematically, the steering level acts as a critical level for the system (see Chapter 6), which reflects vertically propagating wave activity. In the case of the unstable modes, they are 'over reflected', that is, they gain energy from the reflection at the critical level. There is a short wave and a long wave cutoff, and the scale and growth rate of the most unstable mode are similar to those of the most unstable Eady wave. For typical midlatitude values of the various parameters, the steering level is around 70 kPa, and the wavelength of the most unstable mode is around 4000 km. Once again, the source of the instability may be envisaged as an interaction between two wave trains. One is similar to the boundary wave of the Eady model with the upper boundary removed to infinity. The other is a Rossby wave which can propagate as a result of the potential vorticity gradient in the fluid interior.

All the models described in this section succeed in representing, in a qualitative way, the distribution of temperature flux, and hence the primary eddy generating energy conversions, actually observed in the atmosphere. However, none of them in any way represents the observed momentum fluxes. Indeed, a cursory examination of the solutions for the Eady model reveals that there are no horizontal phase tilts of the wave and hence no momentum fluxes. The same is true of the two-level and Charney models. In order to generate unstable modes with horizontal phase tilts, it is necessary to introduce a basic flow $U = U(y, z')$. This apparently trivial generalization in fact leads to insuperable mathematical difficulties and no general analytical solutions are known. The problem is that the linearized equations are no longer separable, that is, the normal modes cannot be represented by some function

$$\psi^* = A(z')B(y)e^{i(kx - \omega t)}, \tag{5.60a}$$

as for the Eady model and the other models discussed above, but rather by

$$\psi^* = A(y, z')e^{i(kx - \omega t)}. \tag{5.60b}$$

The amplitude $A(y, z')$ must be determined by approximate numerical means for any particular $U(y, z')$. Although no general solutions to this problem are known, some general necessary conditions for instability can be derived on energetics grounds. These are related to the conditions needed for phase locking two wavetrains together, and are expressed in terms of the boundary temperature gradients G_L and G_U and the interior potential vorticity gradient

$[q]_y$:

$$[q]_y = \beta - \frac{\partial^2 U}{\partial y^2} - \frac{\partial}{\partial z'}\left(\frac{f^2}{N^2}\frac{\partial U}{\partial z'}\right). \qquad (5.61)$$

These necessary conditions may be summarized:

(i) $[q]_y$ must change sign in the y–z' plane.
(ii) $-[q]_y$ and G_L must have opposite signs.
(iii) $-[q]_y$ and $-G_U$ must have opposite signs.
(iv) G_L and $-G_U$ must have opposite signs.

Condition (ii) applies to the Charney model and condition (iv) applies to the Eady model. Note that these are necessary but not sufficient conditions for instability. The introduction of extra effects such as friction can have an important effect in modifying (sometimes unexpectedly) the instability of the flow. Rather broad bounds on the growth rate and phase speeds of unstable disturbances can be derived. But more specific information about the actual growth rate or structure of the unstable modes in this more general case can only be obtained by numerical means.

Figure 5.24 shows the results of such a calculation for a realistic jet-like basic flow. In fact, this calculation was carried out for an atmosphere on a sphere. The basic wind and temperature fields were discretized on to a grid in ϕ and p, and the rate of change of perturbation velocity, temperature and surface pressure was calculated using a primitive equation numerical model. The normal modes of the system, including the unstable normal modes, were calculated by determining the eigenvalues and eigenvectors for a resulting 595×595 matrix with complex coefficients. This is no small computational problem, and requires a large computer to be practicable. In some respects, the results do not differ greatly from the most unstable normal mode in the corresponding Charney mode. The growth rate is $0.706\,\mathrm{day}^{-1}$ and the phase speed is $11.9\,°$ of longitude per day. The steering level corresponding to this phase speed is sketched on the zonal wind cross section; its height varies with latitude, being nearest to the ground in the centre of the jet. The temperature fluxes are poleward, confirming that the energy conversions are those of baroclinic instability. They are large below the steering level and decrease in the upper troposphere. The new element is the poleward momentum flux. This tends to be poleward south of the jet (where $[u]_y$ is positive) and equatorward to the north of the jet (where $[u]_y$ is negative); consequently, the momentum flux represents a conversion of eddy kinetic energy to zonal kinetic energy, as shown by Fig. 5.14. The effect of the momentum fluxes is to offset the growth of eddy energy.

Although the distribution of eddy momentum flux with latitude is similar to that observed, the maximum is too low in the troposphere and its magnitude is much too small relative to the temperature fluxes. We conclude that the linear theories of this section give a reasonable account of the observed scale, growth rates and temperature flux distribution of the transient eddies. But they do not account very well for the observed momentum fluxes, and they in no way predict the actual levels of eddy activity to be expected. Linear theory has its limitations!

5.5 Baroclinic lifecycles and high frequency transients

Linear theory of baroclinic instability, such as the Eady model, is based on a linearization of the equations. The amplitude of disturbances is assumed to be so small that products of eddy quantities can be neglected. If instability is present, disturbances of certain wavelengths will amplify exponentially, and will dominate the flow after a sufficient time. But as the unstable eddies continue to grow, the assumptions permitting the linearization will eventually break down. In the language of global circulation studies, the eddy fluxes of temperature, momentum, and so on, will become so large that they begin to modify the zonal mean temperature and wind fields. What will happen then? If baroclinic instability is responsible for generating the transients observed in the tropospheric midlatitudes, what happens in the atmosphere?

We might imagine a number of possibilities. The simplest is that as the growing wave modifies the zonal mean field, it should reduce and eventually remove the available potential energy which provides the source of energy for the instability. The wave might then equilibrate with some finite amplitude, with small conversions which enable a balance to be maintained between the generation of AZ by differential solar heating, and the destruction of kinetic energy by friction. Something like this can apparently take place in laboratory systems (see Section 10.5). For a range of parameter settings, experiments with rotating baroclinic fluids exhibit a regime of 'regular wave' behaviour, in which a strong wave with constant or, at most, regularly fluctuating, amplitude propagates around the apparatus. Alternatively, the growing wave might modify the mean flow in such a way as to make it more unstable, thereby accelerating the growth of the wave. Eventually, some catastrophic 'breaking' of the wave field will occur, completely wiping out the wave and the zonal flow upon which it grew. Some fluid instabilities (such as Kelvin–Helmholtz billows on a stratified, sheared flow) do appear to have this character. Synoptic scale disturbances do not. The third possibility is that the wave growth might be episodic, that is, the wave will grow

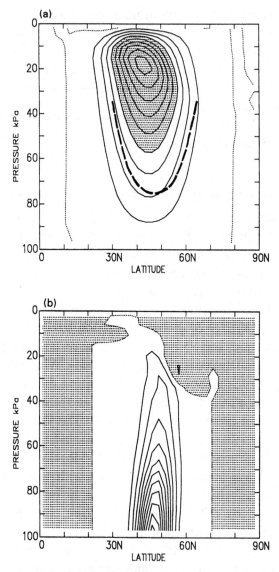

Fig. 5.24. Linear baroclinic instability of a zonal jet. (a) The basic zonal flow, contour interval 5 m s^{-1}. The heavy dashed line indicates the 'steering level' where $[u]$ is equal to the phase speed of the most unstable mode. (b) The distribution of the eddy temperature flux $[v^*T^*]$ of the most unstable normal mode. The amplitude has been arbitrarily scaled so that the maximum value of $[v^*T^*]$ is close to that observed. Contour interval 2 K m s^{-1}.

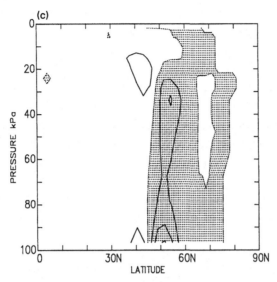

Fig. 5.24 (*cont.*). (c) The distribution of the eddy momentum flux $[u^*v^*]$, scaled by the same factor as in (b), contour interval $5\,\mathrm{m^2\,s^{-2}}$.

until its sources of energy are exhausted. Then friction and other dissipative processes will destroy the eddy energy. Eventually, the zonal flow will become unstable once more, and another round of wave activity can begin. Synoptic experience suggests that this third picture might be nearer the atmospheric situation. We are used to depression systems which form, deepen, and then eventually occlude and decay.

A more precise theoretical discussion of these possibilities is provided by 'weakly nonlinear' theory. This is restricted to situations which are only marginally unstable; the degree of instability is measured by some small parameter ϵ. The streamfunction and other variables are then expanded as a power series in ϵ:

$$\psi^* = \Psi_0 + \epsilon\Psi_1 + \epsilon^2\Psi_2 + \ldots, \text{etc.} \tag{5.62}$$

By substituting in the governing equations and equating coefficients of ϵ, a sequence of problems is obtained. The first equations in the sequence simply describe gesotrophic balance and the linear problem already discussed. But at the next order, an 'amplitude equation' governing the long term behaviour of the wave amplitude is obtained. This treatment has been accorded to the two-layer system and also to the Eady model. The results depend critically on whether, for example, friction is included in the system and, if so, how large the friction is. Without any friction at all, the instability is perfectly reversible. We find that the wave grows to a maximum amplitude, then begins to decline until it returns to its original amplitude. This cycle repeats indefinitely, and its character is determined by the initial conditions. Such a model is not very realistic, since in a real situation, friction will always be present. If friction is substantial, the wave will simply equilibrate to some fixed amplitude such that the generation of eddy kinetic energy exactly balances its dissipation by friction. In the case when the friction is small but not zero (i.e. proportional to some small power of ϵ), more complicated and more interesting behaviour can be found. The wave can settle into a pattern of regular oscillations or 'vacillations' which are independent of the initial conditions. In some cases, these vacillations can be irregular or 'chaotic'.

Such analyses are mathematically rather complex, and since they rest on very artificial assumptions of marginal instability which are perhaps rarely realized in practice, they are not necessarily relevant to the actual evolution of unstable disturbances. Numerical integration provides a more accessible, and perhaps not much less general, approach to the problem.

A classical numerical experiment is to determine an unstable eigenmode of a given zonal flow, such as was shown in Fig. 5.24, and to use this as initial data for a numerical integration of the full nonlinear equations of motion. In the simplest such experiments, there is no friction or heating, so that any chosen initial $[u]$ and the $[\theta]$ field in thermal wind balance with it is an exact solution of the governing equations. The initial jet is chosen to be representative of the observed midlatitude tropospheric jet. Provided the eigenmode has sufficiently small amplitude to begin with, the nonlinear terms are negligible and the wave simply amplifies exponentially. As the wave amplitude becomes finite, the zonal flow begins to change, and the rate of increase of amplitude becomes less rapid. Eventually, the wave reaches a maximum amplitude. Thereafter, for a large number of realistic initial jets, the waves collapse, often very rapidly, leaving a final state of zonal flow, though with the jet profoundly modified by the wave event. Such a sequence has been called a 'baroclinic wave lifecycle', and it has much in common with the observed synoptic evolution of midlatitude disturbances in the atmosphere.

Figure 5.25 shows the energetics of such a lifecycle. In the linear phase, KE increases exponentially, while AZ and KZ remain more or less steady. As the growth of KE levels off in the mature stage, a drop in AZ reveals the modification of the zonal flow. The decay phase is characterized by a rapid drop of KE and a corresponding increase of KZ. During the growth stage, the conversions are dominated by CA and CE, indicating that the temperature fluxes are poleward, upward and large. In the mature stage,

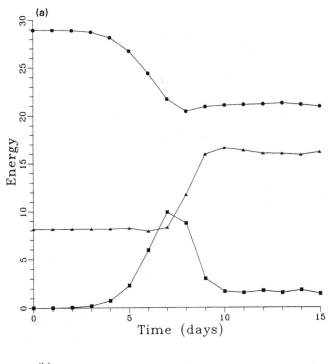

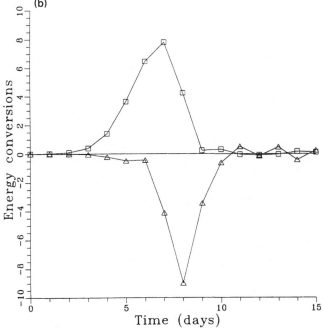

Fig. 5.25. Showing the Lorenz energetics during a lifecycle experiment starting from the initial state shown in Fig. 5.22. (a) Energies *AZ* (filled circles), *KE* (filled squares) and *KZ* (filled triangles), in units of $10^5\,\mathrm{J\,m^{-2}}$. (b) Conversions *CE* (open squares) and *CK* (open triangles), in units of $\mathrm{W\,m^{-2}}$.

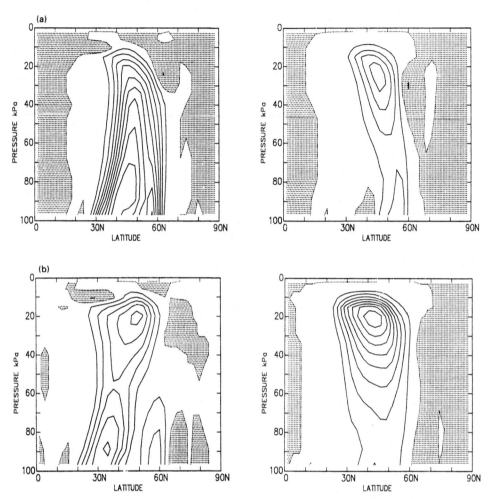

Fig. 5.26. Poleward fluxes of temperature, $[v^*T^*]$ (left), and momentum, $[u^*v^*]$ (right), in a typical baroclinic lifecycle experiment. The initial state was shown in Fig. 5.24: (a) at day 6, at the time when the nonlinear saturation of the wave is taking place; (b) at day 8, when the momentum fluxes are strongly converting eddy kinetic energy to zonal kinetic energy; (c) averaged from day 0 to 15.

these conversions become small, while the decay phase is dominated by CK, indicating that the momentum fluxes are large, accelerating the zonal flow at the expense of the eddy energy. The fact that AZ remains rather constant during this decay phase means that the modification of the zonal wind is barotropic in nature, so that no change in the zonal mean temperature field is required to retain thermal wind balance.

The distributions of the eddy fluxes of temperature and momentum show

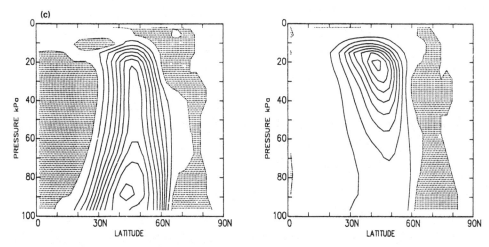

Fig. 5.26 (*cont.*). Contour intervals: in (a) and (b), the contour interval is $10 \, \mathrm{K \, m \, s^{-1}}$ for $[v^* T^*]$ and $50 \, \mathrm{m^2 \, s^{-2}}$ for $[u^* v^*]$; in (c) the contour interval is $2 \, \mathrm{K \, m \, s^{-1}}$ for $[v^* T^*]$ and $10 \, \mathrm{m^2 \, s^{-2}}$ for $[u^* v^*]$. Negative values indicated by shading.

marked changes from one stage of the lifecycle to the next. The patterns associated with the initial linear normal mode have already been given, in Fig. 5.24. The corresponding plots for the mature and decaying stages are shown in Fig. 5.26. At first, the fluxes are concentrated near or below the steering level. As the mode grows, the intensity of the fluxes increases, but the pattern of distribution remains close to that of the linear stage. As the disturbance approaches its maximum amplitude, the momentum fluxes become stronger relative to the temperature fluxes, and their largest values are found at upper tropospheric levels. In the decay phase, the temperature fluxes become quite weak, but the momentum fluxes are strong and concentrated near the tropopause, which acts more or less as an upper boundary to the eddy part of the flow. The fluxes averaged through the lifecycle are shown in Fig. 5.26(c). They are strikingly similar to the temperature and momentum fluxes observed in the winter troposphere, with large temperature fluxes at low levels and large momentum fluxes at upper levels. These results suggest that the observed transients may be regarded as a random superposition of lifecycle events, occurring at different times and at different longitudes throughout the midlatitudes. Further insight into these processes is afforded by relating the pattern of eddy fluxes to the propagation of wave-like disturbances out of the baroclinic regions. This will be treated in Section 6.6.

The zonal flow at the end of the lifecycle is nearly stable. The eddy energy is decaying, although for some zonal flows, a generally weaker 'secondary development' may follow. A numerical eigenvalue calculation shows that, indeed, the final state is at most only weakly unstable. Such unstable modes as there are have narrow and contorted structures, so that there is little possibility of the remaining wavelike disturbances projecting on to and exciting these new normal modes. Yet, according to the general criteria for baroclinic instability developed in the preceding section, the flow should still be unstable. It possesses available potential energy, which has been depleted only marginally by the first baroclinic lifecycle. There are substantial contrasts of potential vorticity gradient and surface temperature gradient, suggesting the possibility of further instability. The resolution of this apparent paradox lies in the form of the final zonal flow. Figure 5.27 shows the difference in the zonal flow between the beginning and end of the lifecycle event. It is nearly barotropic, that is, it is independent of height, which is consistent with the small reduction in AZ. A westerly acceleration has taken place near the latitude where the baroclinic waves were centred, and there has been easterly acceleration to the south and north. Now suppose the surface wind is subtracted from all levels in the model. This barotropic change will leave the wind field still in thermal wind balance with the temperature field, and in fact will make only small changes to the potential vorticity gradients. This is because the term $[u]_{yy}$ makes only a small contribution to the potential vorticity gradient in the midlatitudes. The modified zonal flow is not very different from the starting flow at the start of the lifecycle, and eigenvalue analysis shows that it is very nearly unstable. In other words, the baroclinic lifecycles stabilize the zonal flow not so much by depleting the zonal available potential energy as by introducing a horizontally sheared barotropic component to the flow. This component severely restricts the possible balanced structures which eddies can take, and thereby reduces their instability.

This last result illustrates the limitations of the frictionless adiabatic lifecycle as a model of global circulations. Extremely large surface winds are associated with the final state, with maximum speeds of up to $40\,\mathrm{m\,s^{-1}}$. In the real atmosphere, friction would moderate these winds quickly. At the same time, the introduction of heating would tend to restore any changes made to the initial zonal mean temperature fields. Such a model is really a simple global circulation model, such as has been introduced in Section 2.4 and used for illustration elsewhere in this book. Figure 5.28 shows a low level field of temperature from such a model. A number of cyclonic centres, in different stages of development, can be seen around the globe. The zonal

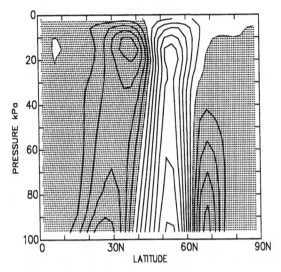

Fig. 5.27. The difference between the final and initial zonal flow during the baroclinic lifecycle shown in Figs. 5.25 and 5.26. Shading indicates regions where easterly acceleration has taken place. Contour interval $5\,\mathrm{m\,s^{-1}}$.

mean temperature and momentum fluxes are quite similar both to those observed and to those resulting from the time average of the lifecycle. No single wavelength dominates, and just as for the observed atmosphere, the levels of eddy energy are continually fluctuating. Experiments with various geometries reveal that these fluctuations are a result of interactions between the different wavelengths of eddies present, as well as between the unstable eddies and the zonal mean flow. But the source of eddy energy on all scales is clearly the result of baroclinic instability in the midlatitudes. The lower frequency transients exhibited by such experiments will be considered further in Chapter 7.

5.6 Problems

5.1 Given that the global average zonal wind $<u>$ is $15\,\mathrm{m\,s^{-1}}$, estimate the typical atmospheric kinetic energy in $\mathrm{J\,kg^{-1}}$. If the friction spins up the atmosphere on a timescale of five days, work out the typical dissipation of kinetic energy in $\mathrm{W\,m^{-2}}$. Compare your value with the global mean insolation.

5.2 Use Fig. 5.7(a) to estimate the divergence of the transient eddy heat flux in DJF, northern hemisphere in the subtropics and the midlatitudes.

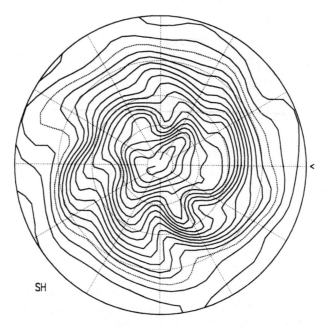

SH

Fig. 5.28. Temperature at 90 kPa at an arbitrary instant in a run of a simple global circulation model of the type described in Section 2.4. Contour interval 4 K. The quasi-horizontal exchange of parcels of warm and cold air at different longitudes is clearly seen, and demonstrates the baroclinic instability mechanism at work.

Hence calculate the heating and cooling rates due to transient eddies in $W\,m^{-2}$.

5.3 Perform a similar calculation for the eddy zonal momentum flux at 50°N in DJF (see Fig. 5.5(a)) to estimate the acceleration of the midlatitude flow due to transient eddies. Estimate the 'spin-up time' τ_D required in order that friction should balance the eddy convergence of momentum flux.

5.4 High frequency transients in the upper troposphere at 45°N are characterized by $\overline{u'^2} = 40\,m^2\,s^{-2}$, $\overline{v'^2} = 80\,m^2\,s^{-2}$, $\overline{Z'^2} = 3600\,m^2$ and $\overline{u'v'} = 20\,m^2\,s^{-2}$. Estimate the typical horizontal scale of eddies, the ratio of their major and minor axes and their tilt.

5.5 A typical geopotential height variance at 70 kPa and 45°N is 30 m, a typical meridional wind variance is $5\,m\,s^{-1}$ and a typical poleward temperature flux is $6\,K\,m\,s^{-1}$. Estimate the horizontal separation between the trough at 90 kPa and at 50 kPa.

5.6 The Brunt–Väisälä frequency is about $10^{-2}\,s^{-1}$ in the troposphere. The available potential energy is observed to be $4 \times 10^6\,J\,m^{-2}$. Estimate the pole–equator temperature difference, and compare your result with values taken from cross sections of the observed (potential) temperature structure. Using values of *AE*, *KE* and *KZ* in Fig. 5.15, estimate representative values of the typical eddy temperature fluctuation, the typical eddy velocity and the typical zonal wind speed in the annual mean.

5.7 Using the cross sections of eddy temperature and momentum fluxes given as Figs. 5.5 and 5.7, together with the sections of $[\bar{u}]$ and $[\bar{\theta}]$ given in Fig. 4.2, estimate the typical magnitudes of the two terms in the eddy energy equation.

6

Wave propagation and steady eddies

6.1 Observations of steady eddies

Despite the eddy–zonal flow partitioning which we have employed in preceding chapters, the seasonal mean flow is very far from being zonally
symmetric. Such departures from symmetry are important in accounting for
regional variations of climate. They also modify the global patterns of heat
and momentum transport, especially in the northern hemisphere winter. In
this chapter, we will discuss some observations of the steady wave pattern,
and show how rather simple theories based on linear wave propagation can
account for some of the gross features of these observations.

The steady waves are most pronounced in the northern hemisphere winter, and have their largest amplitudes in the upper troposphere. In some
circumstances, they also become very important at high levels in the winter
stratosphere, a point that we will return to in Chapter 9. Figure 6.1 shows
the winter mean geopotential height field at 25 kPa in both hemispheres.
The characteristic features of the northern hemisphere picture are the pronounced troughs over Canada and Japan, with ridges over the eastern side
of the two ocean basins. One's subjective impression is of a predominantly
zonal wavenumber 2 pattern. This general pattern is very persistent and
can be seen in individual seasons with only relatively small variations. The
corresponding picture for the southern hemisphere looks, at first sight, much
more axisymmetric. Closer examination reveals that the main vortex is significantly displaced from the pole, while a characteristic three trough pattern
can be seen around the periphery of Antarctica.

The character of the non–zonal disturbances to the upper level flow is
more obvious if the zonal mean is subtracted from diagnostics such as those
shown in Fig. 6.1. Such zonal anomalies are shown in Figs. 6.2 and 6.3. These
diagrams show the zonal anomaly of the time mean streamfunction $\overline{\psi}^*$ at

25 kPa rather than the geopotential height anomaly. This is to emphasize the important zonal anomalies of the flow in the tropics and subtropics as well as at high latitudes. As we shall see in later sections, the influence of the tropics on disturbances in midlatitudes is now believed to be substantial. The presence of waves in the northern hemisphere winter, Fig. 6.2(a), is clear; at high latitudes, wavenumbers 2 and 3 dominate, while a strong wavenumber 1 signature is seen at lower latitudes. Many of the eddies show a systematic phase tilt, with their major axes orientated from south-west to north-east. Such phase tilts indicate that a significant poleward momentum flux must be associated with the steady eddies. The steady eddies are also of substantial amplitude, though with rather longer zonal wavelength, in the southern hemisphere winter, shown in Fig. 6.3(b). However, little systematic phase tilt can be discerned in the southern hemisphere.

Steady waves are also seen in the summer season, though with sharply reduced amplitudes. The major feature in the northern hemisphere summer, Fig. 6.3(a), is an anticyclonic circulation centred over south western Asia; this is associated with the upper level Asian monsoon circulation and is more pronounced at higher levels such as 10 kPa. Apart from this, steady eddies are rather weak and have a smaller scale than those in the winter season. Similar remarks also apply to the southern hemisphere summer, Fig. 6.2(b); it is dominated by a rather weak zonal wavenumber 1 pattern.

A notable feature of all these diagrams is the absence of much flow across the equator. The steady disturbances fill their respective hemispheres, but by and large do not spill over into the opposite hemisphere. The exception is the Asian monsoon circulation, which does involve significant cross equatorial flow (especially at low levels). The reasons for this apparent insulation of the hemispheres from one another will be matters upon which our theories of steady waves will be required to comment.

There is considerable continuity between the upper and lower level steady waves. Rather than present diagrams such as Figs. 6.2 and 6.3 for many levels throughout the troposphere, Fig. 6.4 shows representative longitude–pressure cross sections. Attention is restricted to the winter seasons. In the northern hemisphere winter, the dominant zonal wavenumber 2 can be discerned. More striking is the systematic phase tilt of the waves, with the upper level troughs and ridges well to the west of their low level counterparts. As we saw in Chapter 5, such a tilt is inevitably associated with a poleward transport of temperature by the eddies. In fact, the steady eddies make an important contribution to the poleward transport of heat in the northern hemisphere winter. Although these sections do not resolve the stratosphere adequately, the rapid increase of the steady eddy amplitudes at the uppermost levels is

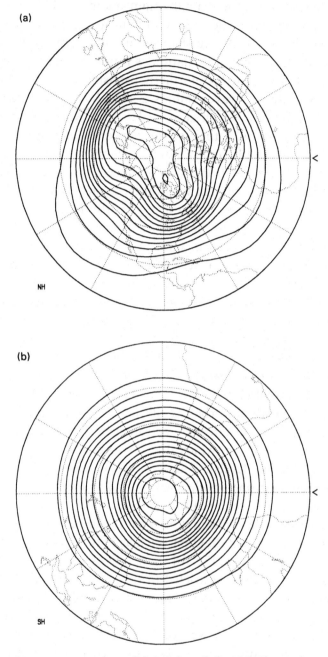

Fig. 6.1. The time mean geopotential height of the 25 kPa surface, based upon ECMWF data for six seasons: (a) Northern hemisphere, DJF. (b) Southern hemisphere, JJA. Contour interval 100 m.

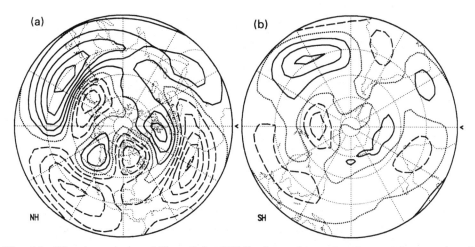

Fig. 6.2. The time mean eddies of the 25 kPa flow, shown by the zonal anomaly of the mean streamfunction, $\overline{\psi}^*$ for the DJF season. Contour interval $5 \times 10^6\,\mathrm{m^2\,s^{-1}}$, negative values dashed. Based upon six years of ECMWF data. (a) Northern hemisphere. (b) Southern hemisphere.

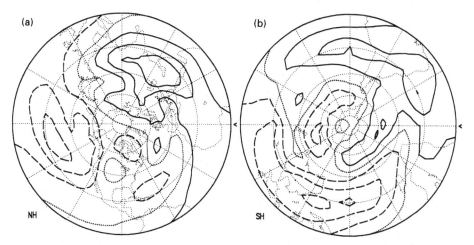

Fig. 6.3. As Fig 6.2, but for the JJA season. (a) Northern hemisphere. (b) Southern hemisphere.

significant. In the southern hemisphere winter, note the somewhat smaller amplitude and longer wavelength of the disturbances. More importantly, the vertical phase tilt is small; there is little systematic poleward temperature flux by the steady eddies in the southern hemisphere.

The total effects of the steady eddies for the transport of heat and mo-

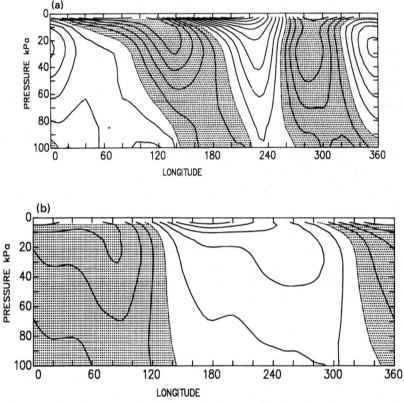

Fig. 6.4. Longitude–pressure sections of the zonal anomaly of geopotential height $\overline{Z}^*$, based on six years of ECMWF data. Contour interval 50 m, negative values shaded. (a) For DJF at 60 °N. (b) For JJA at 60 °S.

mentum are summarized by latitude–pressure sections of the steady poleward temperature flux $[\overline{v}^* \overline{T}^*]$ and momentum flux $[\overline{u}^* \overline{v}^*]$, shown in Figs. 6.5 and 6.6 respectively. The fluxes are significant only for the northern hemisphere winter, when a poleward temperature flux extends throughout the depth of the midlatitude troposphere. The maximum value of more than 12 K m s^{-1} is comparable to the transient eddy temperature flux. At the same time, there are large values of the momentum fluxes in the upper troposphere, with convergence around 50 °N. Once again, the distribution and magnitude is comparable to the fluxes carried by the transients. In contrast, the fluxes in the summer season and for the southern hemisphere are a good deal smaller than the transient fluxes.

The persistence of these steady disturbances throughout an entire season, and their repeatability from one year to another, strongly suggests that they

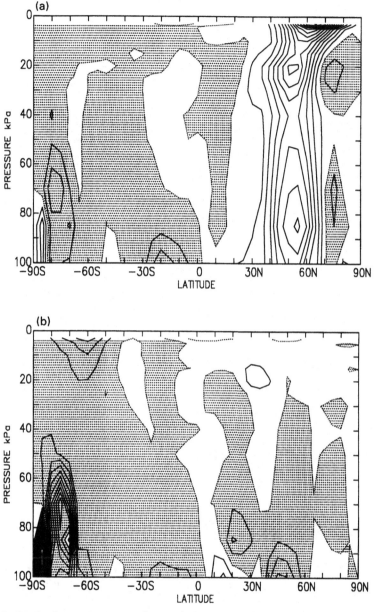

Fig. 6.5. Latitude–pressure sections of the northward temperature flux by steady eddies, $[\overline{v}^* \overline{T}^*]$, based upon six years of ECMWF data. Contour interval $2\,\mathrm{K\,m\,s^{-1}}$, negative values shaded. (a) DJF. (b) JJA.

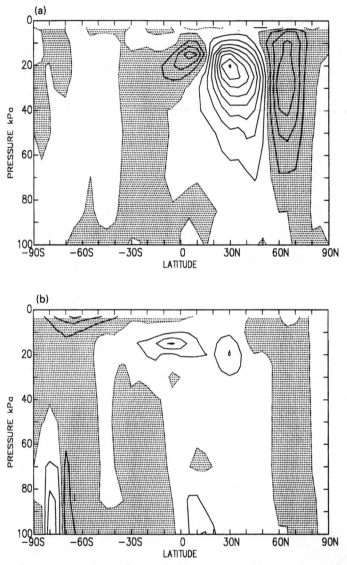

Fig. 6.6. Latitude–pressure sections of the northward momentum flux by steady eddies, $[\bar{u}^*\bar{v}^*]$, based on six years of ECMWF data. Contour interval $5\,\mathrm{m}^2\,\mathrm{s}^{-2}$, negative values shaded. (a) DJF. (b) JJA.

must owe their existence to some permanent forcing. Such forcing might be related to land–sea contrasts or to the disturbance of the zonal flow by features such as mountains. In either case, the lower boundary is implicated. On the global scale, major mountain ranges occupy only relatively small areas of the Earth's surface. The most significant ranges are the Tibetan plateau

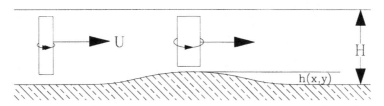

Fig. 6.7. Schematic view of a barotropic model of midlatitude flow over mountains.

with the associated Himalayan chains and the Rockies; in the southern hemisphere, the Andes cordilleras and the great ice sheets of Antarctica are the dominant features. These mountain ranges are quite restricted in their extent. Thermal forcing by varying sea surface temperatures, or by the temperature contrast between land and ocean surface, are again often confined to relatively small areas. For example, large changes in sea surface temperature are associated with narrow boundary currents running along the western sides of ocean basins. In the southern hemisphere, even the contrast between land and sea is localized, since only 20% of the surface is solid land or ice. These considerations raise the question of whether locally forced disturbances could propagate to remote parts of the globe, generating a 'wake' of disturbances which might fill much of the hemisphere. Much of the remainder of this chapter will deal with the horizontal and vertical propagation of eddies; in so doing, some of the gross features of the steady eddies which have been detailed in this section will be accounted for in a straightforward way.

6.2 Barotropic model

A surprising number of the features shown in the diagrams in the preceding section can be reproduced with the very simplest of barotropic models. Such models are of course highly idealized and cannot be used to give accurate simulations of the time mean circulation. But their use helps us to identify fundamental processes rather clearly; they provide conceptual models in terms of which the observations and results of more complicated models can be interpreted.

Consider the situation depicted in Fig. 6.7. A single layer of fluid, with surface pressure p_s, passes over a mountain whose height is $h(x, y)$. This height will be supposed small compared to the atmospheric scale height. Our starting point for the mathematical formulation of this model is the quasi-geostrophic vorticity equation, Eq. (1.62), in the form:

$$\frac{\partial \xi}{\partial t} + \mathbf{v} \cdot \nabla \xi + \beta v = f \frac{\partial \omega}{\partial p}. \tag{6.1}$$

For a barotropic flow, in which there are no horizontal variations of temperature, the flow is independent of pressure (or height). Consequently, the stretching term on the right hand side of the equation is independent of pressure. We will consider three possible contributions to the stretching term. Flow over mountains introduces a lower boundary condition:

$$\omega = \mathbf{v} \cdot \nabla p_s = -\frac{p_R}{H} \mathbf{v} \cdot \nabla h. \tag{6.2}$$

Here, p_R may be thought of as the mean pressure of the lower surface. Consistent with the barotropic model, we assume that ω changes linearly to zero above the boundary. Then $\partial \omega / \partial p = \mathbf{v} \cdot \nabla p_s / p_R$, and a forcing term can be introduced to the right hand side of Eq. (6.1). A second contribution to the stretching is provided by surface friction. This may be modelled by the introduction of an Ekman layer to the lower boundary. The Ekman layer introduces a vertical 'pumping' velocity proportional to the interior relative vorticity at the top of the Ekman layer:

$$\omega = -\frac{p_R}{H}\left(\frac{K}{2|f|}\right)^{1/2} \xi. \tag{6.3}$$

In a laboratory situation, K is simply the kinematic viscosity of the working fluid, and this relation is generally accurate. In the atmosphere, K must be thought of as an empirically determined 'eddy viscosity' which parametrizes the momentum transports by turbulent motion. The Ekman layer model is then highly idealized and not very accurate, but it serves adequately for qualitative purposes. Applying this vertical velocity as a lower boundary condition to the barotropic model, a term $-\xi / \tau_D$, where the 'spin up' time $\tau_D = (2H^2/|f|K)^{1/2}$, is introduced on to the right hand side of Eq. (6.1). A typical atmospheric spin up time is around five days. A third forcing is provided by heating. It can only be introduced into a barotropic model in a rather approximate fashion. Heating may be balanced by ascent in a stably stratified atmosphere. In a barotropic context, assume that this ascent increases linearly with pressure, so that it is zero at $p = 0$. The resulting divergence $D = -\partial \omega / \partial p$ acts as a forcing on the relative vorticity. Such an idealization does represent the typical distribution of heating in the midlatitudes reasonably well. But it is a poor representation of the effect of heating in the tropics, where the ascent is a maximum in the middle troposphere. Thus the vorticity tendency is cyclonic at lower levels and anticyclonic aloft. Such an inherently baroclinic process cannot be modelled properly by a simple barotropic equation set. Nevertheless, the barotropic model is very useful at our present level of discussion. If all these forcing effects are included, the barotropic vorticity equation becomes:

$$\frac{\partial \xi}{\partial t} + \mathbf{v} \cdot \nabla \xi + \beta v = -\frac{f}{H} \mathbf{v} \cdot \nabla h - \frac{\xi}{\tau_D} - fD. \qquad (6.4)$$

Now suppose that the basic zonal flow $[u]$ is some prescribed function of y only, say $U(y)$. The vorticity associated with this basic state is $-\partial U/\partial y$. If we suppose that the magnitude of the eddy vorticity $|\xi^*|$ is small compared to $|U_y|$, then we may linearize the equation about the zonal mean state to yield:

$$\frac{\partial \xi^*}{\partial t} + U\frac{\partial \xi^*}{\partial x} + (\beta - U_{yy})v^* = -\frac{fU}{H}\frac{\partial h}{\partial x} - fD - \frac{\xi^*}{\tau_D}. \qquad (6.5)$$

The term $(\beta - U_{yy})$ represents the poleward gradient of absolute vorticity; it is a fundamental property of the basic state. Ekman pumping acts so that relative vorticity decays exponentially towards zero on a timescale τ_D, thus providing a damping effect on the vorticity field. The orographic and diabatic forcing terms both take the form of simple prescribed forcings, independent of the eddy flow. We anticipate that a steady state can be defined in which the Ekman dissipation balances these vorticity forcings.

In order to appreciate the properties of the model more clearly, assume that the orography is the only forcing acting, and that it is sinusoidal in form:

$$h = h_0 e^{i(kx+ly)}. \qquad (6.6)$$

We may take the flow to be confined to a channel, width π/l in y and periodic in x. We suppose that U is independent of y for the present. Then seeking solutions of the form

$$\xi^* = Z e^{i(kx+ly)}, \qquad (6.7)$$

we find that the amplitude Z is related to the amplitude of the mountain h_0 by

$$Z = \frac{fU\{(U - \beta/K^2) + i/(\tau_D k)\}}{\{(U - \beta/K^2)^2 + 1/(\tau_D k)^2\}} \frac{h_0}{H}, \qquad (6.8)$$

where $K = (k^2 + l^2)^{1/2}$. Thus the response is wavelike, with zonal wavenumber k and meridional wavenumber l. The calculation can be generalized easily to arbitrary orography by assuming that Eq. (6.6) is just one term in a Fourier series representing the mountain. Then the total response is given simply by summing expressions such as Eq. (6.8) over all permitted k and l. The amplitude of the response is

$$|Z| = \frac{fU}{\{(U - \beta/K^2)^2 + 1/(\tau_D k)^2\}^{1/2}} \frac{h_0}{H}, \tag{6.9}$$

while its phase relative to that of the mountain is given by

$$\delta = \tan^{-1}\left\{\frac{(\tau_D k)^{-1}}{U - \beta/K^2}\right\}. \tag{6.10}$$

This represents a classical damped resonant response. The crucial parameter is the dimensionless number $\beta/(UK^2)$. For long waves, with $\beta/(UK^2) \gg 1$, the vorticity is in phase with the mountain, with cyclonic vorticity above its summit. For shorter waves, when $\beta/(UK^2) \ll 1$, the response is in anti-phase with the mountain, with anticyclonic vorticity above the mountain. When $\beta/(UK^2) = 1$, the response is large and limited simply by the Ekman pumping. Taking a typical atmospheric zonal wind of 15 m s^{-1}, the critical total wavenumber $K = 10^{-6}$ m^{-1} at 45° of latitude.

This response is illustrated by Fig. 6.8, which takes a β-plane centred at 45°N and width 5000 km along which blows a basic zonal wind of 15 m s^{-1}. The meridional wavenumber l is taken to be 6.3×10^{-7} m^{-1}, so that a half wavelength in the meridional direction fits into the channel. An Ekman spin up time of five days is assumed. The resonance is quite sharp, and close to zonal wavenumber 4. Only the longest waves can therefore lead to cyclonic vorticity over the mountain; for a narrower mountain, no such waves would be excited. Realistic mountain ranges, in which the important x- and y-scales are fairly short, are likely to be characterized by the response well to the left of the resonance point.

Now suppose that the forcing is confined to fairly concentrated regions, and is small over the rest of the globe. A mountain range, or region of strong heating, may be regarded as a 'wavemaker', exciting waves with a variety of zonal and meridional wavenumbers. Under appropriate conditions, they may propagate as Rossby waves to remote parts of the globe. Much of the remainder of this chapter will be devoted to a discussion of the character of such propagation and of the conditions under which it may occur.

In the absence of any forcing, the linearized vorticity equation is simply:

$$\frac{\partial \xi^*}{\partial t} + U\frac{\partial \xi^*}{\partial x} + (\beta - U_{yy})\frac{\partial \psi^*}{\partial x} = 0. \tag{6.11}$$

By seeking solutions of the form

$$\xi^* = Z e^{i(kx + ly - \omega t)}, \tag{6.12}$$

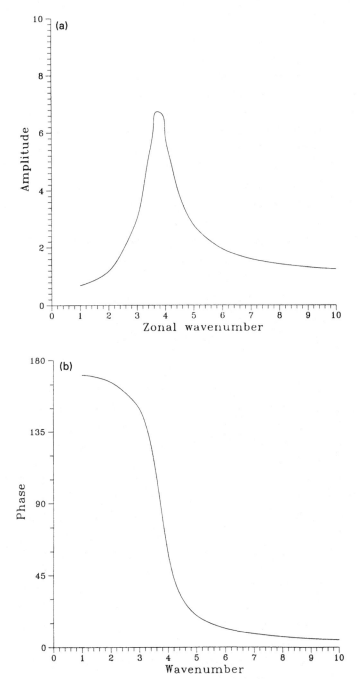

Fig. 6.8. (a) The amplitude and (b) the phase, of the steady vorticity wave in response to orographic forcing as a function of zonal wavenumber. The calculation has been performed for a β channel centred at 45°N, width 5000 km, with a zonal wind of 15 m s⁻¹.

we obtain a dispersion relationship:

$$\omega = Uk - (\beta - U_{yy})k/K^2. \tag{6.13}$$

Such solutions are called Rossby waves. They are low frequency, large scale wavelike disturbances whose phase speed, ω/k, is westward relative to the basic flow U. The longest waves propagate westward most rapidly, while short waves (i.e. those with large K) move at a speed close to U. Compared with more familiar examples of wave motion, such as sound waves or buoyancy waves, Rossby waves are curious in that only westward propagating waves are possible.

For the steady response to forcing, $\omega = 0$ and so the dispersion relationship may be regarded simply as a diagnostic relationship determining the meridional wavenumber l as a function of the zonal wavenumber k:

$$l = \pm \left\{ (\beta - U_{yy})/U - k^2 \right\}^{1/2} \tag{6.14}$$

for steady Rossby waves. If this expression is imaginary, propagation is not possible and disturbances are evanescent. Note that β is of course always positive and its magnitude is generally large compared to that of U_{yy}; hence steady Rossby waves are nearly always evanescent when U is negative. Shorter waves (i.e. waves with large k) are also likely to be evanescent, except for very small but positive U. The existence of propagating waves with a zonal wavenumber k depends upon the total steady wavenumber $K_s = \sqrt{(\beta - U_{yy})/U}$. Propagation is possible if K_s is real and greater than k. A typical tropospheric zonal wind is $15 \, \text{m s}^{-1}$; at 45°N, β is $1.6 \times 10^{-11} \, \text{m}^{-1} \text{s}^{-1}$. Assuming that U_{yy} is small compared to β, we find that the total steady wavenumber is $1.04 \times 10^{-6} \, \text{m}^{-1}$, which corresponds to zonal wavenumber 4 or 5 at this latitude. Near the tropospheric jet core, $-U_{yy}$ can be large, and could increase the total steady wavenumber. Equatorward of the jet, U is smaller and β is larger, so we anticipate a general increase in K_s as the subtropics are approached. Figure 6.9 shows K_s calculated using the 30 kPa zonal wind for the southern hemisphere 1979 winter by way of example. The K_s is imaginary north of 9°S and has very large values immediately poleward of that latitude. It generally falls as higher latitudes are entered, though there is a deep minimum around 40°S. This was a somewhat unusual feature of that particular winter, and was associated with an unusually marked double structure of the tropospheric jet.

When propagation is possible, we are concerned with the rate at which wave packets can be spread to remote parts of the globe. The relevant quantity is the group velocity, which describes the rate at which 'wave activity' or

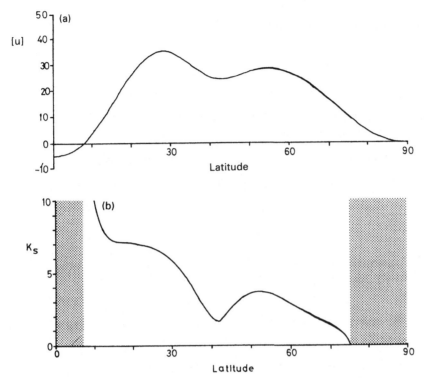

Fig. 6.9. (a) The 30 kPa zonal wind observed during the southern hemisphere FGGE winter. (b) The corresponding total steady wavenumber K_s. (From James 1988.)

'wave energy' is dispersed across the globe. It is given by

$$\mathbf{c}_g = (\partial\omega/\partial k, \partial\omega/\partial l). \tag{6.15}$$

From the dispersion relation, Eq. (6.13), the two components of the group velocity for steady wave packets are:

$$c_{gx} = 2(\beta - U_{yy})k^2/K^4, \quad c_{gy} = 2(\beta - U_{yy})kl/K^4. \tag{6.16}$$

The propagation is in the direction which makes an angle α with the zonal direction, where:

$$\alpha = \tan^{-1}(c_{gy}/c_{gx}) = \tan^{-1}(l/k). \tag{6.17}$$

This equation reveals that the choice of the positive root in Eq. (6.14) corresponds to northward propagating packets, while the negative root corresponds to southward propagation. The magnitude of the group velocity is

$$|\mathbf{c}_g| = 2(\beta - U_{yy})k/K^3, \tag{6.18}$$

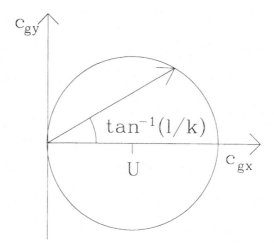

Fig. 6.10. Construction showing the magnitude and direction of the group velocity of steady Rossby waves in a zonal flow U.

which can be rewritten, using the dispersion relation and noting $\omega = 0$, in the convenient form:

$$|\mathbf{c}_g| = 2U \cos \alpha. \qquad (6.19)$$

Figure 6.10 shows a simple construction for the group velocity of steady Rossby waves. When the propagation is purely zonal, the group speed is simply twice the zonal wind. When it is nearly meridional, the group speed becomes small. Returning to the example of steady Rossby waves excited at 45 °N in a zonal wind of 15 m s^{-1}, consider propagation of packets with zonal wavenumber 3, that is, with $k = 6.7 \times 10^{-7}$ m^{-1}. Since K_s was found to be 1.03×10^{-6} m^{-1}, and we must have $l^2 = K_s^2 - k^2$, it follows that $l = 7.8 \times 10^{-7}$ m^{-1}. The packet will propagate in a direction making an angle of 49 ° with the zonal direction, and the group speed in this direction will be 20 m s^{-1}. The reader may quickly verify that such a packet will require around one day to influence a latitude 20 ° further poleward.

This meridional propagation of Rossby waves is a crucial feature of the response of a barotropic model to local forcing. The frequent use of channel models, in which v^* is supposed to be zero at certain latitudes north and south of the source, obscures this aspect of Rossby wave propagation and has encouraged the use of a conceptual model in which Rossby waves propagate zonally. In fact, the $v^* = 0$ boundary condition at the poleward and equatorward boundaries of such a channel implies Rossby wave reflection. The zonal wavetrain predicted by such bounded channel models may be regarded as the result of interference between an equatorward propagating wavetrain and a reflected poleward propagating train. It is difficult to justify the existence of such reflecting regions in more realistic models of the atmosphere.

As the packet propagates in the meridional direction, it will encounter changing local values of parameters such as $\beta - U_{yy}$. Provided the distance over which such changes take place is in some sense long compared to the dimensions of the Rossby wave, it is possible to make some predictions about the evolution of the propagating packet. The technique is called 'ray tracing'; it is frequently employed in many branches of physics such as optics and electromagnetic propagation through an inhomogeneous medium. In such applications, there is frequently a huge scale separation between the wavelength of the waves and the variations of the medium through which they propagate. When this is the case, the slowly varying approximation is extremely good. In our atmospheric application, it is much more doubtful whether the mathematical conditions for the application of ray tracing can be rigorously justified. The application of ray tracing can be defended as providing an initial conceptual model of atmospheric forcing and can be verified by comparison with less restrictive, but correspondingly more complex, models. We may rewrite the dispersion relation, Eq. (6.13), in the form:

$$\omega = \omega(k, \hat{\beta}), \tag{6.20}$$

where $\hat{\beta} = (\beta - U_{yy})$ is the basic parameter controlling the propagation. Then, using the identities $\partial k/\partial y = \partial l/\partial x$, $\partial k/\partial t = -\partial\omega/\partial x$ and $\partial l/\partial t = -\partial\omega/\partial y$, it can be shown that the rate of change of frequency following the wave packet is:

$$\frac{D_p\omega}{Dt} = \left(\frac{\partial\omega}{\partial\hat{\beta}}\right)\frac{\partial\hat{\beta}}{\partial t}, \tag{6.21}$$

where

$$\frac{D_p}{Dt} \equiv \frac{\partial}{\partial t} + c_{gx}\frac{\partial}{\partial x} + c_{gy}\frac{\partial}{\partial y} \tag{6.22}$$

denotes the rate of change following the wave packet. But in our linearized model, $\hat{\beta}$ does not vary in time; it therefore follows from Eq. (6.21) that the packet conserves its frequency as it propagates. For the steady waves we are currently discussing, this is of course simply zero. A similar relationship describing the variation of wavenumber of the packet can also be derived:

$$\frac{D_p k}{Dt} = -\left(\frac{\partial\omega}{\partial\hat{\beta}}\right)\frac{\partial\hat{\beta}}{\partial x}, \quad \frac{D_p l}{Dt} = -\left(\frac{\partial\omega}{\partial\hat{\beta}}\right)\frac{\partial\hat{\beta}}{\partial y}. \tag{6.23}$$

Since the basic state is purely zonal, $\hat{\beta}$ depends upon y only and so the zonal wavenumber of a packet is conserved. In contrast, its meridional wavenumber will evolve as the packet moves from one latitude to another. The meridional wavenumber could in principle be derived by integrating Eq. (6.23) with respect to time. In practice, it is much simpler to use the diagnostic relationship based on the dispersion relation with $\omega = 0$, Eq. (6.14). Since we know ω, k and l as functions of latitude, we can calculate the group velocity at any location. The trajectory followed by the packet of Rossby waves is therefore described by:

$$\frac{\mathrm{d}x}{\mathrm{d}t} = \frac{2\hat{\beta}k^2}{K^4} \tag{6.24a}$$

and

$$\frac{\mathrm{d}y}{\mathrm{d}t} = \frac{2\hat{\beta}kl}{K^4}, \tag{6.24b}$$

where

$$l = \pm\sqrt{K_S^2 - k^2}. \tag{6.14}$$

Consider the propagation of a wave packet with zonal wavenumber k away from some midlatitude source. Two trajectories are possible provided that $k < K_s$; one, corresponding to the negative root in Eq. (6.14), is directed to the south, while the positive root leads to a northward trajectory. In general, we expect that K_s will increase as the subtropics are approached. At some critical latitude, where U changes sign, K_s will first become extremely large and then imaginary. As K_s becomes larger, l must become larger, and so the equatorward propagating ray will turn into a more meridional direction. At the same time, from Eq. (6.19), the group speed will become smaller. As the critical latitude is approached, the packet will propagate extremely slowly in a nearly meridional direction. The meridional scale of the Rossby waves will become extremely small (l large). Indeed, the packet will, according to this linear theory, take an infinite time to reach the critical latitude, which will act as something of a 'black hole' to Rossby wave information approaching it from higher latitudes. Figure 6.11 gives a schematic illustration of the approach to the critical latitude. If the effect of friction were included, we might anticipate that the slowly moving, short length scale packet would be dissipated in the vicinity of the critical line. In a truly inviscid world, the linear theory would break down in the vicinity of the critical latitude. A more sophisticated analysis suggests that in these circumstances, the critical line might partially reflect some of the wave activity incident upon it. Whether or not such reflection is important in the atmosphere is difficult to determine.

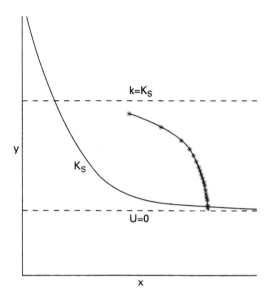

Fig. 6.11. Schematic illustration showing an equatorward Rossby ray approaching a critical latitude where $U = 0$. The variation of K_s is shown, and the crosses indicate the location of the packet after equal intervals of time.

Evidence from the observed steady momentum fluxes (see below) suggests that rather little reflection can be taking place. But the breakdown of zonality and stationarity of the basic flow make it very difficult to interpret the details of behaviour observed near $U = 0$.

For the poleward propagating ray, the packet will in general move into an environment where K_s is smaller. Once more, it may become imaginary at some latitude where β becomes smaller than U_{yy}, though such a latitude is much less pathological than a critical latitude where U changes sign. As K_s becomes smaller, the packet will adjust by acquiring a smaller l, i.e., by becoming more extended in the meridional direction. That is, the ray will turn into a more zonal direction. Eventually, at a latitude where $K_s = k$, it will become completely zonal. The scale assumption of small spatial scale of the packet compared to the scale of variations of K_s becomes invalid as such a latitude is approached and, strictly, a different analysis is needed in this region. For our present purposes, it is enough to note that the meridional wavenumber continues to decrease and becomes negative. The ray is refracted away from the $K_s = k$ latitude and back into lower latitudes, as shown in Fig. 6.12. Eventually it slows down and approaches the critical latitude.

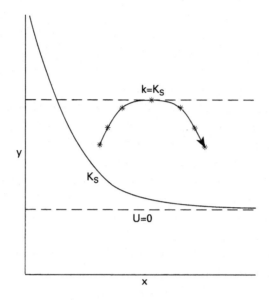

Fig. 6.12. As Fig. 6.11, but illustrating a poleward propagating ray approaching the latitude where $K_s = k$.

A wind profile such as:

$$U = U_E \cos \phi \qquad (6.25)$$

provides a particularly simple example of steady Rossby wave propagation. Such a profile corresponds to uniform superrotation of the atmosphere with angular velocity U_E/a relative to the solid Earth. It is particularly simple because β , U and U_{yy} all have the same cosine dependence on latitude so that the total steady wavenumber K_s is simply given by:

$$K_s^2 = \frac{2\Omega a + U_E}{a^2 U_E}, \qquad (6.26)$$

which is independent of latitude. Rays are straight lines in this flow. The longer waves propagate more meridionally while the shorter waves propagate more zonally. If full spherical geometry is retained, the rays are found to follow great circle tracks. Wave packets originating in one hemisphere will pass into the opposite hemisphere before returning to the region in which they were excited.

But generally, the zonal wind becomes easterly in the tropics. A critical latitude where $U = 0$ is typical, and acts to isolate one hemisphere from another. Let us return to the profile illustrated in Fig. 6.9. We imagine that a wavemaker which excites a wide spectrum of wavenumbers has been

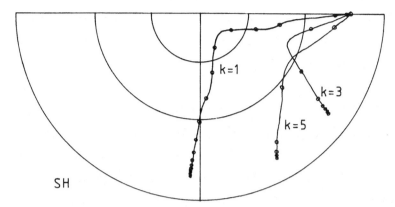

Fig. 6.13. Rays emanating from a wavemaker at 10 °S. The basic zonal flow is based on the 30 kPa southern hemisphere wind for June–July 1979 (the FGGE year, shown in Fig 6.9). The circles mark the daily position of wave packets. (Adapted from James 1988.)

inserted into the flow at 10° of latitude. Ray paths followed by packets with different zonal wavenumbers are shown in Fig. 6.13. Some rays propagate polewards from the source until they reach a latitude where $K_s = k$; they are then refracted back towards the tropics. Other rays propagate equatorwards from the wavemaker. They turn into a more meridional direction and slow down as they approach the critical latitude. The smaller wavenumber packets propagate rather slowly in a more meridional direction, while the larger wavenumber packets propagate more zonally and more rapidly, as we would expect from Eq. (6.19). But, clearly, purely zonal propagation is very unlikely. In no sense do the strong westerlies of midlatitudes act as a zonal wave guide, contradicting the implicit assumption of so many β-channel models.

Thus, the simple ray tracing theory predicts that disturbances originating in the midlatitudes of one hemisphere cannot propagate into the opposite hemisphere. The theory can easily be modified for transient disturbances with phase speed c in the zonal direction. The same results apply, except that the critical latitude beyond which no propagation is possible is where $U = c$. Equally, the theory suggests that a wavemaker in the deep tropics will not produce a response in the midlatitudes. This insulation of one hemisphere from the other, and from the equatorial regions, has been exploited in numerical prediction models and global circulation models which only treat one hemisphere and apply boundary conditions based on symmetry (or, where appropriate, antisymmetry) about the equator.

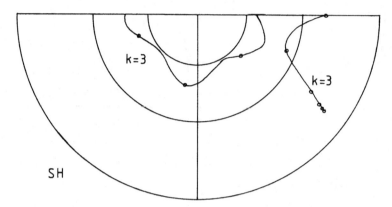

Fig. 6.14. As Fig. 6.13, but for a zonal wavenumber 3 disturbance excited at 20 °S and 55 °S.

In order to produce waveguide behaviour, there would have to be a maximum of K_s in midlatitudes. Short zonal wavelengths excited in such a region would propagate to south and north until refracted back towards their original latitude at latitudes where $K_s = k$. In fact, the example shown in Fig. 6.9 has a local maximum of K_s at 55 °S. It is sufficiently pronounced to act as a waveguide for zonal wavenumber 3. Fig. 6.14 shows how wavenumber 3 could indeed be confined in this case. Of course, the scale separation can hardly be justified in this case, and we must not take such a result too literally. Even if we can accept that the assumptions of ray tracing are not too grossly violated, such a waveguide would be extremely leaky. There is a rather narrow range of latitudes between 50 and 40 °S where wavenumber 3 cannot propagate. Our theory suggests that the disturbance will be evanescent in such a region, dying away exponentially with distance from the regions where propagation can be supported. But it would still have some amplitude at 40 °, where it would excite equatorward propagating packets. So if we were to take the ray tracing literally in this case, we would anticipate that the ray would be weakened each time it was reflected from the equatorward boundary of the waveguide, and would soon have negligible amplitude. The effect is analogous to the 'quantum tunnelling' of elementary particles across some energy barrier.

6.3 Application to observed steady eddies

To begin with in this section, we will calculate the momentum fluxes carried by propagating, steady Rossby waves. As well as considering the direction of propagation, and the strength and nature of the forcing, we will need to take into account the amplitude of the wave packet as it passes into regions of differing K_s.

In the last section, the waves were described by the perturbation stream-function:

$$\psi^* = \Psi e^{i(kx+ly)}, \tag{6.27}$$

where Ψ is a (generally complex) amplitude. Such an expression is always implicitly qualified by 'real part of', so it is more precise to write the perturbation streamfunction as:

$$\psi^* = \frac{1}{2}\left\{\Psi e^{i(kx+ly)} + \tilde{\Psi}e^{-i(kx+ly)}\right\}, \tag{6.28}$$

where $\tilde{\Psi}$ denotes the complex conjugate of Ψ. The two horizontal components of the perturbation velocity field are therefore:

$$u^* = -\frac{\partial \psi^*}{\partial y} = \frac{1}{2}\left\{-il\Psi e^{i(kx+ly)} + il\tilde{\Psi}e^{-i(kx+ly)}\right\}, \tag{6.29a}$$

$$v^* = \frac{\partial \psi^*}{\partial x} = \frac{1}{2}\left\{ik\Psi e^{i(kx+ly)} - ik\tilde{\Psi}e^{-i(kx+ly)}\right\}. \tag{6.29b}$$

Then the poleward momentum flux is written:

$$u^*v^* = \frac{1}{4}\left\{kl\Psi^2 e^{2i(kx+ly)} + kl\tilde{\Psi}^2 e^{-2i(kx+ly)} - 2kl\Psi\tilde{\Psi}\right\}. \tag{6.30}$$

The first pair of terms describes a wavelike variation, with zonal wavenumber $2k$. This contribution will of course average to zero around a latitude circle. The third term is a constant. We conclude that the zonal mean poleward momentum flux is simply:

$$[u^*v^*] = -\frac{kl}{2}|\Psi|^2. \tag{6.31}$$

We will turn to the problem of estimating the variation of the wave amplitude Ψ along the ray path shortly. For the present, we note that the poleward momentum flux takes the opposite sign to l. Thus associated with a poleward propagating packet will be an equatorward momentum flux, and vice versa.

The simplest example to consider is the case of uniform superrotation, with K_s given by Eq. (6.26). Rays are straight lines in our β-plane approximation. A poleward momentum flux is associated with those rays with negative l, which propagate towards the equator. For those poleward propagating rays with positive l, the momentum flux will be equatorward. The net result is rather curious. There will be an eddy momentum flux which converges

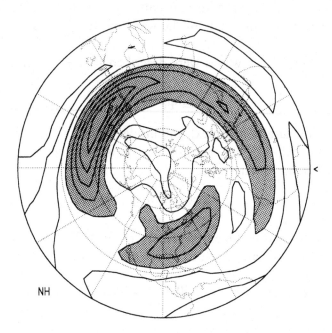

Fig. 6.15. The time mean wind speed at 25 kPa for DJF in the northern hemisphere. Contour interval $10\,\mathrm{m\,s^{-1}}$, with values in excess of $25\,\mathrm{m\,s^{-1}}$ indicated by shading. Note the maxima downstream from the Rockies and the Tibetan plateau.

towards the latitude of the localized forcing region. If a zonal wind were forced to blow over an isolated mountain, the radiated Rossby waves would induce eddy momentum fluxes which would lead to a zonal jet at the latitude of the mountain. One's intuitive reaction might be to expect that the drag exerted by the mountain on the barotropic flow would decelerate the flow at this latitude, rather than accelerate it!

In the northern hemisphere winter, there are two important sources of orographic wave forcing in the midlatitudes, namely the Tibetan plateau and the Rockies. They both have their largest amplitudes between 30 and 40 °N, just polewards of the subtropical jet maximum. The pattern of steady momentum fluxes shown in Section 6.1 is certainly consistent with the idea of orographic forcing of Rossby waves giving rise to the observed steady eddies. There is a poleward steady momentum flux (equatorward propagation) at these latitudes. The equatorward steady momentum flux is found at higher latitudes, north of 45 °N. At the same time, the strongest zonal wind maxima are downstream of the two mountain ranges. Figure 6.15 shows the observed wind speed at 25 kPa.

The amplitude of the waves can be calculated using the slowly varying approximation to obtain an equation describing the slow variation of Ψ with y. An alternative approach is to use the concept of 'wave action'. The quantity:

$$A = \frac{[u^{*2} + v^{*2}]/2}{(\omega - Uk)} \qquad (6.32)$$

is called the 'wave action density'. It can be shown that $c_{gy}A$ is conserved following the ray, provided that the medium varies only in the meridional direction and the slowly varying approximation is valid. From Eq. (6.27) it is easily shown that

$$\frac{1}{2}[u^{*2} + v^{*2}] = \frac{1}{2}K^2|\Psi|^2. \qquad (6.33)$$

For steady waves, the dispersion relation can be used to re-write the poleward component of group velocity as:

$$c_{gy} = \frac{2\hat{\beta}kl}{K^4}. \qquad (6.34)$$

Hence, for steady waves of zonal wavenumber k, we have:

$$c_{gy}A = -l|\Psi|^2 = \text{ constant.} \qquad (6.35)$$

That is, along the ray, the amplitude varies as $l^{-1/2}$. The amplitude of an equatorward propagating ray will become smaller as the packet approaches the tropics where its group velocity is increasingly meridional. Conversely, the amplitude on a ray propagating polewards is expected to increase as the latitude where $k = K_s$ is approached. Of course, as pointed out in the preceding section, the simple, slowly varying sinusoidal solutions break down at this latitude, though a more sophisticated matched asymptotic analysis can be carried out. One consequence of this result is that relatively weak forcing at low latitudes could excite a meridionally propagating wavetrain which would achieve substantial amplitude at higher latitudes. From this idea has come a great deal of work relating anomalous weather patterns in the midlatitudes to unusual forcing of waves in the subtropics. We will return to these ideas in Section 7.2.

The results just derived must be treated with a degree of caution. Strictly, they apply to a highly localized isotropic wavemaker. If an extensive mountain were considered, interference between wavetrains emitted from different parts of the orography could lead to the variation of amplitude with latitude being very different from the $l^{-1/2}$ behaviour discussed here.

It is clear that a number of highly restrictive approximations have been made in deriving the ray tracing theories of the preceding two sections. The reader may well find the continual appeal to the slowly varying approximation unconvincing, especially when it appears that the most important steady waves are of rather long zonal wavelength. Before returning to a consideration of real data, it is worth comparing these results with those from an intermediate model. Figure 6.16 shows a numerical calculation using the linearized barotropic vorticity equation, Eq. (6.11). The zonal flow was based upon the climatological zonal mean flow at 30 kPa observed in the northern hemisphere winter. A spectral representation of the fields was used, and forcing was provided by a single isolated mountain at 30 °N. No slowly varying approximation has been made, and so the calculation is equally valid for all wavenumbers. Two wavetrains, one propagating poleward and one propagating equatorward from the mountain are clearly seen. The amplitude of disturbances on the poleward propagating track in particular show an increase towards higher latitudes. The agreement with the rather simple arguments of the preceding section, despite the lack of formal scale separation, suggests that ray tracing arguments can give at least a qualitative account of the distribution of steady waves for a wide range of conditions.

A second calculation, shown in Fig. 6.17, is very similar. The same zonal mean flow is used. But instead of passing over an isolated circular mountain, it passes over the actual Earth orography, smoothed to match the resolution of the numerical model. The vorticity field reveals the presence of two dominant sets of wavetrains. The larger emanates from the Tibetan plateau, where the equatorward train of waves is especially marked. The other originates from the Rockies, where both poleward and equatorward trains of waves can be seen. The vorticity field is relatively undisturbed over Europe and western parts of Asia. The corresponding streamfunction, Fig. 6.17(b), shows sharp troughs over the eastern coast of North America and over the east Asian coast, with pronounced ridging in the eastern part of both the Pacific and Atlantic Oceans. The pattern should be compared with the observed geopotential height, Fig. 6.1. All these various features are present in the observed fields. Indeed, Fig. 6.16 is as accurate a representation of the observed steady eddy pattern as is produced by many sophisticated global circulation models. Part of the reason for this is that a linear calculation does not permit the eddy fluxes carried by the eddies to change the zonal flow, so that by specifying the zonal mean wind, a large part of the climatology has already been determined. However, the model clearly demonstrates that the radiation of Rossby waves by mountains provides a useful conceptual model of the observed steady eddy pattern.

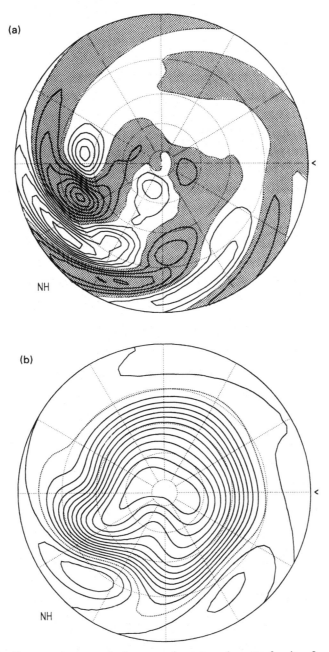

Fig. 6.16. The linear response of a barotropic atmosphere to forcing from an isolated circular mountain at 30°N. The mean flow is for DJF at 30 kPa, averaged from six seasons of ECMWF data. (a) Eddy part of relative vorticity, ξ^*, contour interval $10^{-5}\,\mathrm{s}^{-1}$. Shading indicates positive vorticity anomalies. (b) Streamfunction ψ, contour interval $2 \times 10^7\,\mathrm{m}^2\,\mathrm{s}^{-1}$.

6.4 Vertical propagation of Rossby waves

The theory in the preceding section can be adapted for the case of vertically propagating Rossby waves. The results enable us to extend the discussion of the earlier part of this chapter to a stratified fluid, and are important in helping to clarify the links between the troposphere and the higher regions of the atmosphere. These links will be discussed in Chapter 9. Once again, the approximations made in the theory can easily be criticized. But the simple analysis has great value in establishing a conceptual framework into which the results of more realistic, but consequently less accessible, calculations can be fitted. We will first consider the problem illustrated in Fig. 6.18. An atmosphere is forced by disturbances at its lower boundary; these disturbances may represent flow over mountain ranges, thermal forcings, or (especially in the case of the stratosphere) they may represent disturbances in some lower region of the atmosphere. The problem we will address is whether these disturbances may influence the flow at great heights above the lower boundary, or whether their influence is confined to the lowest levels of the atmospheric layer.

Our starting point will be the quasi-geostrophic potential vorticity equation. Because a particular application of the results is to the stratosphere it is convenient to express that equation in log pressure or 'pseudo-height' coordinates. The pseudo-height is given by $z' = H \ln(p_R/p)$, where the atmospheric scale height $H = RT_0/g$. The quasi-geostrophic potential vorticity equation is written:

$$\frac{D_g}{Dt}\left\{f + \beta y + \nabla^2\psi + \frac{f^2}{\rho_R}\frac{\partial}{\partial z'}\left(\frac{\rho_R}{N^2}\frac{\partial\psi}{\partial z'}\right)\right\} = 0, \qquad (6.36)$$

where ρ_R is the density and N is the Brunt–Väisälä frequency. The density of an isothermal atmosphere varies as

$$\rho_R = \rho_0 e^{-z'/H}. \qquad (6.37)$$

The analysis will be simplified drastically by assuming that the basic state consists of constant zonal wind U and Brunt–Väisälä frequency N. These assumptions can be relaxed, as we shall see later. We partition the streamfunction into zonal mean and eddy parts:

$$\psi = -Uy + \psi^* \qquad (6.38)$$

where $|\psi_y^*|$ is supposed to be small compared with $|U|$. For the present we will assume that the perturbations are steady. Then the quasi-geostrophic

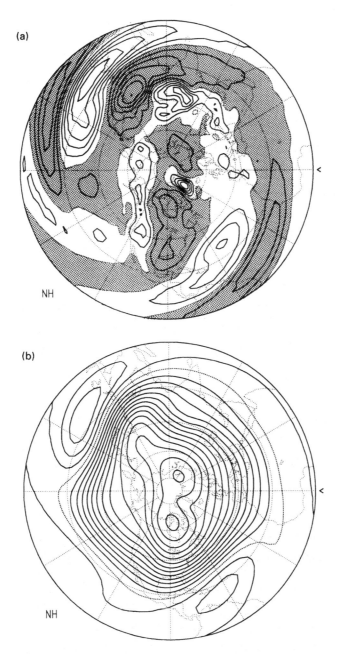

Fig. 6.17. The linear response of a barotropic atmosphere to forcing from the northern hemisphere orography. Mean flow as in Fig. 6.16. This diagram should be compared with the observed field, shown in Fig. 6.2. (a) Eddy part of relative vorticity, contour interval $2 \times 10^{-5}\, \text{s}^{-1}$, positive values shaded. (b) Streamfunction, contour interval $2 \times 10^{7}\, \text{m}^2\, \text{s}^{-1}$.

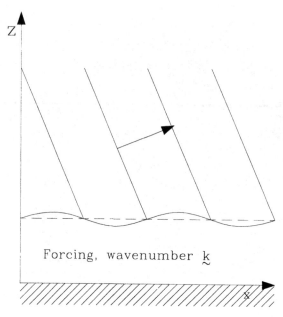

Fig. 6.18. Schematic illustration of the problem to be discussed in this section.

potential vorticity equation can be linearized to give:

$$U \frac{\partial}{\partial x} \left\{ \nabla^2 \psi^* + \frac{f^2}{\rho_R} \frac{\partial}{\partial z'} \left(\frac{\rho_R}{N^2} \frac{\partial \psi^*}{\partial z'} \right) \right\} + \beta \frac{\partial \psi^*}{\partial x} = 0. \qquad (6.39)$$

The forcing at the lower boundary is supposed to be wavelike; then we seek wavelike solutions of the form:

$$\psi^* = \Psi(z') e^{z'/2H} e^{i(kx+ly)}. \qquad (6.40)$$

The factor $\exp(z'/2H)$ is introduced to account for the decrease of density with height. An upward propagating wave must have an increasing amplitude with height if its wave activity is to remain constant. Substitution of this form of solution into Eq. (6.39) yields an equation for the amplitude Ψ:

$$\frac{d^2 \Psi}{dz'^2} + \frac{N^2}{f^2} \left(\frac{\beta}{U} - K^2 - \frac{f^2}{4N^2 H^2} \right) \Psi = 0. \qquad (6.41)$$

As in the preceding section, the total wavenumber K is given by $K = \sqrt{k^2 + l^2}$.

For the stratosphere, N^2 is typically $4 \times 10^{-4}\,\text{s}^{-2}$ and the temperature is around $220\,\text{K}$. Thus the scale height H is $6.4\,\text{km}$, and the term $f/(2NH)$ corresponds to zonal wavenumber 2 in the midlatitudes. Clearly, for negative

U or for large positive U, the coefficient of Ψ is negative, but otherwise it is positive.

These two cases have quite different types of solution. When the coefficient of Ψ is positive, the amplitude relation, Eq. (6.41), may be written:

$$\frac{d^2\Psi}{dz'^2} + m^2\Psi = 0, \tag{6.42}$$

where

$$m = \pm\frac{N}{f}\left(\frac{\beta}{U} - K^2 - \frac{f^2}{4N^2H^2}\right)^{1/2}. \tag{6.43}$$

Equation (6.42) is a wave equation with solutions:

$$\Psi(z') = Ae^{imz'} + Be^{-imz'} \tag{6.44}$$

and m plays the role of a vertical wavenumber. The constants of integration A and B are determined from the condition that the group velocity is upwards, as assumed in Fig. 6.18. Then $B = 0$ and $A = \Psi_0$ is determined by the amplitude of the forcing. The disturbance is vertically propagating and is given by the expression:

$$\psi^* = \Psi_0 e^{z'/2H} e^{i(kx+ly+mz')}. \tag{6.45}$$

That is, the amplitude increases with height as the density decreases. The wave action density, however, remains constant. Such wavelike solutions are favoured for small but positive U or for small total wavenumber K.

In the alternative case, the amplitude equation may be written:

$$\frac{d^2\Psi}{dz'^2} - \mu^2\Psi = 0, \tag{6.46}$$

with

$$\mu = \pm\frac{N}{f}\left(\frac{f^2}{4N^2H^2} + K^2 - \frac{\beta}{U}\right)^{1/2}. \tag{6.47}$$

This has the general solution:

$$\Psi(z') = Ae^{\mu z'} + Be^{-\mu z'}, \tag{6.48}$$

so that μ plays the role of an e-folding height. Note that if a boundary condition of no forcing is applied at $z' \to \infty$, then $A = 0$. The constant B is simply determined by the lower boundary condition, that is, by the forcing imposed at $z' = 0$. The amplitude of the wave may either increase or decrease with height, according to whether μ^{-1} is larger or smaller than $2H$, but the wave action always decreases with height. These evanescent

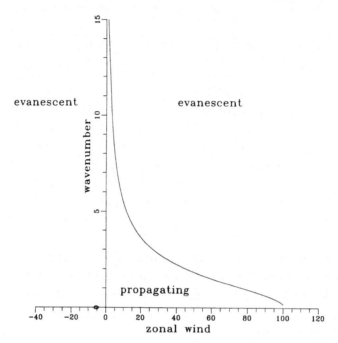

Fig. 6.19. The transition between vertically propagating and evanescent waves as a function of zonal wind U and total wavenumber K. The calculations are for a latitude of $45°$ and $N = 2 \times 10^{-2}\,\mathrm{s}^{-1}$.

solutions are favoured by large wavenumber disturbances, by easterly wind, no matter how weak, or by strong westerly wind.

Figure 6.19 shows the transition from propagation to evanescence as a function of U and K for typical midlatitude stratospheric values of the parameters. In general, only rather long waves are able to propagate vertically. For example, with a zonal wind of $15\,\mathrm{m\ s^{-1}}$, only total wavenumbers of 4 or less (i.e. $K < 9.6 \times 10^{-7}\,\mathrm{m^{-1}}$) can propagate vertically. When propagation is possible, the typical vertical wavelength is rather long. For example, for total wavenumber 3 and U of $15\,\mathrm{m\ s^{-1}}$, the vertical wavelength is 47 km. For the same parameters, but with total wavenumber 7, the e-folding height for the decay of wave activity is 4.2 km.

Associated with the upward propagating Rossby wave is a poleward temperature flux. The hydrostatic relationship leads to

$$T^* = \frac{fT_0}{g}\frac{\partial \psi^*}{\partial z'}, \qquad (6.49)$$

where T_0 is a reference temperature for the level; it can be taken as the mean stratospheric temperature since this varies rather little with height in the lower stratosphere. The perturbation streamfunction for a vertically propagating wave is given by Eq. (6.45), so that the poleward velocity may be written:

$$v^* = \frac{k\Psi_0}{2}e^{z'/2H}\left(ie^{i(kx+ly+mz')} - ie^{-i(kx+ly+mz')}\right) \tag{6.50}$$

and the temperature perturbation is

$$\begin{aligned}
T^* &= \frac{fT_0\Psi_0}{2g}e^{z'/2H}\left\{\left(\frac{1}{2H} + im\right)e^{i(kx+ly+mz')}\right. \\
&\quad \left. + \left(\frac{1}{2H} - im\right)e^{-i(kx+ly+mz')}\right\}. \tag{6.51}
\end{aligned}$$

Thus the poleward temperature flux is

$$\begin{aligned}
v^*T^* &= \frac{fT_0}{4g}\Psi_0^2 e^{z'/H}k\left\{\left(\frac{i}{2H} - m\right)e^{2i(kx+ly+mz')}\right. \\
&\quad \left. - \left(\frac{i}{2H} + m\right)e^{-2i(kx+ly+mz')} + 2m\right\}. \tag{6.52}
\end{aligned}$$

The first two terms in the brackets are wavelike and their zonal mean is zero; the third constant term determines the zonal mean poleward temperature flux, which is

$$[v^*T^*] = \frac{fT_0}{2g}km|\Psi_0|^2 e^{z'/H}. \tag{6.53}$$

As we discussed in Section 5.2, the existence of this temperature flux implies that the upward propagating Rossby wave must show a phase tilt to the west with height. It is easy to use the same arguments to show that if the wave is evanescent in z', then

$$[v^*T^*] = 0. \tag{6.54}$$

In this case, there is no phase tilt, and the troughs and ridges of an evanescent disturbance will be exactly vertical.

Equation (6.53) is at first sight a curious result. Since this model involves no vertical wind shear, the mean temperature gradients are zero, and so there is no energy conversion associated with this flux. In fact, the arguments of Chapter 4 would show that a meridional circulation would develop; provided the Rossby waves are linear, steady and there is no dissipation, the zonal mean heat flux carried by this circulation would exactly balance the eddy heat flux just calculated.

6.5 The Eliassen–Palm flux

The question of how wave activity propagates in both the vertical and poleward directions has now been raised. In this section, we shall see that the propagation of Rossby waves is intimately connected to their interaction with the basic zonal state, thus confirming a result which has already been hinted at in the earlier discussions. The arguments of this section are not restricted simply to the steady disturbances which have been the focus of our consideration throughout the earlier part of the chapter, but, rather, can be applied to propagating Rossby waves of any frequency. The theory that we will develop will bring together the results of Sections 6.2 and 6.3, which were concerned with horizontal propagation and those of Section 6.4, where we discussed vertical propagation.

In this general theory, we will begin with the quasi-geostrophic version of the potential vorticity equation:

$$\frac{Dq}{Dt} = S \tag{6.55}$$

where

$$q = f + \beta y + \xi + \frac{1}{\rho_R} \frac{\partial}{\partial z'} \left(\frac{\rho_R f^2}{N^2} \frac{\partial \psi}{\partial z'} \right) \tag{6.56}$$

and S is a source term which is taken to include all processes such as heating and friction which can alter the potential vorticity of an element of air. In the entirely artificial case where Rayleigh friction and Newtonian cooling are assumed, both with the same time constant τ, S can be written as $(q_e - q)/\tau$, q_e being the equilibrium distribution of q corresponding to zero relative vorticity and radiative equilibrium stratification. But, in general, it is misleading to regard the effects of friction or cooling simply as the dissipation of potential vorticity anomalies.

First, we consider the propagation of disturbances by linearizing the potential vorticity equation about a general zonal flow $U(y, z')$; denoting the perturbation potential vorticity by q^*, the linearized equation is:

$$\frac{\partial q^*}{\partial t} + U \frac{\partial q^*}{\partial x} + [q]_y \frac{\partial \psi^*}{\partial x} = S^*. \tag{6.57}$$

The poleward gradient of potential vorticity of the basic state $[q]_y$ is a crucial parameter. It may be expressed as:

$$[q]_y = \beta - U_{yy} - \frac{1}{\rho_R} \frac{\partial}{\partial z'} \left(\frac{\rho_R f^2}{N^2} \frac{\partial U}{\partial z'} \right). \tag{6.58}$$

Note that there are three contributions to the potential vorticity gradient.

The first term is simply the poleward gradient of planetary vorticity and the second the poleward gradient of relative vorticity. The third term is dominated by $(f/N)^2 \partial^2 U / \partial z'^2$ if the vertical variations of U are significant on a scale smaller than the density scale height. The factor (N/f) serves to determine the natural ratio of horizontal and vertical scales. The perturbation potential vorticity can be expressed in terms of the geostrophic streamfunction:

$$q^* = \nabla^2 \psi^* + \frac{1}{\rho_R} \frac{\partial}{\partial z'} \left(\frac{\rho_R f^2}{N^2} \frac{\partial \psi^*}{\partial z'} \right). \tag{6.59}$$

Now assume that the perturbation is wavelike in x, y and z', and has frequency ω so that it may be written:

$$\psi^* = \Psi_0 e^{z/2H} e^{i(kx+ly+mz'-\omega t)}. \tag{6.60}$$

Substitution in Eq. (6.57) yields the dispersion relation:

$$\omega = Uk - \frac{[q]_y k}{\left\{ K^2 + \frac{f^2}{N^2} \left(m^2 + \frac{1}{4H^2} \right) \right\}}, \tag{6.61}$$

where the source or sink term S^* has been dropped. This is justified provided the frequency and rate of propagation are all fast. The effect of retaining weak dissipation will be examined later. Equation (6.61) is familiar. When U is constant and $\omega = 0$ (steady waves) it reduces to Eq. (6.47). If vertical variations of U and ρ_R are dropped, it reduces to Eq. (6.14). The dispersion relationship enables the group velocity to be calculated, so that the direction of wave activity propagation can be determined. The poleward component of group velocity is:

$$c_{gy} = \frac{\partial \omega}{\partial l} = \frac{2[q]_y kl}{\left\{ K^2 + \frac{f^2}{N^2} \left(m^2 + \frac{1}{4H^2} \right) \right\}^2} = \frac{2[q]_y kl}{K_T^4} \tag{6.62}$$

and the vertical component is:

$$c_{gz} = \frac{\partial \omega}{\partial m} = \frac{2[q]_y f^2 km}{N^2 \left\{ K^2 + \frac{f^2}{N^2} \left(m^2 + \frac{1}{4H^2} \right) \right\}^2} = \frac{2[q]_y f^2 km}{N^2 K_T^4}. \tag{6.63}$$

If a slowly varying approximation were made, these two expressions would form the basis of three-dimensional ray tracing, analogous to the two-dimensional theory developed in Section 6.2. But for the purposes of interpreting atmospheric data, we need not pursue the rigorous development of ray tracing theory.

We may at this point bring together the discussions which led us to Eqs. (6.31) and (6.52). The poleward momentum flux carried by disturbances of the form given in Eq. (6.60) is simply:

$$[u^*v^*] = -\frac{kl}{2}|\Psi_0|^2 e^{z'/H},\qquad(6.64)$$

which is just the result derived earlier but with possible vertical variations included. But comparison with Eq. (6.62) shows that this poleward momentum flux can be related to the poleward group velocity:

$$c_{gy} = -\frac{4[q]_y e^{-z'/H}}{K_T^4|\Psi_0|^2}[u^*v^*].\qquad(6.65)$$

Similarly, the poleward potential temperature flux:

$$[v^*\theta^*] = \frac{f\theta_R}{2g}|\Psi_0|^2 e^{z'/H} km\qquad(6.66)$$

(cf. Eq. 6.52) can be related to the vertical component of the group velocity:

$$c_{gz} = \frac{f}{\theta_{Rz'}}\frac{4[q]_y e^{-z'/H}}{K_T^4|\Psi_0|^2}[v^*\theta^*].\qquad(6.67)$$

From Eqs. (6.65) and (6.67), we see that the direction of the vector

$$\mathbf{F} = \left(-\rho_R[u^*v^*],\quad \rho_R f[v^*\theta^*]/\theta_{Rz'}\right)\qquad(6.68)$$

is parallel to the local group velocity. The vector $\mathbf{F}$ is called the Eliassen–Palm flux, and it is an important diagnostic. The interpretation of the Eliassen–Palm flux as measuring group velocity is robust in situations where group velocity is meaningful. Unlike ray tracing, it does not rely on any assumption about a slowly varying background flow.

But the Eliassen–Palm flux may also be interpreted in other ways. Let us return to the quasi-geostrophic potential vorticity equation, Eq. (6.55), and take its zonal average:

$$[q]_t + [v^*q^*]_y = [S].\qquad(6.69)$$

Consistent with the quasi-geostrophic assumption, the poleward eddy transport of q is the only dynamical term in the zonal mean equation. But the poleward flux of potential vorticity $[v^*q^*]$ is related to the Eliassen–Palm flux. From Eq. (6.59), it follows that:

$$v^*q^* = \left(\frac{v^{*2}-u^{*2}}{2}\right)_x - (v^*u^*)_y + \frac{fv^*}{\rho_R}\frac{\partial}{\partial z'}\left(\frac{\rho_R f\theta_R}{g\theta_{Rz'}}\frac{\partial\psi^*}{\partial z'}\right),\qquad(6.70)$$

where the continuity equation $u_x^* + v_y^* = 0$ has been invoked. The last term can be simplified by relating v_z^* to θ_z^* using thermal wind balance, so that Eq. (6.70) becomes:

$$v^* q^* = + \left(\frac{v^{*2} - u^{*2}}{2} - \frac{g \theta^{*2}}{(\theta_R^2)_z} \right)_x - (u^* v^*)_y + \frac{1}{\rho_R} \frac{\partial}{\partial z'} \left(\frac{\rho_R f v^* \theta^*}{\theta_{Rz'}} \right). \tag{6.71}$$

On taking the zonal averages, the first term, which is a derivative with respect to x, vanishes, and so we may write:

$$[v^* q^*] = \frac{1}{\rho_R} \left\{ \frac{\partial}{\partial y} (-\rho_R [u^* v^*]) + \frac{\partial}{\partial z'} \left(\rho_R f \frac{[v^* \theta^*]}{\theta_{Rz'}} \right) \right\}. \tag{6.72}$$

That is, the modification of the basic state by the eddies is described by the divergence of the Eliassen–Palm flux. This result is quite general; it is equally valid for steady or transient disturbances, provided the quasi-geostrophic approximation can be sustained. The result using pseudo-height $z' = H \ln(p_R/p)$ as the vertical co-ordinate has been given here. An equivalent result using pressure as the vertical coordinate can easily be derived; in this case the Eliassen–Palm flux can be written:

$$\mathbf{F} = \left\{ -[u^* v^*], \quad -\frac{hf}{s^2} [v^* \theta^*] \right\}. \tag{6.73}$$

Note that the vertical component has changed sign, consistent with pressure decreasing upwards. Returning to the linearized potential vorticity equation, Eq. (6.57), in the form:

$$\left(\frac{\partial}{\partial t} + U \frac{\partial}{\partial x} \right) q^* + [q]_y v^* = 0 \tag{6.74}$$

(where friction and diabatic heating are ignored), a further interpretation of the Eliassen–Palm flux can be made. Multiply Eq. (6.74) by $\rho_R q^* / [q]_y$ and take its zonal average; the result of this manipulation is:

$$\frac{\partial}{\partial t} \left(\frac{\rho_R [q^{*2}]/2}{[q]_y} \right) + \rho_R \nabla \cdot \mathbf{F} = 0, \tag{6.75}$$

where the identity Eq. (6.72) has been used. Such a conservation relation is of great interest. It states that the quantity

$$A = \frac{\rho_R [q^{*2}]}{2[q]_y} \tag{6.76}$$

is carried by the propagating disturbances, and is conserved. The flux of A is simply the Eliassen–Palm flux. Indeed, A is a fundamental measure of the vigour of the waves; it is conserved in a way in which, for example, the eddy

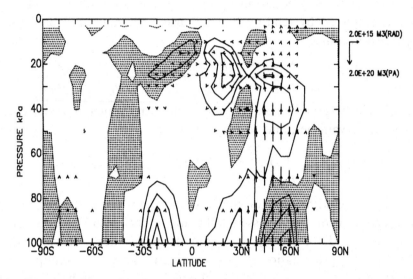

Fig. 6.20. Eliassen–Palm section for the DJF season, based on the steady eddies in six years of ECMWF data. Vectors indicate the magnitude and direction of the Eliassen–Palm flux and contours its divergence, with a contour interval of $10^{15}\,\mathrm{m}^3$; regions of net divergence are shaded. The Eliassen–Palm flux and its divergence has been scaled in a manner appropriate for spherical geometry, as described by Edmon *et al.* (1980).

kinetic energy, is not. It is called the 'wave action density'. Equation (6.76) can be used to deduce the variation of wave amplitude as disturbances propagate into regions of different ρ_R or $[q]_y$. In fact, just such an argument was used in Section 6.3 to consider the momentum flux associated with poleward propagating Rossby waves in a barotropic model. If waves were to propagate into regions of very low density or low potential vorticity gradient, it follows from conservation of wave action density that their amplitude would become very large. The linear assumptions made throughout this section would break down, and the theory would require modification. A plausible scenario is that as the amplitudes and the gradients of vorticity and temperature associated with the waves become large, friction and cooling will become important and the wave activity will be dissipated. But in some circumstances, the nonlinearity could give rise to reflection, or at least to partial reflection. Such conditions are loosely described as 'wave breaking' and could certainly be important in the stratosphere.

The wave activity density has a Lagrangian interpretation in terms of the dispersion of fluid particles initially at the same latitude. From potential vorticity conservation, the poleward displacement of a fluid element is

$$\eta^* = -q^*/[q]_y, \tag{6.77}$$

so that the wave action density is just

$$A = \tfrac{1}{2}\rho_R[\eta^{*2}][q]_y \tag{6.78}$$

That is, the wave action density is proportional to the mean square parcel displacement.

Bearing in mind these theoretical properties of the Eliassen–Palm flux, it is of interest to plot both the flux itself and its divergence when studying the general circulation. A latitude–pressure cross section showing vectors of **F** and contours of $\nabla \cdot \mathbf{F}$ has been advocated as a compact diagnostic of the propagation of disturbances and their interaction with the zonal mean flow. The vectors give information about the propagation of the disturbances, while the divergence gives information about the generation, dissipation and mean flow interaction of the disturbances.

Figure (6.20) shows such a cross section for the steady waves of the troposphere. The Eliassen–Palm flux vectors are vertical in the lower troposphere, but become more horizontal near the tropopause at around 25 kPa. They turn mainly towards the tropics, though some wave activity is seen turning more polewards and propagating through the tropopause into the stratosphere. A partitioning by wavenumber indeed reveals that the shorter waves contribute to the equatorward flux along the tropopause, while only long waves continue upwards and polewards into the stratospheric polar night jet. Convergence of the Eliassen–Palm flux is largest just below the tropopause in the midlatitudes, indicating that the mean flow is modified in this region by the effects of the eddies. The large divergence near the surface demonstrates the generation of steady eddies by various surface forcing processes there.

6.6 Eliassen–Palm fluxes and baroclinic lifecycles

Most of this chapter has been concerned with the steady disturbances seen in the seasonal mean flow, both of the troposphere and of the stratosphere. A final topic in this chapter is to extend our discussion to the transient disturbances of the troposphere. The theory of the preceding section can be applied equally well to propagating transient disturbances; such application leads to helpful insight into the nature of the nonlinear baroclinic lifecycles described in Section 5.5 and leads to an interpretation of the observed transient eddy activity in terms of baroclinic life cycles.

Consider the two paradigms of linear baroclinic instability, the Eady and

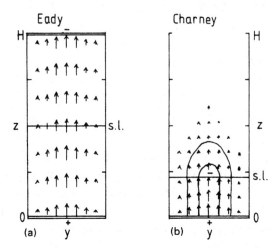

Fig. 6.21. Schematic Eliassen–Palm sections for (a) a linear unstable Eady mode; and (b) a linear unstable Charney mode.

Charney models (see Section 5.4). The linear Eady mode is simplest; it has no horizontal momentum fluxes and the temperature flux is constant with height. The corresponding Eliassen–Palm section is shown schematically in Fig. 6.21. The Eliassen–Palm flux is directed vertically and does not vary with height. We may consider that the divergence is concentrated into a thin layer at the lower boundary and the convergence into a thin layer near the top boundary. The essence of the Eady mode is indeed the interaction between trapped neutral modes on the upper and lower boundaries. A linear Charney mode is only marginally more complicated to describe in terms of its Eliassen–Palm section, despite the considerable mathematical complexity involved in its derivation. Again, the momentum flux is zero and so the Eliassen–Palm fluxes are vertical, with divergence concentrated in an infinitesimal layer at the surface. But the convergence is spread through the lower troposphere, centred on the 'steering level' where $U = c$, c being the linear phase speed of the mode.

Baroclinically unstable linear normal modes can be derived for more realistic jet-like flows on the sphere, as discussed in Section 5.4. The nonseparable nature of the linearized equations makes the problem very intractable to analytic techniques and the normal modes generally have to be computed numerically. Fig. 6.22(a) shows the Eliassen–Palm section for such a mode. In the lower troposphere, the section is quite similar to that of the Charney mode, Fig. 6.21(b), with maximum convergence near 70 kPa. Above this level, momentum fluxes are small but nonzero and some turning of the vectors can be seen. The remainder of the diagram shows Eliassen–Palm sections from various stages of a lifecycle integration, starting from the linear normal mode as initial condition. As the wave saturates, the low level fluxes become smaller, and the maximum wave activity is seen at higher levels. In the decay phase, large, nearly horizontal vectors are seen near the tropopause; wave activity is propagating equatorward in the upper troposphere, and is absorbed before the tropical easterlies are reached. Thus, the baroclinic lifecycle may be described in terms of linear instability generating wave activity at low levels in the midlatitudes. This is followed by upward and, finally, equatorward propagation; at the same time the original seat of the instability is switched off. Finally, wave activity is dissipated in the subtropics, returning energy to the zonal mean flow. This is a typical evolution and the predominantly equatorward propagation is a ubiquitous feature of such waves. Anomalous zonal flows can be produced in which the equatorward propagation is suppressed, and wave activity is focussed towards the pole. In this case, the rapid collapse of the mature disturbance is not seen, and, instead, a slowly decaying, nearly circular, system results.

The final frame of Fig. 6.22, showing the time mean Eliassen–Palm sections for the entire lifecycle, is remarkably similar to the section shown in Fig. 6.23. The latter is based on transient temperature and momentum fluxes observed during the northern hemisphere winter. It shows the same divergence in the midlatitudes and upward propagation through the midtroposphere as does the idealized lifecycle calculation. There is the same strong equatorward propagation along the tropopause towards lower latitudes. In other words, the pattern of transient wave activity in the troposphere is characteristic of nonlinear baroclinic lifecycles. The low level temperature fluxes are described quite well by linear theory, but the momentum fluxes are a feature of the nonlinear mature and decaying phase of the normal modes.

This chapter has taken as its theme the way in which the troposphere can support wave propagation of quasi-geostrophic, balanced disturbances. The insight that such a model gives us helps us to understand many of the features of the distribution of the steady and transient wave activity of the middle latitudes. But these interpretations must not be pushed too far; the assumptions of linearity and slowly varying background states that we have been forced to make do not hold to any great degree of accuracy. But our models do give a vocabulary which enables the results of much more complex, generally numerical calculations to be described.

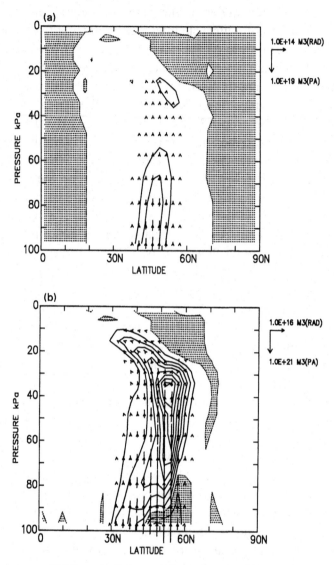

Fig. 6.22. A sequence of Eliassen–Palm sections for the idealized lifecycle of an unstable baroclinic mode: (a) linear normal mode, used as initial data, and scaled as in Fig. 5.24; (b) day 8.

6.7 Problems

6.1 From the longitude–height section given in Fig. 6.4(a), deduce the typical zonal wavelength and vertical phase tilt of midlatitude steady eddies in the northern hemisphere winter. Assuming that the flow is geostrophically

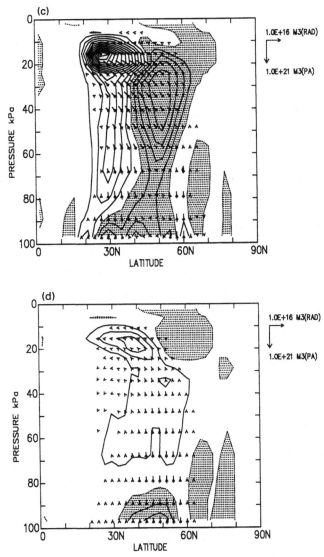

Fig. 6.22 (*cont.*). (c) Day 12; and (d) average through from day 0 to day 15. Contour interval for $\nabla \cdot \mathbf{F}$ is $4 \times 10^{15} \, \mathrm{m}^3$ in (a), (b) and (c) and $2 \times 10^{15} \, \mathrm{m}^3$ in (d).

balanced, deduce the typical poleward temperature flux associated with the steady eddies. Compare your results with observations (see Fig. 6.5(a)).

6.2 Model the Rockies as a ridge 1000 km wide (in the zonal direction), 5000 km long (in the meridional direction), 1.5 km high and centred at 45 °N. Estimate the poleward eddy velocity v^* associated with barotropic flow of

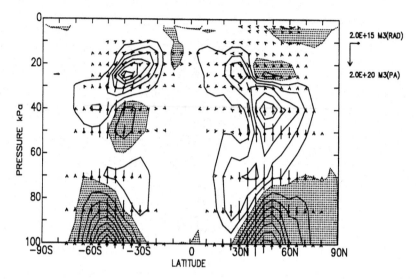

Fig. 6.23. As Fig. 6.20, but for the transient eddies.

$15 \, \text{m s}^{-1}$ over such a ridge. Deduce the sign of the vorticity over the summit of the mountain.

6.3 Repeat the calculations of problem 2, but include the friction term. Assume that the spin-up time τ_D is five days. By how much would this friction modify the value of v^* calculated in problem 2?

6.4 Consider an atmospheric flow of 'uniform super-rotation' in which the zonal wind is given by $U = U_0 \cos \phi$. If a midlatitude β-plane approximation is made, show that Rossby wave packets propagate in straight lines for this particular flow. If U_0 is $20 \, \text{m s}^{-1}$ calculate the maximum zonal wavenumber which can propagate at $45\,°\text{N}$. Calculate the direction in which a zonal wavenumber 4 packet will propagate.

6.5 A zonal wavenumber 3 steady Rossby wave is radiated from a mountain at $45\,°\text{N}$ where the mean zonal wind is a constant $15 \, \text{m s}^{-1}$. Estimate the time taken for a wave packet to reach $65\,°\text{N}$.

6.6 An alternative form of ray tracing considers the distance travelled by the packet, s and the angle the ray makes with the zonal direction α as the fundamental variables. Show that using these variables, Eqs. (6.24a, b) can be rewritten:

$$\frac{ds}{dt} = 2U \cos \alpha, \quad \frac{d\alpha}{dt} = \frac{U \cos^4 \alpha}{k^2} \frac{d}{dy}(K_s^2).$$

Suggest why these equations might be more practical for numerical integration than Eqs. (6.24a, b).

6.7 Take typical values of the poleward temperature flux due to transient eddies in the northern hemisphere winter from Fig. 5.7(a). Hence, estimate the Eliassen–Palm flux divergence in the upper troposphere and the rate of change of zonal wind associated with the transient eddy temperature flux. Compare this with the acceleration associated with the transient eddy momentum flux convergence.

6.8 Starting from Eq. (5.37) which gives the rate of change of total eddy energy, show that the rate of generation of eddy energy is related to the Eliassen–Palm flux by:

$$\frac{dE}{dt} = < \mathbf{F} \cdot \nabla[\bar{u}] > .$$

7

Three-dimensional aspects of the global circulation

7.1 Zonal variations in the tropics

Up to this point, we have followed a traditional exposition of the global circulation by concentrating upon the zonal mean circulation and upon the zonal mean fields of eddy quantities. But the global circulation is far from zonally symmetric. Tropical heating has distinct maxima at particular longitudes. In the midlatitudes, the transient eddies are not distributed uniformly around the latitude circles, but are concentrated into isolated 'storm tracks', especially in the northern hemisphere. This chapter will be devoted to a description of such zonal asymmetries and their consequences.

The various diagnostics of the steady and transient eddy activity which we have considered in earlier chapters become small in the tropics. Eddy kinetic energy is much smaller in the tropics than in the midlatitudes. Similarly, eddy temperature and momentum fluxes, both steady and transient, are much smaller in the tropics. Thus a picture emerges in which heat and momentum are transported, essentially by axisymmetric motions in the tropics, with eddies taking over in the subtropics and midlatitudes.

There is some truth in this picture. But it can also be misleading. First, consider the heating fields shown in Fig. 3.8. The forcing of the circulation is certainly not axisymmetric, especially in the tropics. Rather, there are a small number of centres of intense heating. The most important is over south-east Asia in the JJA season, and over Indonesia in the DJF season. Other maxima are apparently associated with the continental land masses. One matter that we will wish to consider is why such a highly localized distribution of heating leads to a more or less axisymmetric response.

It must also be said that a closer look at the tropical zone shows that the response is not entirely axisymmetric. Figures 7.1 and 7.2 show vectors of the mean wind and near the tropopause and at low levels. The 15 kPa level

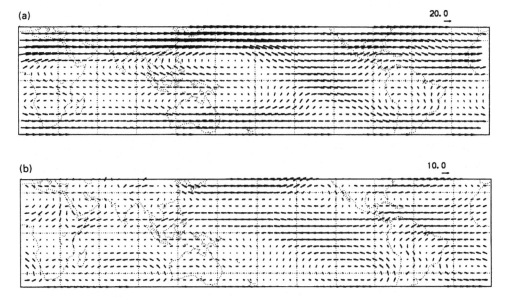

Fig. 7.1. Vectors of the time mean horizontal wind, $\bar{v}$, for the DJF season. Based on six years of ECMWF data for the DJF season. (a) At 15 kPa, near the tropical tropopause: the sample vector represents 20 m s^{-1}. (b) At 85 kPa, just above the atmospheric boundary layer. The sample vector represents 10 m s^{-1}.

is near the tropopause in the deep tropics. There, essentially zonal winds are dominant in both seasons. This is one immediate reason why the eddies do not contribute significantly to meridional transports in the tropics: the meridional part of the eddy wind field is very small. However, the eddy zonal wind is by no means small; there are a number of significant extrema of $\bar{u}$ distributed around the globe. In the DJF season, there are strong easterlies over Indonesia and strong westerlies over the eastern Pacific and over the Atlantic. The zonal mean zonal wind therefore involves a good deal of cancellation; in fact, at these levels, the mean wind is westerly (see Fig. 4.1). In the JJA season, the zonal component again dominates, but the pattern of maxima is quite different. There is a region of strong easterly flow stretching from Africa in the west to Indonesia in the east. Elsewhere, there are weaker easterlies, with very little trace of the patches of westerly flow seen in DJF.

Turning now to the low level wind fields, the same dominance of the zonal over the meridional components of the wind is clear. In DJF, the wind is mainly easterly, with regions of convergence more or less corresponding to regions of divergence at 15 kPa and vice versa. In JJA, a similar pattern is found over most of the tropics. However, dramatically different fields occur over the Indian Ocean. Here we see strong meridional flow, in a broad

(a)   20.0

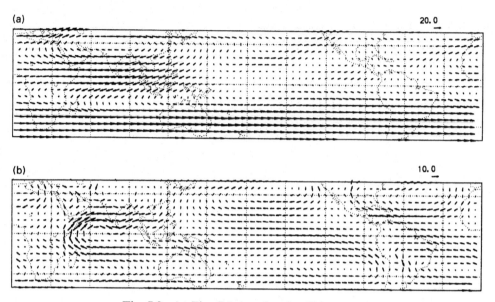

(b)   10.0

Fig. 7.2. As Fig. 7.1, but for the JJA season.

band parallel to the African coast, which links easterly flow in the southern subtropics and westerly flow over India. This flow is part of the Asian monsoon system, a planetary scale system which is sufficiently important to merit separate discussion in the next section.

Comparing the flows at 85 and 15 kPa, there is a strong tendency for low level convergence to be associated with upper level divergence, and vice versa. This is consistent with midlevel ascent and descent linking the two levels, and suggests that the tropical response to localized heating consists of a set of overturning circulations in the longitude–height plane. Such circulations are sometimes called 'Walker circulations', and were first identified by comparing the surface meteorological records at Darwin, north-east Australia and Tahiti in the mid-Pacific.

In attempting to account for these observations, it has to be conceded that elegant theories of tropical dynamics comparable to the quasi-geostrophic and related formulations for midlatitudes do not exist. A group of theories, based on a linearization around a resting atmosphere, represents the simplest approach, and will be outlined in this section. But it is not straightforward to relax the very restrictive assumptions of such a theory. The alternatives soon involve rather elaborate numerical solutions of the full dynamical equations.

Scaling arguments show that temperature fluctuations will be very much smaller in the tropics than in midlatitudes; in fact

$$\frac{\Delta\theta}{\theta} \approx \frac{U^2}{gH} \text{ (tropics)}, \quad \frac{\Delta\theta}{\theta} \approx \frac{fUL}{gH} \text{(midlatitudes)}, \tag{7.1}$$

where H is a typical height scale and L is a typical horizontal scale; $\Delta\theta/\theta$ is around 10^{-3} for the tropics and an order of magnitude larger in the midlatitudes. This means that horizontal temperature gradients in the tropics are small so that horizontal advection cannot balance the large heating observed in convecting regions. Instead, that heating must be balanced by vertical advection, so that:

$$w = \frac{g}{\theta_R N^2}\mathcal{Q}. \tag{7.2}$$

Strong ascent, of around $3\,\mathrm{cm\,s^{-1}}$, is expected in the convecting regions, where the heating may be as large as $5\,\mathrm{K\,day^{-1}}$. Throughout the rest of the tropics, weak descent of around $0.3\,\mathrm{cm\,s^{-1}}$ is needed to balance radiative cooling. Continuity implies that these vertical motions will in turn excite horizontal velocities. The character of the horizontal velocity field is elucidated by considering the momentum equations.

The Coriolis force is zero on the equator, but varies rapidly with latitude. In the dynamical equations, the Coriolis parameter can be approximated by $f = \beta y$ where $\beta = 2\Omega/a$. This approximation is sometimes called the 'equatorial β plane'. Making this approximation, the flow is linearized about a motionless state in which atmospheric variables change only in the vertical direction. The solution can then be separated into a height dependent part and a part which depends upon the horizontal coordinates and time. The horizontal part is governed by the linearized 'shallow water' equations, with a suitable 'equivalent depth'. These equations may be written:

$$\frac{\partial u}{\partial t} - \beta y v = -g\frac{\partial h'}{\partial x} - \frac{u}{\tau_D}, \tag{7.3a}$$

$$\frac{\partial v}{\partial t} + \beta y u = -g\frac{\partial h'}{\partial y} - \frac{v}{\tau_D}, \tag{7.3b}$$

$$\frac{\partial h'}{\partial t} = -h_0\nabla \cdot \mathbf{v} + \mathcal{Q} - \frac{h'}{\tau_D}. \tag{7.3c}$$

Here, Eq. (7.3c) is the linearized continuity equation for a shallow layer of incompressible fluid of depth h_0; $\mathcal{Q}$ represents a forcing which can be related to the heating. Dissipation in the form of Rayleigh friction or Newtonian cooling acting on perturbations has been included. The simplest interpretation of the equivalent depth h_0 is provided by considering an incompressible atmosphere with constant Brunt–Väisälä frequency N and a rigid lid at height H. Then the gravest vertical modes have pressure and horizontal velocity perturbations which vary as $\cos(\pi z/H)$ and a perturbation vertical velocity which varies as $\sin(\pi z/H)$. The equivalent depth is given by:

$$h_0 = \frac{N^2 H^2}{\pi^2 g}. \tag{7.4}$$

Setting H to the depth of the tropical troposphere, around $18\,\mathrm{km}$, a typical equivalent depth is around $400\,\mathrm{m}$. A more satisfactory way of obtaining the same equations, but for a compressible atmosphere with variations of N, etc., in the vertical, is to solve an eigenvalue problem for the vertical structure; similar equivalent depths for the gravest vertical modes emerge. We will restrict our attention to the gravest vertical modes which have a single maximum of vertical velocity at midtropospheric levels, and zero vertical velocity at top and bottom. This is justified because the vertical distribution of tropical heating is expected to excite disturbances which project principally on to such a structure.

Before considering the full forcing problem, it is worth considering some of the wave motions which can take place in the tropical atmosphere. The shallow water equations support gravity waves which propagate at speed $(gh_0)^{1/2}$. But further large scale tropical modes are also possible, and these dominate the large scale response to isolated heating maxima. The simplest solutions to Eqs. (7.3a–c) are obtained when friction and heating are ignored, and the meridional velocity is set to zero. The equations become:

$$\frac{\partial u}{\partial t} = -g\frac{\partial h'}{\partial x}, \tag{7.5a}$$

$$\beta y u = -g\frac{\partial h'}{\partial y}, \tag{7.5b}$$

$$\frac{\partial h'}{\partial t} = -h_0\frac{\partial u}{\partial x}. \tag{7.5c}$$

This set is satisfied by solutions of the form:

$$u = \mathcal{U}(y)f(x - c_0 t), \quad h' = \alpha \mathcal{U}(y)f(x - c_0 t). \tag{7.6}$$

Substituting in, we find:

$$c_0 = \pm(gh_0)^{1/2}, \quad \alpha = (h_0/g)^{1/2} \tag{7.7}$$

and:

$$\frac{\partial \mathcal{U}}{\partial y} = -\frac{\beta}{c_0} y \mathcal{U}. \tag{7.8}$$

Equation (7.8) is simple to integrate, giving:

$$\mathcal{U}(y) = u_0 \exp\left(-\frac{\beta}{2c_0} y^2\right). \tag{7.9}$$

The negative root for c_0 is not physically reasonable, since it would lead to exponentially increasing zonal velocity perturbations away from the equator. The choice of the positive root leads to nondispersive, eastward propagating waves, which are confined to the region of the equator. The meridional scale of these waves is $(2c_0/\beta)^{1/2}$ or around 2000 km when the equivalent depth is 400 m. Their eastward phase speed is about $60 \, \mathrm{m \, s^{-1}}$. These trapped disturbances are called 'equatorial Kelvin waves' and are an important component of the response of the tropical atmosphere to localized thermal forcing.

A variety of further trapped equatorial wave solutions to Eqs. (7.3a–c) can be found, with $v \neq 0$ in general. The general dispersion relationship is:

$$\left(\frac{\omega}{c_0}\right)^2 - k^2 - \frac{\beta k}{\omega} = \frac{(2n+1)\beta}{c_0}, \tag{7.10}$$

where n is an integer. The meridional scale for all these trapped equatorial waves is of order $(c_0/2\beta)^{1/2}$, but their frequencies, and hence their phase speeds, vary substantially. The dispersion relationships are illustrated in Fig. 7.3. The Kelvin wave dispersion relationship is also consistent with Eq. (7.10) when $n = -1$, and this is also plotted on Fig. 7.3. For $n \geq 1$, there exist high frequency equatorial gravity waves which propagate either to the west or to the east. But there is also a group of low frequency 'planetary waves' which propagate to the west; a typical phase speed is around $20 \, \mathrm{m \, s^{-1}}$ for $h_0 = 400$ m. These latter, along with the Kelvin wave, are an important component of the tropical response to localized heating. The $n = 0$ mode is known as the mixed Rossby–gravity wave. For large positive k, these waves are similar in structure to the eastward propagating gravity waves. But for negative k, they are slowly westward propagating modes, similar to the planetary waves.

Now consider the flow forced by a localized heating maximum in terms of these equatorially trapped waves. The problem is rather different from the midlatitude problem discussed in Section 6.2. There, the response to forcing was discussed in terms of Rossby waves which had zero phase speed with respect to the forcing. In the equatorial case, the phase speeds of all the waves shown in Fig. 7.3 are large compared to the typical tropospheric zonal

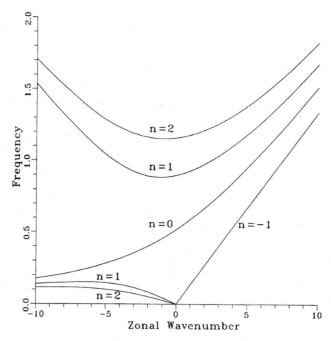

Fig. 7.3. The dispersion relationship, Eq. (7.10), for equatorially trapped wavelike solutions to the linear shallow water equations on an equatorial β-plane. The frequency is given in units of Ω and the wavenumber in units of a^{-1}. The family of curves is governed by the single parameter $c_0/(\Omega a)$ which took the value 0.134 for this calculation.

flow speeds, and so it is pointless to seek wave solutions which are stationary with respect to the forcing. Instead, solutions can be constructed in which the zonal flow is taken to be zero, but in which a Kelvin wave propagates eastward from the forcing region, while planetary waves propagate westward. Steady solutions can be constructed if there is some dissipation in the system, so that the u, v and h' perturbations decay on a timescale τ_D. Then the disturbance decays away from the forcing region on a spatial scale $L = (c_{gx}\tau_D)$ where $c_{gx} = \partial\omega/\partial k$ is the zonal component of the group velocity of the waves. For the Kelvin waves, this scale is large. Taking τ_D as five days, we have $L = 26\,000\,\mathrm{km}$. Hence the Kelvin wave emanating from a single heating maximum would fill much of the tropics. The westward extensions of the response would decay more rapidly; a typical L would be around 8000 km.

A typical solution to the forced problem is shown in Fig. 7.4. As one would expect, there is low level convergence into the heating region. But

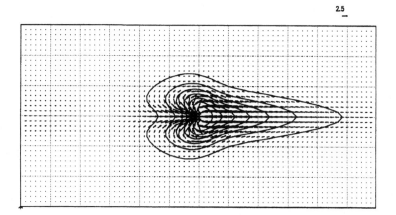

Fig. 7.4. The linear response to a localized heating maximum centred on the equator, showing the low level wind vectors and contours of the pressure perturbation. The upper level fields are, of course, simply the same but with the signs reversed. The plot is based on a numerical solution of the linear shallow water equations; the wind vectors have been arbitrarily scaled so the maximum is $10\,\mathrm{m\,s^{-1}}$.

this convergence is dominated by $\partial u/\partial x$. The meridional winds are virtually zero to the east of the heating region, consistent with the response there being a Kelvin wave, and are generally small to the west of the heating. Thus the major response consists of a zonally overturning 'Walker' cell, with two cyclones on the north-west and south-west flanks of the heating region. Any meridional temperature flux will be small, and will have opposite signs at upper and lower levels. The important conclusion is that the localized heating maxima noted in Fig. 3.8 will have little impact on the meridional heat transport out of the tropics. Rather, the meridional heat tranports are dominated by the mean heat transports. Localized heating will lead to zonal overturning and associated zonal heat transports.

A similar solution can be constructed when the heating maximum is centred away from the equator. The heating can be decomposed into a part which is symmetric about the equator (as in Fig. 7.4) and a part which is antisymmetric, and the response to each forcing superposed. Figure 7.5 illustrates such a solution. To the east of the forcing region, only the symmetric part of the heating has any effect, and the solution is identical to that of Fig. 7.3. To the west of the heating maximum, a large asymmetry in the cyclones either side of the heating region develops, with the cyclone in the same hemisphere as the heating maximum quickly dominating. This solution has similarities to the JJA circulation in the Indian region, a point to which

2.5

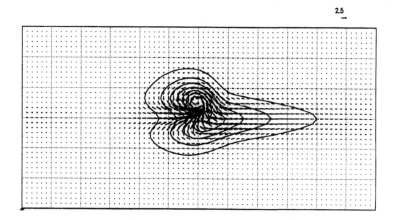

Fig. 7.5. As Fig. 7.4, but showing the surface pressure and surface wind vectors for the case of a heating maximum located 10° to the north of the equator.

we will return in the next section. The meridional winds are stronger, with the result that the zonal mean of the solution shows a significant Hadley cell straddling the equator. The strongest ascent is in the vicinity of the heating maximum, north of the equator, and the descent is largely to the south of the equator in the 'winter' hemisphere. The results have the same general pattern as the axisymmetric theory of Section 4.2, although, in this linearized problem, the strong subtropical zonal winds associated with the Hadley circulation are absent because there is no meridional advection of angular momentum.

It is as well to conclude this section with a word of warning. The vertical modal decomposition on which the theory of this section is based involves linearizing around a state of zero motion. This is a very restrictive assumption which, strictly, can only be justified for very light winds and weak forcings. Such solutions do indeed have much the same qualitative appearance as large scale disturbances observed in the tropical circulation. But more elaborate nonlinear (and generally numerical) models have to be used to establish the quantitative details of the response to large amplitude forcing with general vertical structure. Although waves with structure like the Kelvin and planetary waves discussed in this section are observed in the troposphere, their phase speeds and detailed structure do not accord with the simple theory. The effects of feedbacks between the large scale flow and moist convection, and more realistic boundary layer processes, have to be included. These matters are more properly left to a text on tropical atmospheric dynamics for their fuller discussion.

7.2 Monsoon circulations

One of the largest and most dramatic steady departures from zonal symmetry in the tropics is the summer Asian monsoon circulation. The term 'monsoon' strictly means any seasonal reversal in the circulation. However, the summer monsoon over India and south-east Asia is such a dominant component of the circulation, and is of such enormous human and economic importance that it is often referred to simply as 'the monsoon'. The circulation is primarily associated with changes in the distribution of heating between summer and winter. But other effects, such as the feedbacks between large scale circulation and the release of latent heat in cumulus scale convection, and the influence of mountains also play an important role, making the Asian monsoon a complex and still imperfectly understood component of the JJA circulation.

The essential character of the monsoon is well illustrated by Fig. 7.2. Throughout the JJA season, a strong anticyclone at 15 kPa is centred near India. The zonal extent of this anticyclone is truly enormous. Its influence is clear over more than 90° of longitude, from North Africa to the western parts of the Pacific Ocean. Its meridional extent is around 3 000 km. To its south, strong easterly winds are located near the equator; these make a very significant contribution to the zonal mean easterlies during this season. The low level winds, at 85 kPa for example, tend to exhibit a cyclonic circulation. But the most prominent low level feature is the jet which crosses the equator along the east African coast and, from there, across the Arabian Sea into India. The heating was shown in Fig. 3.8(b); the intense maximum over south-east Asia is one of the most prominent features of this plot. It is largely associated with the seasonal precipitation in this part of the world which lends the monsoon its significance for human activities.

The summer monsoon circulation is a thermally driven circulation which arises primarily from the temperature differences between the hot continental areas in the summer northern hemisphere and the colder oceans in the southern hemisphere. This is something of an oversimplification, though, as is clear from the observations that comparable strong monsoonal circulations do not feature over other subtropical continental areas such as northern Australia or South America. There is a complex feedback between the flow field and the heating, especially through the interaction between moist convection and large scale flow, which is poorly understood but which is undoubtedly involved in ensuring that the Asian monsoon is so much more prominent than the monsoonal circulations over other continents. The special orography of that particular area also modifies the circulation considerably.

The result is that the monsoon is a particularly complex weather system. Its quantitative description is beyond the scope of simple models and probably requires the use of a full global circulation model to do it justice. As a consequence, the brief discussion in this section will be mainly descriptive.

Consider, first, the situation shown in Fig. 7.6, in which a hot land mass lies to the north of the equator, and cooler ocean lies to the south. The asymmetric Hadley cell discussed in Section 4.2 provides a zonally symmetric example of such an idealized monsoonal circulation. Low level cross equatorial flow is set up by the heating distribution. Under the combined influence of pressure gradient forces, friction and Coriolis accelerations, air parcels will have consistently negative relative vorticity, i.e. cyclonic in the southern hemisphere and anticyclonic in the northern hemisphere. The convergence of moist air at low levels in the heating region will lead to convection and release of latent heat, thus giving a positive feedback which will reinforce the monsoonal circulation. At the same time, the warm anomaly over the continent implies, by thermal wind balance, increasing anticyclonic vorticity with height, so that a strong upper anticyclone will overlie the low level 'heat low'.

Now consider the effect on this circulation of a high mountain barrier lying across the equator. Such a barrier is provided by the highlands of east Africa, where the mean height of the continent rises to $1 - 2$ km within a short distance of the coast. The low level barrier provided by these mountains means that any zonal component of the low level flow is blocked at the equator. In an effect somewhat analogous to the formation of western boundary currents in the ocean (see Section 10.5), the northward flow is concentrated in a low level jet along the foot of the orography. The low level east African jet is a prominent feature of Fig. 7.2(b). It dominates the low level flux of mass across the equator in the JJA season. The strong low level winds are also effective at evaporating water from the Arabian Sea, strengthening the convection over the northern continent. This provides yet another positive feedback which amplifies the monsoon circulation.

The Tibetan plateau and the Himalayan mountains also provide an important element in the observed monsoon, in the form of a total barrier to low level meridional winds. The ascent and rainfall caused as the low level flow encounters the mountain barrier both increases the strength of the heating and confines it to the region south of the Tibetan plateau. At the same time, it has been suggested that the sensible heating, caused as sunlight is absorbed by the elevated Tibetan plateau, provides a high level heat source which strengthens the upper level anticyclone.

The qualitative arguments for the existence of the monsoonal circulation

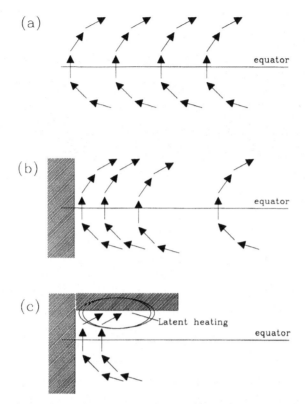

Fig. 7.6. Schematic illustration of idealized monsoonal circulations: (a) flow induced by a hot continent north of the equator and cold ocean south of the equator; (b) effect of orography across the equator; (c) effect of mountains to the north of the heating region.

just given are mainly in terms of the feedbacks between the large scale circulation and the heat source over south-east Asia. However, comparison of Fig. 3.8(b) with Fig. 7.2 shows that the region of strong heating is quite localized compared to the extent of the upper level monsoonal anticyclone. The question arises: what determines the spatial scale of the monsoonal circulation? The linear arguments of the preceding section provide the most accessible simple theoretical prediction. The monsoonal circulation is not unlike the sort of solution shown in Fig. 7.5 which shows the response of the tropical atmosphere to a heating maximum located to the north of the equator. The meridional scale is essentially the equatorial Rossby radius, $(2c_0/\beta)^{1/2}$. The zonal scale according to this theory is related to the friction timescale and the zonal components of the group velocity of waves, i.e., $L = c_{gx}\tau_D$. These ideas certainly suggest the observed zonal elongation of

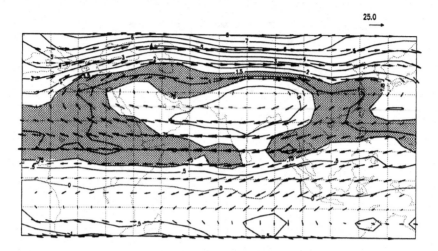

Fig. 7.7. Ertel's potential vorticity on the 360 K surface in the monsoon region during July 1990, computed from ECMWF data by M. Masutani. Shading indicates values between 0.75 and 1.5 PVU. The region covered by the plot extends from $0\,°$E to $140\,°$E and from $20\,°$S to $50\,°$N.

the upper monsoonal circulation, but are probably too crude to give very useful numerical estimates of the scales. It must be recognized that the monsoonal circulation is highly nonlinear, and so the applicability of linear theory is dubious. This is illustrated by Fig. 7.7, which shows the Ertel potential vorticity associated with the upper monsoonal circulation. The nonlinearity is expressed by the large distortion of the potential vorticity contours, and the significant angle between the contours and the velocity vectors.

7.3 Midlatitude storm zones and jets

In the midlatitudes, the distribution of transient eddy kinetic energy, and other eddy quantities, is not uniform. Rather, the transient activity at higher frequencies is concentrated in localized regions. Partitioning the data according to the frequencies of the transients is simply accomplished by the application of a numerical filter, such as that described in Section 5.1. We shall separate high frequency transients, whose periods are less than 6–10 days, from low frequency transients with longer periods. Figure 7.8(a) shows the high frequency eddy kinetic energy for DJF. The field is dominated by two elongated maxima at midlatitudes, one located over the Atlantic and one over the Pacific. Much the same pattern is found when other measures of eddy activity are considered. Some examples of these are shown in the other parts of Fig. 7.8. A similar pair of maxima are seen in the geopotential height variance. The temperature fluxes suggest that the maxima in kinetic energy are closely associated with baroclinic instability. For example, Fig. 7.8(c) shows the vertical temperature flux at 70 kPa. This is upwards nearly everywhere, but has maximum values near the western end of the kinetic energy maxima. As we saw in Section 5.3, such upward temperature fluxes are a necessary condition for the conversions of available potential energy into kinetic energy. Finally, Fig. 7.8(d) shows the poleward momentum flux, $\overline{u'v'}$ at 25 kPa. The largest values are again found close to the maxima in eddy kinetic energy. This time, however, the pattern is rather more complex. Large poleward momentum flux is found to the south and near the eastern end of the kinetic energy maxima. Poleward of this location, there is a weaker maximum of equatorward momentum flux.

The association between baroclinic energy conversions and the maxima of transient eddy activity suggests that there may be a link between synoptic weather systems and the distribution of eddy activity. Indeed, this turns out to be the case. Figure 7.9 shows the tracks of the surface centres of major depression systems observed during a particular winter over the Atlantic. Superimposed is the 90 m variance contour of the 25 kPa geopotential height. The cyclogenesis region lies towards the westwards extremity of the high pass eddy activity maximum, close to the eastern coast of North America. Thereafter, the tracks of the depressions tend to run close to the major axis of the geopotential height variance. The cyclones are observed to decay and fill mainly towards the eastern side of the Atlantic, where the levels of eddy activity are observed to decline. The close association between depression tracks and the high pass filtered eddy variances have led to the latter being called 'storm tracks'. A similar association is found for the Pacific storm track.

The occurrence of discrete storm tracks is presumably related in some way to the strong zonal variations of the surface properties in the northern hemisphere. We will consider exactly how this might operate in the next section. The southern hemisphere is more uniform in its surface properties, especially at latitudes south of 40 °S, where most of the baroclinic energy conversion takes place. Yet a structure analogous to the northern hemisphere storm tracks is still observed. Figure 7.10 is similar to Fig. 7.8, except that it shows the southern hemisphere during the winter (JJA) season. There is a distinct maximum in the high pass filtered eddy kinetic energy in the southern Atlantic and Indian Oceans, and a relative minimum in the southern Pacific. Other high pass eddy statistics show much the same variation along this

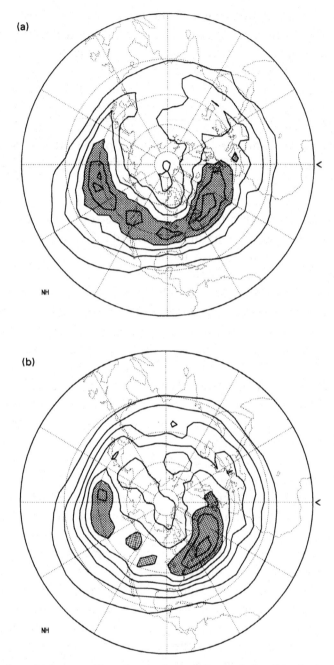

Fig. 7.8. Various high pass filtered transient eddy statistics for the northern hemisphere, DJF period. Based on six years of ECMWF data. (a) Eddy kinetic energy, $\overline{(u'^2 + v'^2)}/2$. Contour interval $25\,\text{m}^2\,\text{s}^{-2}$, values in excess of $100\,\text{m}^2\,\text{s}^{-2}$ shaded. (b) Geopotential height variance, $\overline{Z'^2}^{1/2}$. Contour interval $15\,\text{m}$, values in excess of $90\,\text{m}$ shaded.

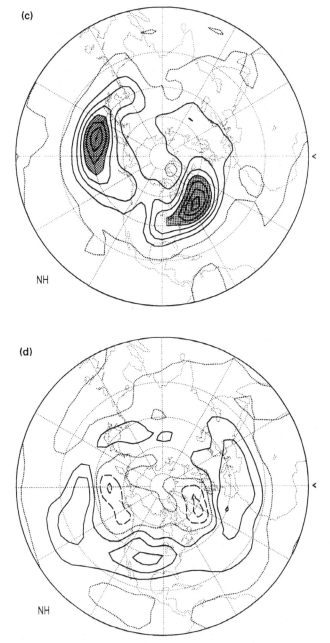

Fig. 7.8 (*cont.*). (c) Vertical eddy temperature flux, $\overline{\omega'T'}$. Contour interval $0.05\,\mathrm{K\,Pa\,s^{-1}}$, values less than $-0.2\,\mathrm{K\,Pa\,s^{-1}}$ shaded. (d) Northward momentum flux, $\overline{u'v'}$. Contour interval $10\,\mathrm{m^2\,s^{-2}}$. Dashed contours indicate equatorward momentum flux.

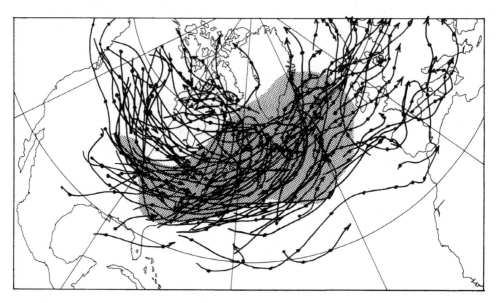

Fig. 7.9. The tracks of low pressure centres over the North Atlantic for the period December 1985 to February 1986. The shading indicates the region where the high frequency $\overline{Z'^2}^{1/2}$ exceeded 90 m in the ECMWF analyses for the same period.

'storm track' as observed in the northern hemisphere, namely, the elongated maximum in geopotential height variance, the large vertical temperature flux at low levels, and the dipolar structure of the poleward momentum flux towards the downstream end of the storm track. Attempts to correlate the tracks of synoptic systems with the variance maximum are less successful than in the northern hemisphere. There is a tendency for the cyclonic systems to spiral polewards from cyclogenesis regions on the equatorward flank of the 'storm track' to decay regions in the 'circumpolar trough', the region of low pressure around the Antarctic coast. Partly at least, this is the result of attempting to identify the centres of synoptic weather systems by means of extrema in the surface pressure field. Because the surface wind field is strong around the southern hemisphere baroclinic zone, there is a natural tendency for centres of low pressure to be displaced poleward of the vortex centre, and for centres of high pressure to be displaced equatorward. But, partly, it seems that this spiral trajectory of weather systems is real. Perhaps it would be better to describe the regions of large high frequency variance as 'storm zones' rather than 'storm tracks'. However, the latter nomenclature is in general use despite being rather misleading.

Each of the three major storm zones has a distinctive seasonal behaviour.

The Atlantic storm track is most pronounced during DJF and least pronounced in JJA. Its actual location varies very little. The Pacific storm track is at its most intense during the transition seasons, MAM and SON, is rather weaker during DJF and is much weaker during JJA. Thus, there is a clear semi-annual as well as annual cycle in its behaviour. The southern hemisphere storm track shows rather little seasonal variation, though it is at somewhat lower latitudes during DJF than the remaining seasons, and is marginally more intense during MAM. Once again, the seasonal cycle has a semi-annual as well as an annual component. The most significant feature is that, for all three storm zones, the location of the beginning of the zone seems to be independent of season, even though the intensity and length of the storm zone varies.

The storm zones occur in association with major jet streams in the troposphere. Particularly in the northern hemisphere winter, the zonal flow varies considerably with longitude. Fig. 7.11 shows the mean wind speed, averaged with respect to pressure, in both hemispheres for DJF and JJA. In the northern hemisphere winter, the most prominent jet is located in the western Pacific, with a shorter and weaker jet over the east coast of North America and the west Atlantic. A third jet is located over Arabia. In the summer, the jets are much weaker, but occur in much the same places. The southern hemisphere jet has a much smaller seasonal cycle. It is strongest over the southern Atlantic and Indian Oceans. Particularly in winter (JJA), there is a minimum of wind speed in the vicinity of New Zealand, with the depth averaged flow splitting to north and south. Comparison between Fig. 7.11 and the various transient eddy quantities shown in Figs 7.8 and 7.9 reveals that the northern hemisphere storm zones are most intense near the longitude of the jet exits, but rather on their poleward flanks. The Atlantic storm zone conforms most closely to this pattern. In the southern hemisphere, the storm zone is much longer, and it begins roughly in phase with the acceleration of the depth average wind. If anything, the strongest eddy activity is slightly towards the equatorward side of the jet. Because the Atlantic storm zone is rather better observed, and certainly more fully documented than the other major storm zones, writers have a tendency to regard it as the 'normal' storm zone from which the Pacific and southern hemisphere storm zones deviate to some extent. But it could equally be that the Atlantic storm zone is the anomalous one.

In Section 5.5, the lifecycle of a single idealized baroclinically unstable disturbance was described. The growing phase of the wave was characterized by large poleward and upward temperature fluxes, associated with the conversion of available potential energy to eddy kinetic energy. As the

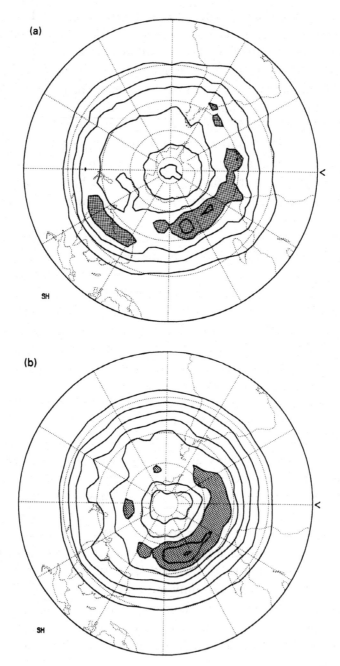

Fig. 7.10. As Fig. 7.8, but showing the southern hemisphere during the JJA season: (a) Eddy kinetic energy, $\overline{(u'^2 + v'^2)}/2$; (b) Geopotential height variance, $\overline{Z'^2}^{1/2}$.

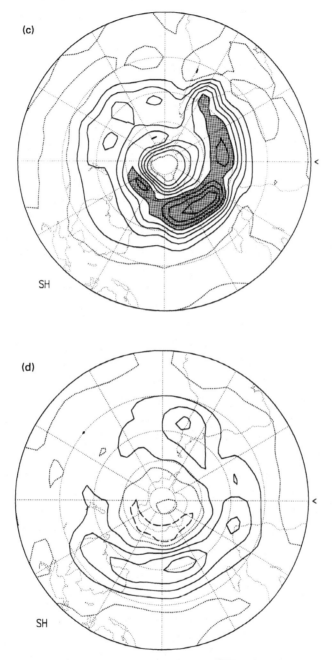

Fig. 7.10 (*cont.*). (c) Vertical eddy temperature flux, $\overline{\omega'T'}$; (d) Northward momentum flux, $\overline{u'v'}$. Plotting conventions as for Fig. 7.8 except for (c), where the contour interval is $0.02 \, \text{K Pa s}^{-1}$ and values less than $0.1 \, \text{K Pa s}^{-1}$ are shaded.

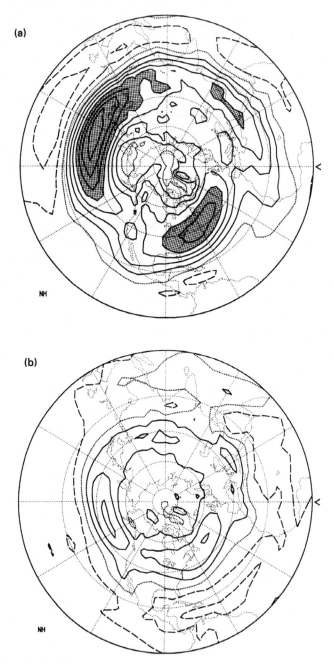

Fig. 7.11. The zonal wind speed, averaged with respect to time and pressure between 100 kPa and 15 kPa. Contour interval 5 m s^{-1}. Shading indicates values in excess of 20 m s^{-1}: (a) northern hemisphere, DJF; (b) northern hemisphere, JJA.

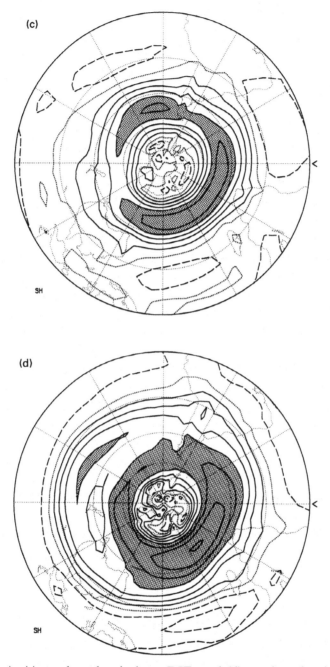

Fig. 7.11 (*cont.*). (c) southern hemisphere, DJF; and (d) southern hemisphere, JJA. Based on six years of ECMWF data.

wave saturated and entered its decay phase, these temperature fluxes became relatively small, but 'barotropic decay' involving strong conversions from eddy to zonal kinetic energy became important. These conversions require large poleward temperature fluxes, with convergence of zonal momentum into the latitude of the main tropospheric jet. The reader will realize that this description of the variations of eddy fluxes and variances in time through a baroclinic lifecycle is very similar to the variations of the eddy fluxes and variances in space as one passes along any one of the three major tropospheric storm zones. For the short Atlantic storm zone, the disposition of cyclone tracks, with cyclogenesis concentrated near the western end of the zone, and cyclolysis towards the eastern end, makes this interpretation straightforward. The length of the storm zone represents the distance across which a typical North Atlantic depression develops and decays. The much longer southern hemisphere storm zone must be interpreted rather differently. Cyclones tend to spiral polewards across this storm zone; there is a greater preponderance of developing cyclones in the section through the Atlantic and western Indian Oceans and a preponderance of decaying systems at the eastern end of the zone.

The concentration of active, growing eddies in the storm track regions suggests that one might expect these regions to be more baroclinically unstable. Linear theory leads to some possible measures of baroclinic instability. For example, the growth rate of the most unstable baroclinic mode is, according to Eady's theory:

$$\sigma = 0.31 \frac{f}{N} \frac{\partial U}{\partial z} \tag{7.11}$$

(see Eq. (5.52)) where U is the basic state zonal wind field. Figure (7.12) shows a plot of this quantity, with $\partial U / \partial z$ taken as $\partial \mid \mathbf{v} \mid / \partial z$, evaluated between 70 kPa and 85 kPa during the northern hemisphere winter. Maxima in the growth rate are seen near the start of the Atlantic and Pacific storm tracks.

7.4 Interactions between transient and steady eddies

The aim of this section is to seek an explanation of why the midlatitude cyclone belt is broken into discrete storm tracks. In fact, much of the discussion will concern consistency relationships between the various circulation statistics; it is much more difficult to make firm statements about causal relationships. This requires more complex analyses of data, or, increasingly, numerical experimentation.

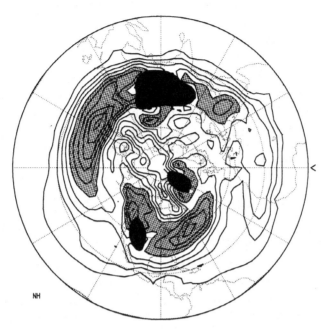

Fig. 7.12. Contours of the Eady baroclinic growth rate at 77.5 kPa, as defined by Eq. (7.11), for the DJF season. Regions which are beneath the ground surface have been blacked out. Contour interval 0.1 day^{-1}, values greater than 0.5 day^{-1} shaded. Based on six years of ECMWF data. (courtesy P.J. Valdes.)

We saw in the preceding section how the northern hemisphere storm tracks are associated with the major tropospheric jet streams. This relationship is indeed consistent with the distribution of cyclogenetic regions around a jet stream, and is not merely a matter of chance association. At the level of quasi-geostrophic theory, the time mean zonal momentum equations can be written:

$$\overline{\mathbf{v} \cdot \nabla u} = f \overline{v}_a \qquad (7.12a)$$

$$\overline{\mathbf{v} \cdot \nabla v} = -f \overline{u}_a \qquad (7.12b)$$

These equations may be thought of as describing the variations of the ageostrophic part of the wind as one follows the time mean track of fluid parcels. For the purposes of the present argument, variations of f are unimportant compared to variations of the wind speed. The jet streams have a predominantly zonal orientation, are generally more or less straight and are most intense near the tropopause. Consider Eq. (7.12a). This relationship suggests that there is likely to be poleward ageostrophic flow in the entrance to jets, and equatorward ageostrophic flow in the exit regions. The ageostrophic wind will be small away from the jet. Figure 7.13 shows the wind speed and ageostrophic wind vectors for the upper level northern hemisphere troposphere in DJF. It confirms this general pattern rather nicely.

This ageostrophic wind field is divergent on the equatorward flank of the jet entrance and on the poleward flank of the jet exit. Continuity requires that there must be ascent at midtropospheric levels below these divergence regions. (The high static stability of the levels above the jet core means that the vertical velocities induced in the stratosphere will be much smaller.) Thus meridional circulations will be set up in the jet entrance and exit regions, with midtropospheric ascent on the equatorward flank of the jet entrances and the poleward flank of the jet exit. In both these regions, there will be generation of cyclonic vorticity by vortex stretching at low tropospheric levels, and hence a tendency to cyclogenesis. The Eady model of baroclinic instability, discussed in Section 5, suggests that the growth rate of the most unstable baroclinic disturbances is proportional to the Coriolis parameter f (see Eq. (5.52)). So the cyclogenetic effect is likely to be particularly pronounced at higher latitudes, that is, on the poleward jet exit.

These arguments will be familiar to synoptic meteorologists. They are an application to the time mean flow of 'development theories' which were elaborated in the pre-numerical weather forecasting era to identify regions where generation or deepening of cyclone systems was likely to occur. They suggest that the meridional circulation associated with the major tropospheric jet streams will favour baroclinic instability in the jet exits, where the observed storm tracks are located. Whether this effect alone is capable of generating the observed storm tracks is another matter. Indeed, one could just as well argue that the existence of a region of enhanced baroclinic instability would reduce the vertical wind shear, and hence lead to a jet exit at upper levels.

Let us now consider the properties of the transient eddies themselves. We will treat the upper tropospheric flow as approximately two-dimensional and nondivergent. The correlations between the two fluctuating components of the wind can be written in tensor notation as:

$$\mathbf{C} = \begin{pmatrix} \overline{u'^2} & \overline{u'v'} \\ \overline{u'v'} & \overline{v'^2} \end{pmatrix}, \tag{7.13}$$

which can be written as the sum of a diagonal and a symmetric tensor:

$$\mathbf{C} = \begin{pmatrix} K & 0 \\ 0 & K \end{pmatrix} + \begin{pmatrix} M & N \\ N & -M \end{pmatrix}, \tag{7.14}$$

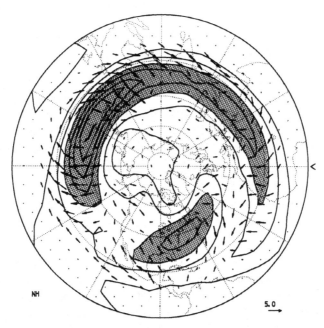

Fig. 7.13. Isotachs of the mean wind at 25 kPa for the DJF period, northern hemisphere, together with vectors of the ageostrophic wind at the same level. Contour interval $10\,\mathrm{m\,s^{-1}}$, with values greater than $30\,\mathrm{m\,s^{-1}}$ shaded. The sample vector represents $5\,\mathrm{m\,s^{-1}}$. Based on six years of ECMWF data.

where $K = \overline{(u'^2 + v'^2)}/2$, $M = \overline{(u'^2 - v'^2)}/2$ and $N = \overline{u'v'}$. The first tensor measures the kinetic energy of the transient eddies. The second also has an immediate physical interpretation: it yields information about the shape of eddies which can be derived using the formulae of Section 5.2. We will refer to this second tensor as the 'anisotropy' tensor. The tilt of the eddies is given by Eq. (5.10). If the velocities are referred to axes rotated through the angle:

$$\varphi = \frac{1}{2}\tan^{-1}\left(\frac{N}{M}\right), \tag{7.15}$$

that is, to axes parallel to the major and minor axes of the eddies themselves, then the anisotropy tensor is diagonalized. The component of eddy velocity parallel to the major axis of the eddy is denoted $\hat{u}'$, while that parallel to the minor axis is $\hat{v}'$. The mean departure of the eddies from circularity is measured by these diagonal elements, which have magnitude $\hat{M} = (M^2 + N^2)^{1/2}$. The quantity $\hat{M}/K$ may be thought of as a dimensionless measure of the eddy anisotropy, varying from 0 when the eddies are exactly circular, and tending towards 1 when the eddies are so elongated that $\overline{v'^2}$ tends to zero.

The properties of the velocity correlation tensor can be summarized in a single plot, in which contours of K are plotted on top of vectors showing the direction of the principal axis of extension of the mean transients, and whose length is proportional to $\hat{M}$. Figure 7.14 compares the results for the high frequency transients with those for the low frequency transients for the northern hemisphere winter. The high frequency eddies are strongest in the two storm track regions, over the Pacific and Atlantic Oceans. They are extended in the meridional direction, but towards the end and to the south of the storm tracks, they develop stronger tilts. The low frequency eddies are mainly zonally elongated; there is a tendency for the largest values to be in the jet exit regions. We will return to the topic of low frequency variability in Chapter 8.

A primary concern of all these studies is the way in which the mean flow $\bar{\mathbf{v}}$ is modified by the eddies and their transports. This becomes more complicated in the case of a flow which varies in the zonal as well as in the meridional direction. A possible approach would start with Eqs. (7.12a, b), ignoring friction, with the winds are expressed as a sum of mean and transient eddy parts:

$$\bar{\mathbf{v}} \cdot \nabla \bar{u} + (\overline{u'^2}/2)_x + (\overline{u'v'})_y = f\bar{v}_a, \tag{7.16a}$$

$$\bar{\mathbf{v}} \cdot \nabla \bar{v} + (\overline{u'v'})_x + (\overline{v'^2})_y = -f\bar{u}_a. \tag{7.16b}$$

A given transient eddy forcing might be balanced either by accelerations of the mean flow (the first terms on the left hand side of these equations) or by an ageostrophic flow. Indeed, the largest eddy term in these equations proves to be $(\overline{v'^2})_y$, and this will principally be balanced by $f\bar{u}_a$, not by variations of the mean flow. The ageostrophic wind can be removed from the balance by combining Eqs. (7.12a, b) to give a vorticity equation. The changes of the mean vorticity would be balanced by the divergence of the eddy vorticity fluxes. This is not a very practical solution, since such vorticity flux divergences involve taking very high derivatives of a field which is likely to be noisy anyway. The components of the eddy anisotropy tensor provide a way round these difficulties.

The time mean vorticity equation without friction can be written:

$$\bar{\mathbf{v}} \cdot \nabla \bar{\zeta} + \nabla \cdot (\overline{\mathbf{v}'\zeta'}) = 0. \tag{7.17}$$

After a certain amount of manipulation using $\zeta' = v'_x - u'_y$, the transient

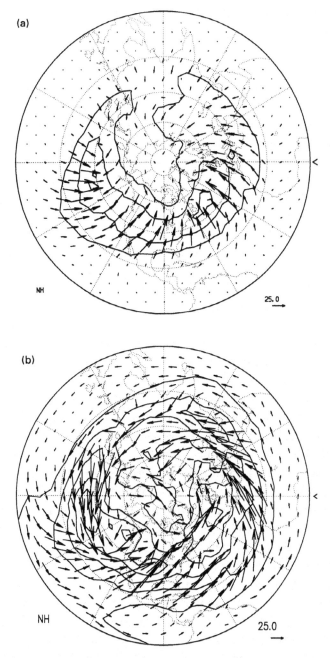

Fig. 7.14. The velocity correlation tensor for the DJF season at the 25 kPa level: (a) high frequency eddies; (b) low frequency eddies. Contours show K, contour interval $50 \, \mathrm{m^2 \, s^{-2}}$, while the vectors are parallel to the major axis of $\mathbf{A}$ and with length proportional to $\hat{M}$; the sample vector has length $25 \, \mathrm{m^2 \, s^{-2}}$. Based on six years of ECMWF data.

eddy vorticity flux divergence can be written:

$$\overline{\mathbf{v}'\zeta'} = (-M_y + N_x, -M_x - N_y), \tag{7.18}$$

so that:

$$\nabla \cdot (\overline{\mathbf{v}'\zeta'}) = -2M_{xy} + N_{xx} - N_{yy}. \tag{7.19}$$

Now it is seen from plots of the variances and covariances of u' and v' presented in the last section that $|N| \ll |M|$, and, furthermore, that the maxima in the high frequency eddy statistics tend to be elongated in the zonal direction. From this, it follows that $|N_{xx}| \ll |N_{yy}|$. Neglecting the N_{xx} term in Eq. (7.19) enables the eddy vorticity flux divergence to be rewritten:

$$\nabla \cdot (\overline{\mathbf{v}'\zeta'}) \simeq \frac{\partial}{\partial y} (\nabla \cdot \mathbf{E}), \tag{7.20a}$$

where the 'vector' $\mathbf{E}$ is defined as

$$\mathbf{E} = (-2M, -N) = (\overline{v'^2 - u'^2}, -\overline{u'v'}). \tag{7.20b}$$

Figure 7.15 illustrates the pattern of mean flow forcing associated with the distribution of $\mathbf{E}$. A local region where $\mathbf{E}$ converges will be characterized by the forcing of anticyclonic vorticity to the north of the convergence region and cyclonic vorticity forcing to the south. The signs are reversed for a region of divergence. The net effect is that convergence is associated with easterly acceleration of the mean flow and divergence with westerly acceleration. The components of $\mathbf{E}$ are reasonably smooth, so plotting vectors of $\mathbf{E}$ gives a fairly clear picture of its pattern of convergence or divergence.

The orientation of the $\mathbf{E}$ vector is closely related to the orientation of the eddies themselves. The direction which the $\mathbf{E}$ vector makes with the x-axis is $\tan^{-1}(N/2M)$, while the angle made by the major or minor axis of the eddies is $\frac{1}{2}\tan^{-1}(N/M)$. When $|N| \ll |M|$, these angles are equal, and, in fact, the difference between them is only 5% when N is as large as $M/2$.

One further interpretation of the components of the $\mathbf{E}$-vector will be discussed before turning to an examination of the observed data. They can be related to the group velocity of the transient disturbances. The arguments are essentially those already developed in Section 6.5, especially Eq. (6.65). The group velocity of Rossby waves in a flow which varies in both the x- and y-directions can be written:

$$c_{gx} = \bar{u} + \frac{\left((k^2 - l^2)\bar{\zeta}_y - 2kl\bar{\zeta}_x\right)}{(k^2 + l^2)^2}, \tag{7.21a}$$

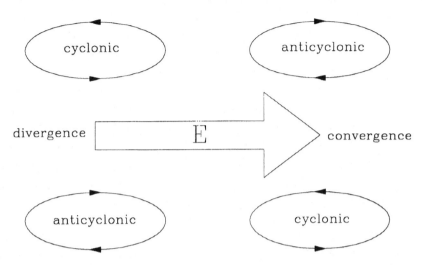

Fig. 7.15. The forcing of the mean flow by convergence and divergence of the E-vector.

$$c_{gy} = \bar{v} + \frac{\left(2kl\bar{\zeta}_y + (k^2 - l^2)\bar{\zeta}_x\right)}{(k^2 + l^2)^2}. \tag{7.21b}$$

Then, assuming that the scale of variations of the time mean flow is large compared to the scale of individual disturbances, and assuming that the eddies can be represented by sinusoidal disturbances, we may write

$$\frac{M}{\overline{\zeta'^2}} = \frac{(l^2 - k^2)}{(k^2 + l^2)^2}, \quad \frac{N}{\overline{\zeta'^2}} = -\frac{2kl}{(k^2 + l^2)^2}. \tag{7.22}$$

It then follows that:

$$\frac{1}{2}\overline{\zeta'^2}(\mathbf{c}_g - \bar{\mathbf{v}}) = (-M\bar{\zeta}_y + N\bar{\zeta}_x, -M\bar{\zeta}_x - N\bar{\zeta}_y). \tag{7.23}$$

Choose local coordinates in which the x-axis is parallel to contours of the time mean absolute vorticity, $\bar{\zeta}$. Generally, the $\bar{\zeta}$ contours will only make a small angle θ with the latitude circles. The transformed M and N are given by

$$(\tilde{M}, \tilde{N}) = (M \cos 2\theta + N \sin 2\theta, -M \sin 2\theta + N \cos 2\theta) \tag{7.24}$$

(see Eqs. (5.8a, b) and (5.9)). In these rotated coordinates,

$$\frac{\overline{\zeta'^2}}{2 \,|\, \nabla\bar{\zeta} \,|}(\mathbf{c}_g - \bar{\mathbf{v}}) = (-\tilde{M}, -\tilde{N}). \tag{7.25}$$

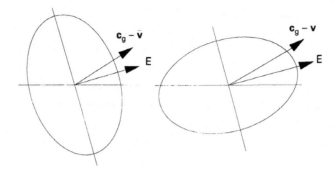

Fig. 7.16. The directions of the eddy axes, the E-vector and the group velocity relative to the mean flow for some typical eddies.

That is, the angle between the group velocity relative to the mean flow and the $\bar{\zeta}$ contours is approximately twice the angle $\tan^{-1}(\tilde{N}/2\tilde{M})$ between the $\bar{\zeta}$ contours and the E-vectors. Figure 7.16 illustrates the relationship between some typical eddies and the directions of the E-vector and the relative group velocity vector. We conclude that, knowing the E-vector, we can summarize information about the typical shape of the eddies, the interaction between the eddies and the mean flow and the direction of the group velocity vector. The latter is perhaps the least satisfactory application of the E-vector, but it is still of qualitative value. Other, more exact diagnostics, have been devised, but they are generally much more difficult to calculate.

Figure 7.17 shows an example of the E-vector, for the high frequency eddies in the northern hemisphere winter. The fields have been averaged in the vertical. The vectors are orientated more or less from west to east, with largest magnitude downstream and slightly poleward of the main jet core. At the end of the storm tracks, the vectors develop a more meridional component, particularly to the south. These features should be compared with the maps of the various transient velocity variances and covariances and with the maps of the mean major axes of the eddies shown in Fig. 7.14. In terms of the forcing of the mean flow, the eddies appear to be reinforcing the existing jet streams. The E-vector shows divergence, implying westerly acceleration, in the region of the jet cores. The maximum divergence is in fact rather poleward of the jet axis, so it is clear that other processes, manifested by time mean ageostrophic circulations, must also act to generate the observed jets. In the eastern part of the ocean basins there is convergence of **E**, or easterly acceleration. These regions are where the westerly jet is weaker. Indeed, these regions are characterized by blocking episodes, when the westerly flow can reverse for periods of several days to some weeks. The

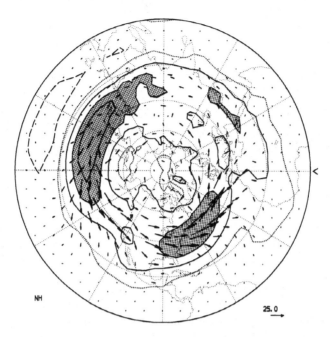

Fig. 7.17. The mean DJF high frequency **E**-vectors for the northern hemisphere, averaged with respect to pressure. The sample vector represents $25 \, \text{m}^2 \, \text{s}^{-2}$. The contours show $\bar{u}$, also averaged with respect to pressure, contour interval $10 \, \text{m} \, \text{s}^{-1}$, with values in excess of $20 \, \text{m} \, \text{s}^{-1}$ shaded. Based on six years of ECMWF data.

E-vector diagnostics suggest that interactions with transient eddies are an important element in initiating and maintaining such blocking episodes.

The two-dimensional theory outlined above is useful in providing a conceptual framework for the general interaction between eddies and a zonally varying flow. Furthermore, if the flow is integrated in the vertical, in order to isolate the barotropic part of the circulation, the thermal effects of the eddies integrate to zero, leaving just the effects of these mechanical forcings by the eddies. However, if one is concerned with a particular level in the atmosphere, thermal effects may be large, and the two-dimensional theory must be extended. Within the quasi-geostrophic system, such an extension is provided by considering the forcing of quasi-geostrophic potential vorticity, rather than simply the vorticity, as was done above. Similarly, the quasi-geostrophic potential vorticity equation yields formulae for both the vertical and horizontal components of group velocity, thus enabling relationships between the eddy statistics and the propagation of wave activity to be derived.

Only the main results of this line of argument are summarized here. The three-dimensional generalization of the **E**-vector is:

$$\mathbf{E} = \left(\overline{v'^2 - u'^2}, \ -\overline{u'v'}, \ -\frac{hf}{s^2}\overline{v'\theta'} \right). \tag{7.26}$$

The meridional and vertical components of this vector are simply the two components of the Eliassen–Palm flux, introduced in Section 6.5. The zonal and meridional components are the components of the barotropic **E**-vector defined in Eq. (7.20b). We conclude that the **E**-vector is a generalization of the Eliassen–Palm flux to three-dimensional flow. The interaction between the three-dimensional **E**-vector and the mean flow is given by:

$$\nabla \cdot (\overline{\mathbf{v}'q'}) \simeq \frac{\partial}{\partial y}(\nabla \cdot \mathbf{E}), \tag{7.27}$$

where q is the quasi-geostrophic potential vorticity. Clearly, this represents a generalization of Eq. (7.20a). A similar parallel exists with Eq. (7.25):

$$\frac{\overline{q'^2}}{2\,|\nabla \overline{q}\,|}(\mathbf{c}_g - \overline{\mathbf{v}}) = \left(-\tilde{M} - \frac{\overline{\theta'^2}}{2s^2}, \ -\tilde{N}, \ -\frac{hf}{s^2}\overline{v'\theta'} \right). \tag{7.28}$$

Here the tilde means that M and N are referred to axes which are parallel and normal to the $\overline{q}$ contours. If the term $\overline{\theta'^2}/(2s^2)$ in the zonal component of the right hand side can be neglected, then this reduces to the three-dimensional **E**-vector. In fact, $\overline{\theta'^2}/(2s^2)$ is not much less than $\tilde{M}$ in the troposphere; nevertheless, Eq. (7.28) suffices to give at least a qualitative description of the group velocity vector. Now consider the typical distribution of **E** in a storm track. The vertical component is largest at low levels towards the westward end of the storm track. The horizontal components are largest at upper tropospheric levels in the middle and eastward sections of the storm track. The meridional component is particularly large towards the eastern end. It follows that a fairly full description of the three-dimensional **E**-vector can be obtained by plotting contours of $hf(\overline{v'\theta'})/s^2$ at low levels superimposed upon vectors showing the horizontal components of **E** in the upper troposphere. Such a plot for the Atlantic storm track is shown in Fig. 7.18.

All this discussion does not take us much closer to an explanation of why storm tracks are seen in their observed locations. Indeed, the poleward heat flux carried by baroclinic waves suggests that the temperature gradients should be weakened after the passage of such a system, making further baroclinic developments in the same region less likely. Yet the opposite is clearly the case: systems tend to follow one another along the same 'storm track'. The **E**-vector suggests that the large scale barotropic circulation induced by a maximum of the transient eddy activity, illustrated in Fig. 7.15,

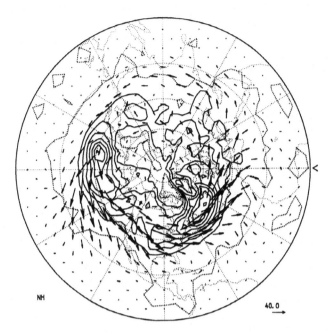

Fig. 7.18. The three-dimensional E-vector for the DJF Atlantic storm track. Contours show $hf(\overline{v'\theta'})/s^2$ at 70 kPa (contour interval 0.5 Pa m s^{-2}), while the vectors show $(\overline{v'^2 - u'^2}, -\overline{u'v'})$ at 25 kPa; the sample vector represents 40 m^2 s^{-2}. Based on six years of ECMWF data.

will tend to tighten the gradients of potential temperature (or any other conserved tracer) upstream of the maximum and to weaken it downstream of the maximum. This effect could offset the direct temperature transports by the eddies themselves, and suggests that any zonally uniform baroclinic zone might tend to break into isolated patches of more intense eddy activity. But, appealing purely to interactions between the transient eddies and the zonal mean flow, there is no reason why these patches should be stationary with respect to the Earth's surface.

It seems that further interactions between the steady waves forced by orography, land–sea contrasts, etc., and the transients must ultimately be responsible for locking the storm tracks to particular geographical locations. As we saw in Chapter 6, the main processes which generate stationary waves in the atmosphere are the forcing of Rossby waves by mountains and by isolated heat sources. In Chapter 6, we largely restricted our discussion to the response of a barotropic atmosphere to such forcings, a restriction justified by the remarkable similarity between the stationary wave pattern forced in that simple model and the observed steady waves. But to discuss the zonal variations of baroclinicity, measured, for example, by the Eady growth rate, Eq. (7.11), the baroclinic response of the atmosphere to forcing must be considered. This is a more difficult technical problem because such an atmosphere is dominated by baroclinic instability; in any time integration of a linear baroclinic model, the unstable modes will quickly swamp the neutral stationary modes. However, such calculations that isolate only the steady response to forcing have been carried out. They reveal that the zonal variations of baroclinicity are not particularly large when the forcing of stationary waves is just due to orography. Similarly, the mechanical forcing of the mean flow by the baroclinic eddies has only a weak effect in modifying the baroclinicity, although it does increase the baroclinicity at the start of the storm track slightly. This confirms the feedback between the transients and the mean flow mentioned above, but suggests that it is fairly weak. A much more important effect comes from the forcing of three-dimensional steady waves by the midlatitude distribution of heating. This has a strong maximum in the storm track regions, as shown in Fig. 3.8(a). It is largest at lower tropospheric levels, with weak cooling above 50 kPa and to the north of the storm tracks.

Figure 7.19 shows the result of one such linear calculation. Here, the time and zonal mean DJF northern hemisphere flow has been forced by the observed heating in the sector 80 °W to 20 °E, shown in Fig. 7.19(a). The upper level streamfunction, Fig. 7.19(b), shows a train of Rossby waves propagating out of the storm track into the tropics, in accordance with the ideas of Section 6.2. But this structure is truly baroclinic; the low level stream function perturbation has a rather different pattern, implying that there is a significant temperature perturbation associated with the propagating wave. As a result, the low level baroclinicity, Fig. 7.19(c) varies sharply across the storm track region, reproducing the observed pattern shown in Fig. 7.12 rather well.

In itself, this result does not really explain the location of the storm tracks. The pattern of heating used to perturb the zonal flow derives largely from the release of latent heat in vigorously developing baroclinic systems. Thus there is a strong feedback between the developing systems and the low level baroclinicity which tends to break the midlatitude baroclinic zone into discrete storm tracks. The question is why this feedback should be so marked in the observed locations and not elsewhere. Various possibilities exist which are connected to the contrast between continents and oceans. One is simply the reduced surface drag over the oceans compared to that over the continents. In a region of uniform baroclinicity, such a reduction in drag could lead to significantly faster growth rates. The other is the forcing by the ocean surface temperatures. The start of the Atlantic storm track is coincident with the confluence of warm water from the Caribbean, with much colder water which has flowed south from the Arctic waters around Greenland. These warm subtropical waters provide a major source of water vapour which can release latent heat when it is condensed in frontal regions. But, as we shall see in Section 10.5, the currents in the upper ocean are largely driven by wind stresses imposed by the overlying atmosphere. The wind stresses represented schematically by the eddy induced circulations shown in Fig. 7.15 are in fact those needed to drive the warm Gulf Stream in the Atlantic and the warm Kuroshio current in the Pacific. It seems that the tropospheric storm tracks are indeed self-sustaining. They owe their permanence to feedbacks involving moist processes in frontal regions, and to the feedbacks between the ocean surface temperature and the surface wind stress. The continental boundaries serve to fix the geographical locations of these feedbacks.

7.5 The global transport of water vapour

Water vapour is the most important variable constituent of the Earth's atmosphere. Its distribution and transport determines rainfall, and therefore it is of great practical concern for agriculture and other human activities. From a scientific point of view, huge amounts of heat energy go to evaporating water which can be released when condensation of liquid water occurs. It is this process which is largely responsible for the highly localized pattern of heating in the tropics, and it is also at least an element in the dynamics of the midlatitude storm tracks. Furthermore, water vapour and clouds profoundly modify the radiative transfer properties of the atmosphere. The resulting feedbacks represent perhaps the greatest uncertainty in modern global circulation modelling.

The concentration of water vapour is measured by the humidity mixing ratio r, defined in Section 1.1 as the ratio of the mass of water vapour in an air sample to the mass of dry air. In the absence of precipitation or evaporation, the humidity mixing ratio of a sample of air is conserved. More generally, a suitable equation for the humidity mixing ratio is:

$$\frac{\partial r}{\partial t} + \mathbf{v} \cdot \nabla r = E - P, \tag{7.29}$$

where E is the rate of evaporation and P is the rate of precipitation of water vapour. For some purposes, this equation might be refined by treating the

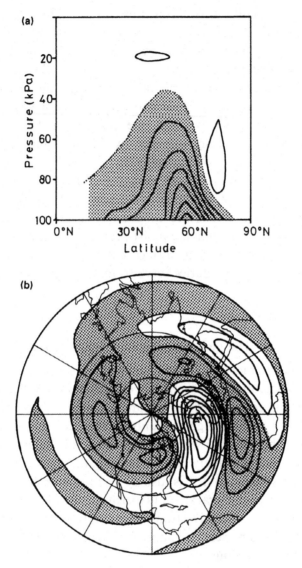

Fig. 7.19. The steady linear response of the DJF northern hemisphere circulation to forcing by the observed heating within the Atlantic storm track region. (a) Cross section of the heating averaged through the sector from 80°W to 20°E, contour interval 0.25 K day^{-1}. (b) Streamfunction perturbation ψ^* at 20 kPa, contour interval 1.5×10^6 m^2 s^{-1}.

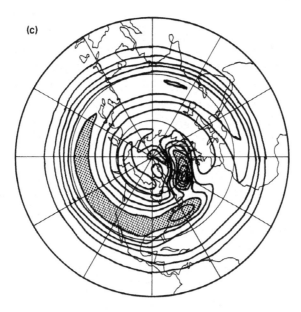

Fig. 7.19 (*cont*.). (c) The Eady growth rate, Eq. (7.11), for the perturbed flow. Contour interval 0.1 day^{-1}. (From Hoskins & Valdes 1990.)

liquid water suspended in cloud droplets separately. A representation of the microphysical processes whereby droplets coalesce and grow, leading to their eventual precipitation, may then be included among the governing equations. Such schemes are attempted in some numerical models of the atmosphere. But for our purposes, it will be sufficient to suppose that water is either in the vapour state or else it is precipitated out of the atmosphere. Thus E is small except near the Earth's surface. On the other hand, P is large at mid-tropospheric levels. Thus it is clear that one effect of the global circulation must be to transport water vapour away from the Earth's surface. It must also transport water vapour away from the principal evaporation regions over warm ocean surfaces to cooler and drier regions.

The time and zonal mean distribution of r, $[\bar{r}]$, is shown in Fig. 7.20. There is considerable variation of $[\bar{r}]$ over the meridional plane, with large values of as much as 1.8×10^{-2} at lower levels in the tropics. At higher levels and latitudes, the concentrations of water vapour are very much smaller, dropping below 10^{-3} at the 50 kPa level, and falling as low as 10^{-6} in the stratosphere. These large variations of humidity mixing ratio largely reflect the distribution of $[\bar{T}]$. The saturation vapour pressure of water vapour, e_s, is a function of temperature only, and is determined by the Clausius–Clapeyron equation:

$$\frac{de_s}{dT} = \frac{Le_s}{R_v T^2},$$ (7.30)

where L is the latent heat of evaporation of water and R_v is the gas constant for water vapour. If L is assumed independent of temperature, which is an approximation which is adequate for our present purposes, Eq. (7.30) is straightforwardly integrated to give:

$$e_s(T) = e_{s0} \exp\left\{ \frac{L}{R_v} \left(\frac{1}{T_0} - \frac{1}{T} \right) \right\}.$$ (7.31)

Here, e_{s0} is the saturated vapour pressure at some reference temperature T_0. For $T_0 = 273$ K, $e_{s0} = 611$ Pa. Thus the saturated vapour pressure increases very roughly exponentially with temperature, doubling every 10 K or so. The saturated humidity mixing ratio r_s is then given by:

$$r_s = \frac{Re_s}{R_v(p - e_s)}.$$ (7.32)

Comparison of Fig. 7.20 with a cross section of $[\bar{T}\,]$ shows that despite its very low humidity mixing ratio, air in the upper tropical troposphere is nearly saturated. As we shall see in Chapter 9, the low humidity of the stratosphere is because the primary transport of air from the troposphere into the stratosphere is across the tropical tropopause, where the temperature is a minimum. Thus, the typical stratospheric r is controlled by r_s at the tropical tropopause. The seasonal variations of $[\bar{r}]$ at lower levels largely reflect the seasonal variations of temperature. The maximum $[\bar{r}]$ is at low levels in the summer hemisphere. The seasonal cycle is more pronounced over the northern hemisphere, a result of the very large seasonal cycle of low level temperature, and hence humidity mixing ratio, over the continental interiors, particularly over central Asia.

Before proceeding further with an analysis of the distribution and transport of water vapour, we should review the errors encountered in measuring the global water vapour field. Routine measurements of r are almost entirely based upon the radiosonde network. Although satellite radiometers do detect emission bands of water vapour, these only give information about the temperature and concentration of water vapour in the upper troposphere. Figure 7.20 shows that more than half the water vapour is below 85 kPa. A new generation of microwave sounders should give information about the vertically integrated wave vapour abundance, though information about its vertical distribution will not be available. These techniques are now operational, but it will be several years before they can contribute to a

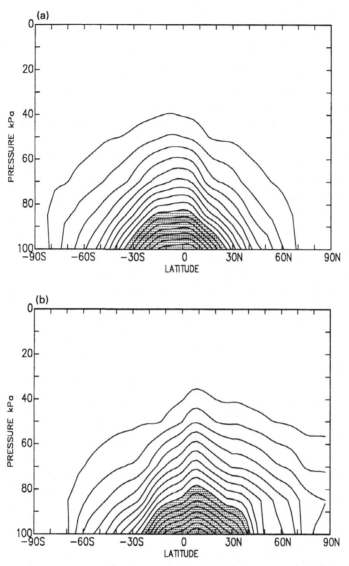

Fig. 7.20. Cross section of humidity mixing ratio $[\bar{r}]$ for (a) December 1991 –
February 1992; and (b) June – August 1991. The contour interval is 10^{-3}, and
shading indicates values in excess of 10^{-2}. (Based on an unpublished analysis of
ECMWF data by J. Dodd.)

detailed climatology of water vapour. As far as the radiosonde network
itself is concerned, each sonde will observe a detailed and accurate profile
of r during its ascent through the lower troposphere; the accuracy of the
measurement declines in the upper troposphere and stratosphere. But the
radiosonde network is highly biased to northern hemisphere land areas. The

water vapour field over the oceans and in the tropics is much more poorly observed. Even in regions where the network is good, problems are still encountered as to how representative a given ascent might be. Water vapour concentrations vary considerably on short spatial scales, and there is no guarantee that a single ascent will be typical of a broader area. Indeed, it is clear that an ascent which passes through a cloud layer will give a very different humidity profile from a neighbouring ascent which passes through cloudless air.

The analysis phase of a numerical weather prediction cycle will generate a detailed moisture field. Over ocean areas, this will consist of almost undiluted background field, and so it should perhaps have more the status of a numerical simulation than of an analysis of observations. Indeed, experiments with numerical weather prediction models suggest that the patterns of rainfall produced during the forecast are not very sensitive to the initial moisture field. For example, simply setting the initial relative humidity to be a constant 70% everywhere does not lead to much poorer rainfall forecasts, at least after an initial period of adjustment, than does attempting a full analysis of r.

Direct measurements of the E and P terms in Eq. (7.29) are also limited in coverage and accuracy. Precipitation is monitored in detail over land areas, but is extremely difficult to estimate over the oceans. Even over land, local effects of orography, etc., make for for very large variations of P over small distances, so there are problems with establishing how representative particular measurements might be. Only relatively few stations are equipped to make direct measurements of E; because of the very strong dependence of E on surface vegetation and other properties, it is again difficult to say how representative such measurements might be. Indeed, E is probably best inferred as a residual between the transport and precipitation terms, rather as was done in Section 3.4 with the thermodynamic equation in determining the heating.

With these reservations in mind, we turn now to consider the three-dimensional distribution of water vapour in the atmosphere. Fig. 7.21 shows the time mean, vertically integrated water vapour abundances for DJF and JJA. The dominant variation is between equator and pole, as suggested by Fig. 7.20. But there are also substantial variations around latitude circles. Some of these are related to orography; there is much less water vapour above a mountain than above a lowland area because of the rapid decrease of $\bar{r}$ with height. But there is a general tendency for the water loading to be lower over the continental interiors than over the oceans in the middle and higher latitudes. In the tropics, there are important variations related to

(a)

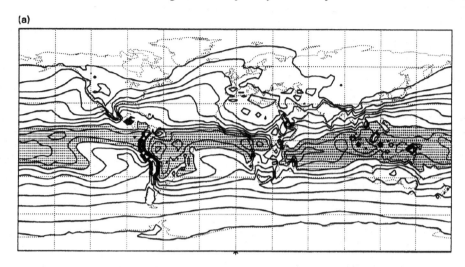

(b)

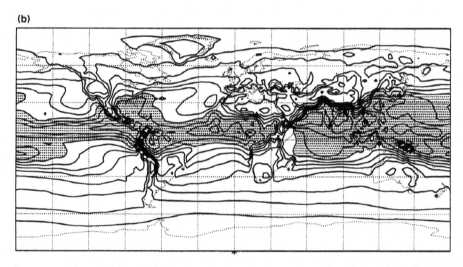

Fig. 7.21. Vertically integrated water vapour abundance for (a) December 1991 – February 1992; and (b) June, July and August 1991. Contour interval $5\,\mathrm{kg\,m^{-2}}$, with shading denoting values in excess of $40\,\mathrm{kg\,m^{-2}}$. (Based on an unpublished analysis of ECMWF data by J. Dodd.)

the fluctuations of sea surface temperature, and hence to E. As mentioned above, this effect is especially marked over central and north-eastern Asia, which is extremely dry in winter.

A major feature is seen in JJA over south-east Asia. The largest values of $\bar{r}$ are found over the Bay of Bengal, with an extended region where it exceeds $50\,\mathrm{kg\,m^{-2}}$ over India and south-east Asia. This maximum is an important part of the monsoon circulation. It reflects a mean convergent flux of low level moisture into this region from the surrounding ocean areas, especially the Arabian Sea. The availability of this moisture leads to convection and latent heat release. A marked maximum in the atmospheric heating rate was shown in Fig. 3.8(b); such heating in turn drives the motions responsible for the monsoon. These processes were described in Section 7.2. Such a feedback between latent heat release and the large scale motion field is a crucial aspect of many tropical circulation systems.

The fluxes of atmospheric water vapour are shown in Figs. 7.22 and 7.23. The first shows the mean vertically integrated flux of water vapour by the atmospheric flow, that is:

$$\mathbf{F}_m = \int_0^{p_s} \overline{\mathbf{v}r}\,\frac{dp}{g}. \tag{7.33}$$

The transport of water vapour by the zonal winds is dominant, with a strong westward transport throughout the tropics and an eastward transport in the midlatitudes that is only about half as strong as that in the tropics. The zonal flow is stronger in the midlatitudes, but the typical values of $\bar{r}$ are considerably smaller. Throughout the tropics, there is a general divergence of $\mathbf{F}_m$ over the oceans and convergence over the continents. In the midlatitudes, maxima of the eastward and poleward transport of water vapour are associated with the storm tracks. The total flux is roughly parallel to the storm track axis. Especially notable is the strong flux of water vapour in the JJA season from the Indian Ocean into south-east Asia, reflecting the low level wind pattern associated with the monsoonal circulation.

The depth integrated transient flux $\mathbf{F}_t$, defined as

$$\mathbf{F}_t = \int_0^{p_s} \overline{\mathbf{v}'r'}\,\frac{dp}{g}, \tag{7.34}$$

is shown in Fig. 7.23. This is dominated by the midlatitude storm tracks. The transient fluxes in the midlatitudes are in a quite different direction from the total fluxes. Instead of being roughly parallel to the storm track axes, they are nearly at right angles to it, with strong poleward and westward fluxes along the storm tracks. Although the transient fluxes are nearly an order of magnitude smaller than the mean fluxes, the divergence of the transient water vapour fluxes is only slightly less than that of the mean fluxes in the midlatitude storm track regions. The moisture flux by the baroclinic eddies at the start of the storm tracks tends to import moisture to the continental interiors of North America and eastern Asia from the warm oceans to the south and east. These transient fluxes are more than offset by the mean fluxes, so that there is net divergence of water vapour out of

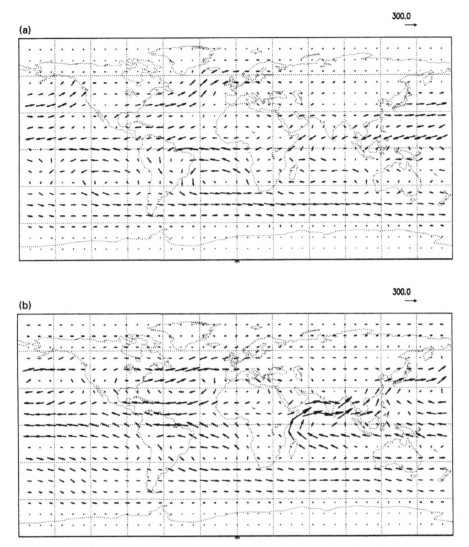

Fig. 7.22. The total fluxes of water vapour, $\mathbf{F}_m$ (see Eq.(7.33)) for (a) December 1991 – February 1992; and (b) June, July and August 1991. The sample arrow represents $300\,\mathrm{kg\,m^{-1}\,s^{-1}}$. (Based on an unpublished analysis of ECMWF data by J. Dodd.)

northern Canada and central Asia. Such a mean circulation is consistent with the mean circulations induced by the concentration of transient activity in the storm tracks, as shown in Fig. 7.15, and undoubtedly plays a role in supplying heat, in the form of latent heat, which is needed to maintain the storm tracks.

Finally, notice that these generalizations do not extend to the southern

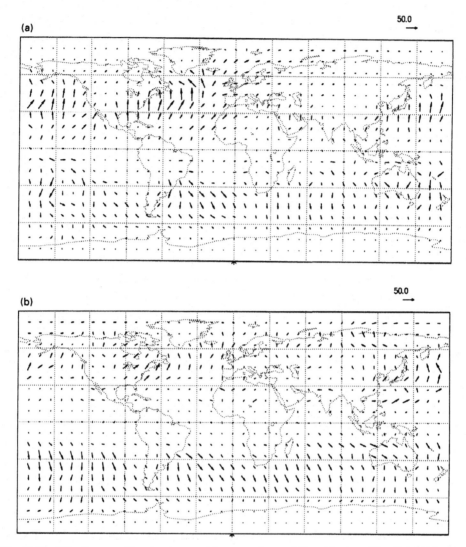

Fig. 7.23. The fluxes of water vapour by transient eddies, $\mathbf{F}_t$ (see Eq.(7.34)) for (a) December 1991–February 1992; and (b) June, July and August 1991. The sample arrow represents $10 \, \text{kg} \, \text{m}^{-1} \, \text{s}^{-1}$. (Based on an unpublished analysis of ECMWF data by J. Dodd.)

hemisphere storm track. The transient fluxes are indeed a maximum in the region of the southern hemisphere storm track, but they are eastwards and polewards throughout. There is some evidence of a flux of moisture out of South America due to the midlatitude transients. This is fed by a mean flux convergence, chiefly from the tropics. It may well be that the southern hemisphere storm track is formed by rather different physical mechanisms from the two northern hemisphere storm tracks. On the other hand, we should recall that the accuracy of these moisture fluxes over the southern hemisphere oceans is probably very poor, and better observations are urgently needed before such conclusions can be established with any certainty.

7.6 Problems

7.1 The forced solution to the shallow water equations shown in Fig. 7.4 is linear, and therefore there is no interaction between the forced disturbance and the mean flow. Use the diagram to discuss qualitatively the momentum flux associated with the forced pattern, and hence deduce the zonal flow that you might expect to be forced.

7.2 From Fig. 7.3, estimate the group velocities of Kelvin waves and planetary waves. If the drag timescale is taken to be five days, estimate the eastward and westward scales of disturbances induced by a tropical heating anomaly situated on the equator.

7.3 Use Fig. 7.17 to estimate the typical divergence of $\mathbf{E}$ near the start and near the end of the Atlantic storm track. Hence, estimate the typical acceleration of the zonal wind in these regions (in units of $\mathrm{m\,s^{-1}\,day^{-1}}$).

7.4 Using the information of Fig. 7.18, estimate the typical gradient of the three-dimensional $\mathbf{E}$-vector in (a) the zonal direction, (b) the meridional direction.

7.5 Suppose that temperature varies sinusoidally around a latitude circle between temperature $T_0 - \Delta T$ and $T_0 + \Delta T$, but that the relative humidity is a constant R_0. Show that the apparent zonal mean relative humidity R_a, based on the zonal mean temperature and specific humidity, can exceed 100%.

7.6 Estimate typical midlatitude values of

$$\frac{\partial \bar{r}}{\partial y}, \frac{\partial \bar{r}}{\partial p}.$$

The typical variance of humidity mixing ratio, $\overline{[r'^2]}^{1/2}$, is 10^{-3} in the midlatitudes. Use this value to estimate the typical displacement of air masses if the fluctuations of water vapour are due solely to: (a) poleward motions; (b) vertical motions. Discuss the relationship between your values and the typical properties of midlatitude weather systems.

7.7 It is believed that air is mostly exchanged between the troposphere and stratosphere in the tropics, where the tropopause is extremely cold, perhaps 200 K. Estimate the saturated vapour pressure at 200 K and hence predict typical values of r in the stratosphere.

8

Low frequency variability of the circulation

8.1 Low frequency transients

In the preceding chapter, Fig. 7.14 compared the eddy correlation tensor for the higher frequency eddies, whose periods are less than around ten days, with the lower frequency transients. The high frequency eddy statistics have a well-defined structure in the midlatitudes, with maxima in the storm track regions. The high pass filter used in Chapter 7 served to isolate a specific family of dynamical processes, namely, those associated with baroclinic instability and the subsequent evolution of baroclinically unstable waves. The low frequency eddy kinetic energy is much less clearly structured. Figure 7.14 shows some evidence of maxima downstream of the high frequency maxima, as well as some correlation between the jet centres and maxima of low frequency variability. But none of these patterns is especially marked.

One reason for this is that the low frequency band covers a very wide range of frequencies. There are disturbances whose periods are very little longer than those of the baroclinic disturbances; indeed, the maxima downstream of the storm track centres are at least partly due to the occlusion and decay of midlatitude cyclones, which become slower moving as they fill. But there are also transients of significant amplitude with very much longer periods. Indeed, a spectral analysis of any long sequence of atmospheric data reveals that variability is observed on as long a period as one cares to specify. Furthermore, the spectrum tends to be red, with amplitude increasing as frequency decreases down to very low frequencies. Within such a range of frequencies, many different physical mechanisms are operating. It is little wonder that their combined signal is rather confused. In studying low frequency variability, it is not enough simply to look at the total low frequency signal. We will need to isolate particular frequency ranges and spatial structures.

One way to do this would be to use spectral analysis or time series filtering to isolate particular narrow ranges of frequencies. This could then lead to the spatial structures of oscillations of different frequency. A difficulty with this approach is the amount of data required to draw reliable conclusions. More seriously, a characteristic of many low frequency phenomena is that they are only quasi-periodic. This means that although they recur, their period can vary, and their structure can vary between one maximum and another. In spectral terms, periodic fluctuations are characterized by broad, rather than sharp, peaks. These features indicate that we are generally dealing with highly nonlinear phenomena, rather than normal modes of oscillation which would emerge from some linear analysis of atmospheric motions.

In this chapter, we will discuss some empirical techniques which have been used to isolate particular low frequency components of the atmospheric circulation. These structures are generally understood to result from localized anomalies of heating or other types of forcing which in favourable conditions can then propagate over substantial parts of the globe. The ray tracing theories of Section 6.2 give a simple description of the ways in which this can happen. We will also look at various quasi-periodic oscillations of the stratospheric circulation and at an important oscillation in the tropical Pacific which involves interactions between the atmospheric and ocean circulations. Finally, we will stress that the atmospheric system alone is sufficiently complex and nonlinear to generate unexpectedly low frequency transients without recourse to any forcing from external systems.

8.2 Teleconnection patterns

Certain analyses of long time series of atmospheric circulation data reveal large scale correlations between the flow at remote locations. These fluctuations belong in the low frequency range of timescales, and they have been dubbed 'teleconnections' to stress the correlation-at-a-distance aspect of their nature. Teleconnections are located in particular places, and take the form of 'standing waves', with fixed nodes and antinodes of low frequency oscillation. They are often orientated in a such a way as to indicate connections between the tropical and midlatitude low frequency transients. The theory of such teleconnections is incomplete, though we will relate them to meridionally propagating Rossby waves. Much of this section will be concerned with the description of empirical statistical techniques which are frequently used to detect teleconnection patterns.

The most straightforward procedure is called correlation analysis. Consider some meteorological field Q defined on a grid of N discrete points;

denote its value on the ith grid point $Q_i(t)$. In many analyses, Q is the field of 50 kPa geopotential height; this field is observed reasonably accurately and long time series of it exist, at least for the northern hemisphere extra-tropics. It is usual to emphasize the low frequency teleconnection patterns by filtering Q_i in some way to remove the higher frequency fluctuations. At its simplest, such a filtering is accomplished by taking monthly mean values of Q_i. The heart of the method is to calculate, for each gridpoint, a correlation with the values of Q at every other gridpoint. Such a correlation is:

$$r_{ij} = \frac{\overline{Q'_i Q'_j}}{\overline{Q_i'^2}^{1/2} \, \overline{Q_j'^2}^{1/2}}. \tag{8.1}$$

Point i is called the 'base point'. The correlation r_{ij} will be large in the neighbourhood of the base point, with values tending towards 1 as j approaches i. If no teleconnection patterns exist, r_{ij} will drop away towards zero when points i, j are separated by a distance greater than the typical horizontal scale of coherent circulation systems. Large positive or negative values of r_{ij} when the points are well separated indicates some kind of teleconnection.

Some typical examples for the northern hemisphere winter are shown in Fig. 8.1. These calculations were based on the mean 50 kPa geopotential height for 45 December, January and February months, beginning with the 1962–3 winter and finishing with the 1976–7 winter. The base point in Fig. 8.1(a) is 55 °N, 20 °W, in the mid-Atlantic. An elliptical region of large correlation, with a meridional width of some 2000 km, is centred upon the base point. The correlations over the Pacific and south-east Asia are small. But large correlations are seen roughly strung along a great circle passing through the base point, with alternating positive and negative correlations. An even more impressive pattern is seen in Fig. 8.1(b). Here the base point is at 20 °N, 160 °W in the central Pacific Ocean. A train of alternating positive and negative centres spreads north from the base point into North America. In contrast, the correlations between this base point and points over the Atlantic and Asia are generally small.

These particular base points were chosen because they lead to especially well-defined teleconnection patterns. Other base points have correlations which fall uniformly away to small values at any significant distance from the base point. These patterns do not deserve to be distinguished as teleconnections. Figure 8.2 summarizes the most significant teleconnection patterns identified in northern hemisphere winter data. The most well defined is the Pacific–North American (PNA) pattern, which is epitomized by Fig. 8.1(b).

Another major pattern is the North Atlantic oscillation (NAO, sometimes

(a)

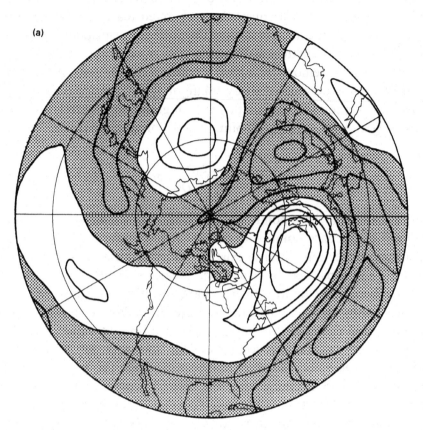

Fig. 8.1. Correlation maps based on 50 kPa monthly mean geopotential height data for 45 winter (DJF) months from 1962/3 to 1976/7: (a) base point 55°N, 20°W.

subdivided into the west Atlantic and east Atlantic patterns). Other, weaker, patterns are sometimes identified, though these are probably not statistically significant. The character of the PNA and NAO patterns is essentially a meridional connection between the tropics or subtropics and the midlatitudes. They are remarkably similar to the Rossby wavetrains discussed in Section 6.2, suggesting that the teleconnection patterns may be thought of as a Rossby train emitted from some anomalous forcing region in the tropics.

The correlation analysis gives information about the spatial structure of teleconnections, but does not give much information about the timescales of teleconnection events. More sophisticated filtering of the time series before carrying out the analysis might help, but the method rapidly becomes cumbersome and the conclusions statistically dubious. A more elegant ap-

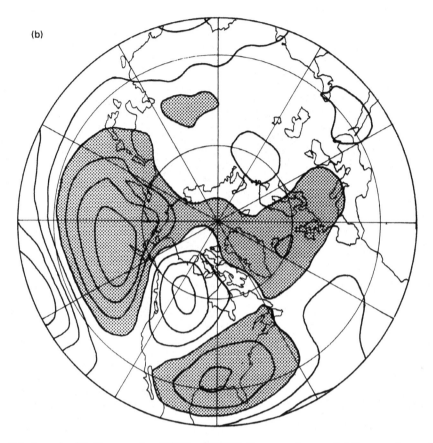

Fig. 8.1 (*cont.*). (b) Base point 20 °N, 160 °W. Contour interval 0.2, negative values shaded. (From Wallace & Gutzler 1981.)

proach which yields such information is called 'empirical orthogonal function analysis' or EOF analysis. It has become a widely used tool in analysing both observations and long numerical simulations of the global circulation.

The basic principle of EOF analysis is as follows. The global field of our general meteorological variable Q_i and its time mean $\overline{Q}_i$, analysed at each of the N grid points, may be thought of as vectors in an N-dimension space. The anomaly Q'_i points from $\overline{Q}_i$ to Q_i. As the system evolves, the vector Q'_i waves about in a generally irregular fashion, oscillating around $\overline{Q}_i$. The question is: are the Q' vectors clustered in certain directions from $\overline{Q}_i$? Can we identify those directions, and attach physical significance to them? The EOF analysis provides a general way of generating a set of orthogonal basis vectors in N-space, in such a way that they best represent any such clusters

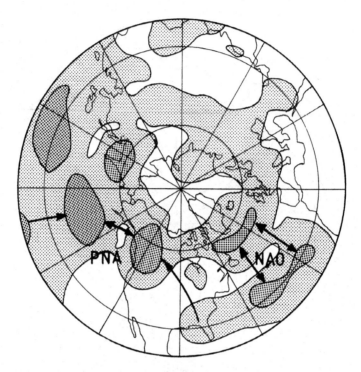

Fig. 8.2. Summary of the main teleconnection patterns for the northern hemisphere winter. The heavy lines denote the 0.6 correlation contours, and the letters indicate the positive and negative centres of the principal patterns. (After Wallace & Gutzler 1981.)

of vectors. Figure 8.3 provides a schematic illustration of what we are trying to do.

The method involves computing the $N \times N$ matrix of covariances of the fluctuations of Q_i' at different points. Denoting the covariance matrix by $\mathbf{C}$ and its elements by $\mathbf{C}_{ij}$, we have:

$$C_{ij} = \overline{Q_i' Q_j'}. \tag{8.2}$$

We then seek the eigenvalues and eigenvectors of $\mathbf{C}$, that is, we solve the eigenvalue problem:

$$\mathbf{C}\mathbf{e}_j = \lambda_j \mathbf{e}_j, \; j = 1, N. \tag{8.3}$$

Since $\overline{Q_i' Q_j'} = \overline{Q_j' Q_i'}$, $\mathbf{C}$ is a symmetric matrix with real elements. This means that its eigenvalues are real and positive. The corresponding eigenvectors form an orthogonal set, and so are the required set of basis vectors or EOFs.

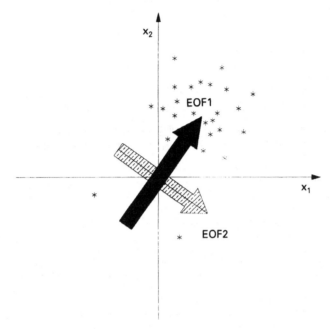

Fig. 8.3. Schematic illustration of the principles of EOF analysis, applied to a system of just two variables. The state of the system at any instant is described by a vector **Q**. They cluster in a particular direction. A compact way of describing the system is in terms of two orthogonal basis vectors, one pointing towards the maximum concentration of **Q**, and the other at right angles to it.

The eigenvalues λ_j are proportional to the fraction of the variance accounted for by each of the eigenvectors. It is conventional to order the eigenvalues in decreasing order of magnitude, so the first EOF explains the largest fraction of the variance of the data, the second EOF explains the second largest fraction of the variance, and so on.

Each EOF can be plotted as a field. The amplitude is arbitrary and is generally normalized in some way. But the spatial structure of the first few EOFs is generally smooth and indicates the most commonly found large scale structures in the field. For a linear system, the EOFs can be shown to correspond to the linear normal modes of the system. For a more realistic system, their significance lies in picking out the populated and unpopulated directions in N-space. Since the time series of Q_i is finite, only the first few EOFs can be regarded as statistically significant. If the time series includes M realizations of the field, then the number of significant EOFs is likely to be $O(M^{1/2})$.

The field itself can be represented very compactly in terms of its EOFs.

Since the EOFs are orthogonal, we may write

$$Q_i(t) = \sum_{j=1}^{N} p_j(t)e_{ij}. \tag{8.4}$$

The time series $p_j(t)$ is called the 'jth principal component' of the series of data. It represents the time series of projections of the data onto the jth EOF. Thus, the principal components yield information about the time behaviour of the spatial structures identified by the EOFs.

Figure 8.4 shows the result of carrying out an EOF analysis of a long set of DJF 50 kPa geopotential height fields. The first two EOFs are shown; between them, they account for 58% of the total variance of the data. These can be related to the teleconnection patterns; the main differences lie over Eurasia, where there is a large degree of overlap (i.e., non-orthogonality) between the various teleconnection patterns shown in Fig. 8.2. Elsewhere, there is a good relationship between the teleconnection patterns and the EOFs. For example, the first EOF is dominated by the PNA pattern, while the largest features in the second EOF correspond to the NAO pattern. Linear combinations of the first few EOFs produce more localized patterns which correspond even more closely to the teleconnection patterns identified by correlation analysis. Such a linear combination of EOFs amounts to a rotation of the vector denoting the EOF in the N-dimension space of Q_i', and so these combinations are called 'rotated EOFs'. Figures 8.4(c) and (d) show the first two rotated EOFs for the DJF 50 kPa geopotential height. They correspond very closely with the PNA and NAO patterns.

At the beginning of this section, teleconnections were described as 'standing oscillations', with antinodal regions, where the correlation with the base point is high, and nodal regions, where the correlation with the base point is close to zero. An inspection of the timeseries of principal components shows that the oscillation is far from sinusoidal. Indeed, some authors have described the oscillations as more similar to an irregular switching between two circulation states, corresponding to the positive and negative signs of the principal components. Each state is envisaged as metastable, so that the circulation remains in that state for some time, before undergoing a rapid but essentially unpredictable transition to the opposite state. Such 'persistent anomalies' or 'multiple flow regimes' are a fascinating possibility which have a long history in synoptic climatology. There have been numerous attempts to identify patterns of 'index cycles' on a regional scale, during which the flow oscillates between a state of strong but zonal flow and one of weaker flow with large amplitude eddies. But at other periods, the principal components oscillate continuously if erratically. During these periods, the oscillation is more similar to a standing oscillation, though with an ill-defined period.

The statistical techniques described in this section serve simply to isolate large scale smooth patterns in the fields of geopotential height or other variables. Of themselves, they do not identify any physical mechanisms for the correlation of fields over large distances. To do this, it is necessary to examine the various terms in the dynamical equations governing the flow, and this requires numerical experimentation with a global circulation model. However, the similarity between teleconnection patterns such as the PNA pattern, and the trains of steady Rossby waves discussed in Chapter 6 suggests that propagation of waves with zero, or small, phase speed is a likely mechanism which could give rise to teleconnection patterns. The pattern could be excited when, for example, a local anomaly in the tropical heating generates a vorticity anomaly at low latitudes. Long wavelength, steady Rossby waves will propagate roughly meridionally away from such a source region to affect higher latitudes. Indeed, the theory of Chapter 6 suggests that the amplitude of the response in the geopotential height field increases with latitude, so that a modest disturbance in the subtropics could lead to a very significant response in the higher latitudes. Heating anomalies, caused, for example, by strong convection over patches of anomalously warm tropical ocean, could then lead to circulation anomalies at large distances from the sea temperature anomaly. An example of just such a connection will be described in Section 8.5. The lifetime of the anomaly is essentially determined by the lifetime of the sea temperature anomaly, that is, by the timescale of circulations in the upper part of the ocean rather than in the atmosphere. Timescales of weeks rather than days are anticipated. It is now generally accepted that many persistent changes of the midlatitude flow are related to the tropical ocean in such ways, so that the problem of forecasting on the seasonal or longer timescale is perceived as being as much to do with forecasting ocean circulations as it is with deterministic forecasting of the atmospheric circulation.

To see how events in the tropics can excite a poleward train of Rossby waves, consider the vorticity equation for a single level in the atmosphere. This may be written:

$$\frac{\partial \zeta}{\partial t} + \mathbf{v} \cdot \nabla \zeta = -\zeta D \tag{8.5}$$

(see Eq. 1.50) where ζ is the absolute vorticity $f + \xi$ and D is the horizontal divergence $\partial u/\partial x + \partial v/\partial y$. From continuity, $D = -\partial \omega/\partial p$. We will suppose that friction is small at upper tropospheric levels, and so no friction term

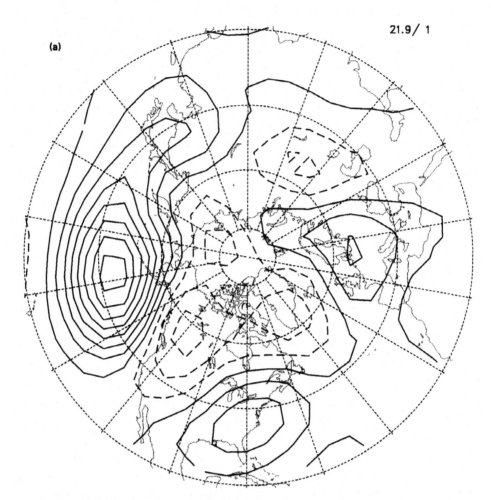

21.9/ 1

(a)

Fig. 8.4. EOFs of the DJF 50 kPa geopotential height field, based on the same 45 month dataset as Fig. 8.1. (a) and (b) show the 1st and 2nd EOFs respectively. (c) and (d) are the 1st and 2nd rotated EOFs, which may be compared with Figs. 8.1(a, b). (Courtesy X. Cheng and J.M. Wallace.)

has been included in Eq. (8.5). At a superficial level, one might think that the left hand side of Eq. (8.5) describes the propagation of Rossby waves, as discussed in Section 6.2, while the term on the right hand side represents the forcing of such waves. In the tropics, the vertical velocity largely balances the heating, so that regions of large latent heat release associated with tropical convection will force large ascent at midlevels, and hence divergent flow at upper tropospheric levels. But in the tropics, ζ changes sign somewhere near the equator, and is generally small throughout the tropics. Indeed, in the

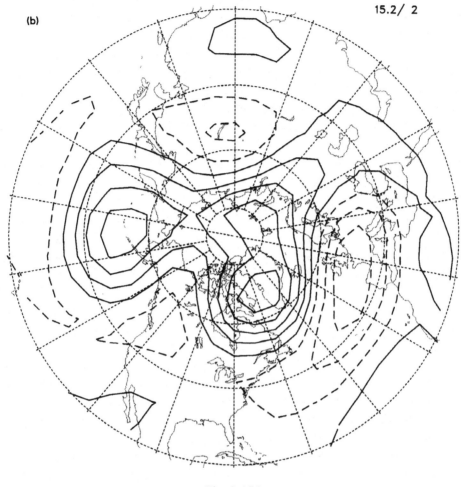

Fig. 8.4(b)

idealized Held–Hou model of the Hadley circulation, ζ is zero right across the Hadley cells. So it appears, despite the clear evidence of the observed teleconnection patterns, that anomalies of heating in the tropics should be ineffective in exciting waves which can provide teleconnections with higher latitudes.

The resolution of this difficulty involves a more careful look at the vorticity equation, Eq. (8.5). Now the velocity field **v** may be decomposed into a purely rotational part $\mathbf{v}_\psi$ and a purely divergent part $\mathbf{v}_\chi$ (a general result known as Helmholtz's theorem). The velocity field can then be described in terms of

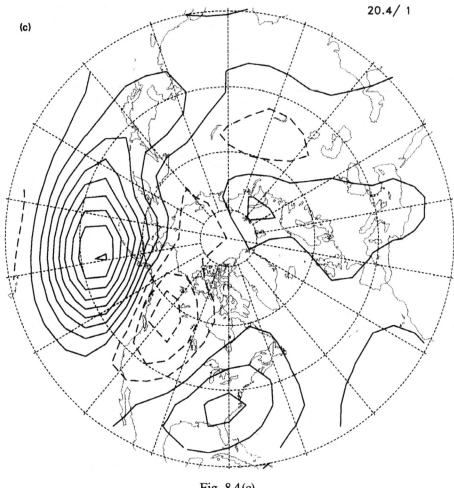

(c)

Fig. 8.4(c)

two scalar fields, the streamfunction ψ and the velocity potential χ since:

$$\mathbf{v}_\psi = \mathbf{k} \times \nabla\psi, \ \mathbf{v}_\chi = \nabla\chi. \tag{8.6}$$

From these definitions, it follows that the relative vorticity $\xi = \nabla^2\psi$ and the divergence $D = \nabla^2\chi$. Now substitute for $\mathbf{v}$ in Eq. (8.5) and rewrite the equation in the form:

$$\frac{\partial \zeta}{\partial t} + \mathbf{v}_\psi \cdot \nabla\zeta = -\zeta D - \mathbf{v}_\chi \cdot \nabla\zeta. \tag{8.7}$$

This equation represents the correct partitioning between the Rossby wave propagation terms, which involve just the rotational part of the wind field,

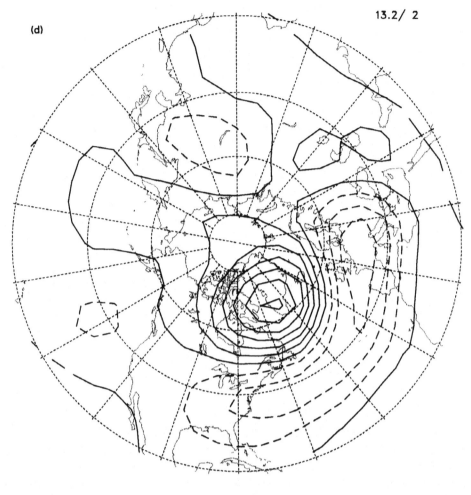

Fig. 8.4(d)

on its left hand side, and forcing terms, involving the divergent part of the wind, on the right hand side. The propagation of Rossby waves is a result of advection of the absolute vorticity by the rotational, not the divergent, part of the wind field. The extra term, representing the advection of absolute vorticity by the divergent part of the wind is not necessarily small, even though $|\mathbf{v}_\chi|$ is generally small compared to $|\mathbf{v}_\psi|$. This is because the rotational wind is generally roughly parallel to contours of constant ζ, while the smaller divergent wind may make a large angle with them. The forcing terms are jointly referred to simply as the 'Rossby source term' S which may

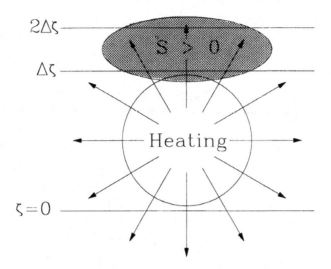

Fig. 8.5. Schematic illustration of the forcing of a train of Rossby waves by a maximum of tropical heating. The hatched region indicates the region where Rossby wave source function is large.

be reduced to the compact form:

$$S = -\nabla \cdot (\mathbf{v}_\chi \zeta). \tag{8.8}$$

We can now see how Rossby waves can be excited by tropical heating, even though ζ is often small in the vicinity of the heating. The divergent flow will be largest around the edge of the heating region, outside the region where D is large. The gradients of ζ become large as one approaches the subtropics, and so S can be large in the subtropics, either side of a tropical heating maximum. Figure 8.5 illustrates this schematically.

These ideas are demonstrated by Fig. 8.6, which shows the Rossby wave source function for the northern hemisphere winter. The heating has a large maximum over the Indonesian region (see Fig. 3.8), although this is in a region of small ζ and $\nabla\zeta$. However, the divergent wind is strongest north of this region, around the south-east coast of Asia. The forcing of the Rossby waves is largest in the region of the Asian jet maximum, some three thousand kilometres away from the heating maximum.

The search for large scale, coherent, low frequency structures in the southern hemisphere circulation has been hampered by the limited data and the rather smaller amplitude of stationary low frequency disturbances. The quality of individual analyses is generally poorer than in the northern hemisphere, because of the dearth of upper air stations. What is more,

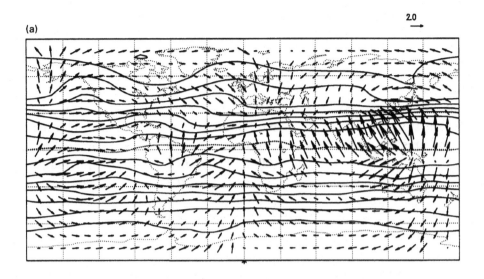

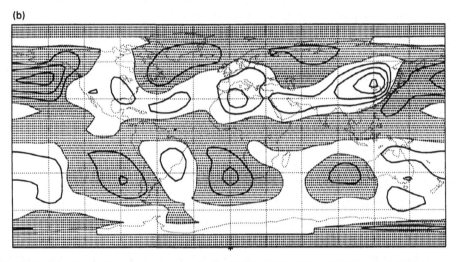

Fig. 8.6. The Rossby wave source S for DJF, 1979–89 at 15 kPa. (a) Contours of absolute vorticity ζ, contour interval $2 \times 10^{-5}\,\mathrm{s}^{-1}$, and vectors of the divergent wind $\mathbf{v}_\chi$. The sample vector indicates $2\,\mathrm{m\,s}^{-1}$. (b) S, contour interval $5 \times 10^{-11}\,\mathrm{s}^{-2}$, with shading indicating negative values.

only short time series of analyses are available, making a statistically secure identification of low frequency phenomena more difficult. This is exacerbated since any southern hemisphere teleconnections have smaller amplitudes, and do not stand out from the irregular background noise provided by synoptic scale fluctuations.

So, although a number of attempts have been made to identify recurrent teleconnection patterns in recent years, the various studies are not consistent in detail. However, two results do appear to be robust, and to be common to several independent studies. Here, a few results from a study by Mo and White (1985), based on monthly mean 50 kPa height and surface pressure fields, are summarized.

The first major feature is a general negative correlation between either the surface pressure or the 50 kPa geopotential height in the polar regions and the tropics, with a node around 60 °S. This implies a related fluctuation in the strength of the midlatitude westerlies. The other major feature is a rather regular zonal wavenumber 3 pattern. Such a pattern is a feature of the mean fields (see Fig. 6.1(b)), but there appears to be an index cycle whereby midlatitude flow switches between a more zonal state and a state with a pronounced wavenumber 3 steady wave. Correlation analysis reveals the presence of three maxima of correlation around the southern midlatitudes. The anomaly at the centre of these maxima can be used to construct an index of the strength of this three-wave pattern. Figure 8.7 illustrates the wave 3 pattern. It is based on a composite of fields for which the wave 3 index is largest, and a similar composite of fields with small wave three index. The diagram shows the difference between these two extremes. The diagram also suggests that the pressure correlation between high and low latitudes may also be at least partly related to this wave 3 oscillation, since it shows generally higher pressure over Antarctica than over the subtropics.

EOF analyses based on monthly mean fields also pick out these patterns. They dominate the first two EOFs, which account for 37% of the total variance in the monthly means.

This section has emphasized the propagation of Rossby waves from the tropics into the midlatitudes. A significant part of the low frequency variance, including persistent flow anomalies in the midlatitudes, can be related to such propagation, and hence to anomalies in the tropical ocean circulation. However, it should be remembered that most of the observed low frequency and steady Rossby wave activity originates in the midlatitudes and propagates mainly towards the tropics. This is clearly demonstrated by the steady momentum fluxes which are predominantly poleward. The midlatitude sources of Rossby waves are orography and ocean–land contrasts, as well as excitation of Rossby waves by maturing transient baroclinic systems.

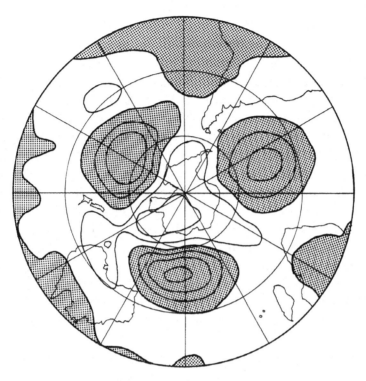

Fig. 8.7. The difference between composites of 50 kPa height fields for the southern hemisphere with maximum wave 3 index, and composites with minimum wave 3 index. Contour interval 40 m, with shading indicating negative values. (After Mo and White 1985.)

8.3 Stratospheric oscillations

A number of rather regular oscillations of the zonal wind are observed in the tropical stratosphere. They provide examples of some of the most regular fluctuations of the atmospheric circulation and are generally thought to be generated internally by the intrinsic dynamics of the atmosphere rather than by some external forcing. We will discuss two such oscillations in this section. One is called the 'quasi-biennial oscillation' (or 'QBO'); the other is the 'semi-annual oscillation' (or 'SAO'). The QBO is observed in the lower and middle stratosphere and can be detected by conventional radiosonde observations. The SAO occurs higher in the atmosphere and is much more difficult to observe. Above the ceiling of radiosondes, the main data sources are satellite soundings of temperature. But near the equator there is no suitable balance condition which can relate these temperature observations to the wind field. Most of the observations of the SAO come from rather limited series of rocket borne measurements, launched from certain tropical stations.

The QBO consists of an oscillation of the zonal wind, from easterly to westerly, at upper levels in the deep tropics. The wind oscillates with a maximum amplitude of around $20\text{--}30\,\mathrm{m\,s^{-1}}$ near $2\,\mathrm{kPa}$, and the oscillations become small below $5\,\mathrm{kPa}$; the easterlies are generally rather stronger than the westerlies. The latitudinal variation is roughly Gaussian with a half width of $12°$ of latitude. The period is rather irregular, being in the range of 22 to 34 months; over a long period, the mean period of the oscillation is around 27 months. This indicates that the oscillation is definitely not related to the annual cycle, though there is some evidence that there is a tendency for the wind reversal to take place preferentially in the northern hemisphere summer. During the westerly phase, the strongest zonal westerlies are actually on the equator; this is a clear sign that eddy fluxes of angular momentum are implicated in the QBO, since this westerly air current has greater angular momentum per unit mass than any part of the Earth's surface. Such a maximum could not be acheived by purely axisymmetric, inviscid circulations. Further discussion of this point will be given in Section 10.3.

An important clue about the nature of the mechanisms driving the QBO is contained in Fig. 8.8. This shows a height–time plot of the zonal component of the wind observed at a number of tropical stations. The oscillation originates at high levels, and propagates slowly downwards. There are some significant asymmetries. For example, the switch from easterly to westerly flow propagates downwards more quickly than does the switch from westerly to easterly. It is easy to see that the QBO cannot simply be a downward propagating wave. Dissipation would quickly damp it, and the exponential decrease of density with height would demand that the amplitude should be largest in the forcing region, but reduce exponentially as the wave propagated downwards. In any case, it is difficult to envisage what mechanisms could excite such a low frequency wave at high levels in the stratosphere. The currently accepted theories of the QBO suggest that the flow is forced from the troposphere, and that it involves interactions between upward propagating waves and the mean flow.

A simulation of the way this might work was provided in an elegant laboratory experiment by Plumb and McEwan (1978). Their apparatus, illustrated in Fig. 8.9, consisted of a cylindrical tank containing stratified brine, with a mechanism for exciting a standing wave, frequency ω and zonal wavenumber k, on the lower boundary. Such a standing wave can be decomposed into two travelling waves, of equal amplitude, but travelling in

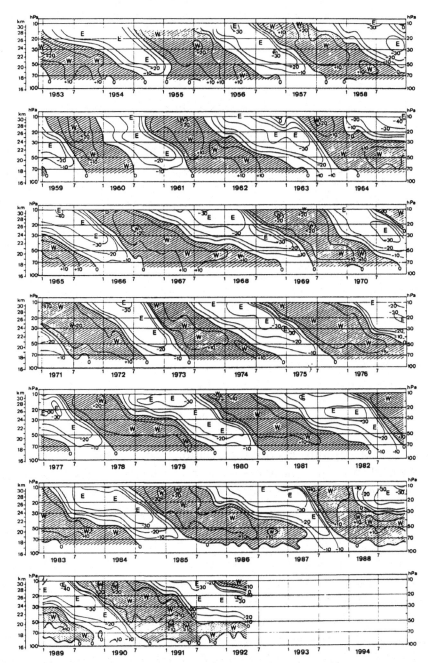

Fig. 8.8. Time–height plot of the monthly mean zonal wind based on observations from various equatorial stations, namely, Canton Island (January 1953 to August 1967), Maldive Islands (September 1967 to December 1975) and Singapore (January 1976 to May 1992). Contour interval 10 m s^{-1}, westerlies shaded. (Updated from Naujokat 1986.)

opposite directions with phase speeds $c = \pm\omega/k$. The brine solution supports internal gravity waves, with zonal wavenumber k and vertical wavenumber m. The dispersion relation for vertically propagating internal gravity waves is:

$$\omega = Uk \pm \frac{Nk}{(k^2 + m^2)^{1/2}}, \tag{8.9}$$

where U is the zonal wind, presumed constant, or at least only a slowly varying function of height z, and N is the Brunt–Väisälä frequency. A theory for the vertical propagation of internal gravity waves through a medium where the zonal wind changes slowly with height can be developed, based on this dispersion relation. The steps are the same as those developed in Section 6.2 for the horizontal propagation of Rossby waves. The frequency and zonal wavenumber are conserved by the propagating wave packet, while Eq. (8.9) can be rearranged to give a diagnostic relationship between m and the local value of the zonal wind U:

$$m^2 = \frac{N^2}{(c - U)^2} - k^2. \tag{8.10}$$

The sign of m is determined by the condition that the group velocity of gravity waves be upward. From the dispersion relation, the vertical component of the group velocity, $\partial\omega/\partial m$, may be written:

$$c_{gz} = \frac{(c - U)^3 km}{N^2}. \tag{8.11}$$

Thus, to ensure upward energy propagation, the negative root for m must be chosen when $U > c$, and the positive root when $U < c$. Equation (8.10) also shows that there must be restrictions on the value of U for there to be any vertical propagation. The vertical wavenumber becomes imaginary, i.e. the waves are evanescent in the vertical, unless:

$$(c - U)^2 \leq N^2/k^2. \tag{8.12}$$

Vertical propagation is possible only for a range of flow speeds centred upon $\pm c$. The size of this range depends upon the selected values of N and k.

An illustration of these restrictions on the value of U is shown in Fig. 8.10, which gives the vertical component of group velocity as a function of the flow speed U. The values of N, k and ω are those used by Plumb and McEwan in their experiment. Vertical propagation is possible only for $|U|$ less than 57 mm s^{-1}. Suppose U is westerly, and of such a magnitude that vertical propagation can take place. The slope of the lines of constant phase is simply m/k; the requirement that c_{gz} be upward means that m takes the

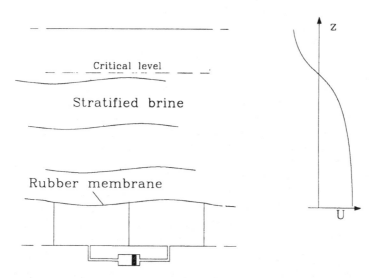

Fig. 8.9. The apparatus used by Plumb and McEwan as a model of the QBO.

opposite sign from $c - U$. So the phase tilt is westward with height when $U > c$. This implies that $[u^*w^*]$ is negative, that is, there must be an upward flux of easterly momentum. Near levels where the waves are dissipated, by internal diffusion, for example, there must be a convergence of this easterly momentum flux, and so an easterly acceleration will be imparted to the flow. Around these levels, U will gradually be reduced, and will eventually approach the value at which vertical propagation ceases. If the forcing at the base is maintained, the zero wind level will therefore gradually descend through the fluid. Eventually, easterlies will reach the base. Then waves for which $U < c$ can propagate in the vertical. All the above arguments reverse, so that these waves will carry westerly momentum aloft, and bring about a flow reversal, first at upper levels, then at lower levels. The apparatus exhibits a low frequency oscillation of its zonal wind, driven by nonlinear interactions between the upward propagating gravity waves and the zonal flow.

It is now generally accepted that the atmospheric QBO is driven in this kind of way. Motions in the troposphere, particularly those associated with clusters of cumulus convection in the tropics, can force a spectrum of disturbances at the tropical tropopause. Vertically propagating gravity waves are probably much less important to the QBO than larger scale vertically propagating waves; the prime candidates are likely to be the planetary scale Kelvin waves, which can propagate into the stratosphere in easterly

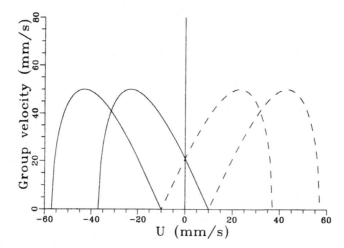

Fig. 8.10. The dependence of the vertical components of the group velocity on zonal flow U for the Plumb–McEwan laboratory model of the QBO. The solid curves show cases where $U < c$, and hence for which $[\overline{u^* w^*}] > 0$. The dashed curves show cases where $U > c$ and $[\overline{u^* w^*}] < 0$.

conditions, and the mixed Rossby–gravity waves which can propagate in westerly conditions. Section 7.1 gives a brief account of equatorially trapped planetary scale waves. The different characteristics of these waves help to account for the observed asymmetry between the easterly and westerly phases of the QBO; the chaotic nature of the forcing, which itself has considerable low frequency variability, explains why the QBO is only quasi-periodic. The confinement of the QBO to tropical latitudes reflects the meridional confinement of the equatorial modes which drive it.

The semi-annual oscillation (SAO) has many features in common with the QBO. Like the QBO, it consists of an oscillation of the zonal wind in the tropics. The oscillation has maximum amplitude near the stratopause and the mesopause; it seems likely that the mesopause oscillation is rather distinct from the stratopause oscillation, and may be driven by different mechanisms. Below 40 km, the oscillation becomes weaker and is swamped by the QBO. The stratopause oscillation has an amplitude of $30 \, \mathrm{m \, s^{-1}}$ and a half width of $25°$ of latitude. The maximum of westerly wind at the tropopause occurs just after the equinox, and the maximum of easterly wind just after the solstice. It is sometimes said that the SAO is global in extent, not merely confined to the tropics. However, it is likely that the extratropical SAO is simply a first harmonic of the annual cycle of zonal wind.

The westerly acceleration of the stratospheric SAO appears to start at the stratopause and to propagate downwards, at a rate of around 10 km per month. This is quite similar to the QBO, and there is strong evidence that vertically propagating Kelvin waves are associated with the westerly acceleration. However, there does not appear to be sufficient Kelvin wave activity to account for the magnitude of the acceleration; it may be that gravity waves also play a role. The easterly acceleration is quite different. It occurs almost simultaneously at all levels, and so vertically propagating waves must play, at most, a subsidiary role. Horizontal transports of angular momentum are needed to provide a simultaneous acceleration at all levels. The zonal mean advection of the summer easterlies across the equator probably accounts for a large fraction of the easterly acceleration, with horizontal momentum transports associated with breaking Rossby waves in the winter hemisphere possibly playing a secondary role. More details about these aspects of the stratospheric circulation are discussed in Chapter 9.

The SAO at the mesopause is not well observed. It has been suggested that it is primarily driven by upward propagating gravity waves, as in the Plumb–McEwan analogue of the QBO.

8.4 Intraseasonal oscillation

Apart from the regular seasonal cycle, quasi-periodic fluctuations of the tropospheric circulation are rather unusual. So it came as a considerable surprise in the early 1970s when a quasi-periodic oscillation of the tropical zonal wind was discovered, with a period of between 40 and 50 days. Since then, the oscillation has been well documented in studies of a number of tropospheric variables. The period is quite variable, and the range has now been extended to 30–60 days. Variously called the 30–60 day oscillation, the 40–50 day oscillation and Madden–Julian oscillation, the best name is perhaps the 'intraseasonal oscillation', that is, a somewhat irregular oscillation whose timescale is long compared to synoptic timescales, but shorter than the seasonal timescale. Figure 8.11 shows the surface pressure record from a single station, Canton Island (3 °S, 172 °W); the oscillation shows even in the raw station data, and is very clear when a suitable band pass filter is employed to remove high frequencies and the seasonal cycle.

Comparison of the records from many tropical observing stations suggests that the oscillation is an eastward moving, zonal wavenumber 1 disturbance, centred on the equator with a half width of about 10 ° of latitude. It is most pronounced in the surface pressure, tropospheric temperature and zonal wind fields, and more recently has also been identified in the patterns of convective cloud distribution and in the distribution of rainfall. The oscilla-

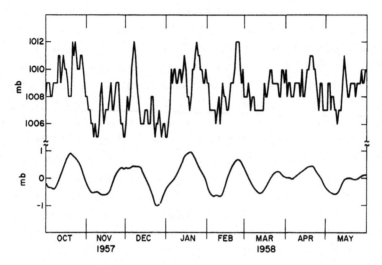

Fig. 8.11. The surface pressure record from Canton (3 °S, 172 °W). The upper curve is the raw data and the lower curve the result of applying a band pass filter centred around 45 days to the data. (From Madden & Julian, 1972.)

tion varies both in intensity and phase speed. It is strongest in DJF and weakest in JJA, but can be detected at all times of the year. Longitude–time plots show that intraseasonal disturbances move eastwards over the Indian Ocean, and reach their largest amplitude over Indonesia. Over the central and western Pacific, they become rather weaker, and take on the character of a standing oscillation. Figure 8.12 gives an indication of the structure of the intraseasonal oscillation, based upon eight DJF periods from the ECMWF archived analyses. EOF analysis has been used to isolate the oscillation in the field of the velocity potential. A numerical filter which isolates periods of 30–60 days was applied to a time series of the velocity potential at 15 kPa and 85 kPa. The resulting band pass filtered time series was used as the basis of an EOF analysis. Figure 8.12(a,b) show the first two EOFs at 15 kPa. The most intense feature of EOF1 is over Indonesia, with upper level divergence over Indonesia and low level convergence. EOF2 is almost out of phase with EOF1, and the corresponding principal component time series (Fig. 8.12(c)) fluctuate in quadrature. Combining the fields of EOF1 and EOF2 gives a generally eastward propagating disturbance with largest amplitude over Indonesia.

The velocity potential is often used, as it was in Fig. 8.12, to diagnose such convectively driven motions in the tropics. It provides a graphic picture of the outflow from strongly convecting regions, and suggests that the global flow responds to such forcing. But this can be very misleading. For example, suppose the wind field associated with an equatorial Kelvin wave is split into its rotational and divergent parts, which may be represented by a stream-function and velocity potential respectively. Both fields will be global. Yet the Kelvin wave is essentially a trapped equatorial disturbance which dies away rapidly at latitudes greater than about 20°. In fact, the rotational and divergent wind vectors for a pure Kelvin wave cancel out in the midlatitudes, but reinforce one another in the tropics. The Helmholtz partitioning of the wind field, while mathematically perfectly valid, is physically misleading in this case. The same is true, at least to a certain extent, of the flow associated with the intraseasonal oscillation.

In the longitude–height plane, the oscillation of the zonal wind is strongest in the upper troposphere, between 15 and 20 kPa, and it has the familiar baroclinic structure of tropical disturbances, with low level easterly anomalies beneath upper level westerly anomalies, and vice versa. The relationship between the wind, surface pressure fields and the distribution of regions of enhanced convection is shown schematically in Fig. 8.13. The cycle is dominated by enhanced convection above an anomalously large low convergence. This appears over Indonesia and moves eastwards across the Pacific before dying away near 180 °W. The anomalous convection is part of a zonal convection cell, or 'Walker circulation', very similar to the Walker circulations of the time mean flow described in Section 8.1.

This structure, together with the eastward phase speed, has suggested that the intraseasonal oscillation may be interpreted in terms of a propagating Kelvin wave. The difficulty with this interpretation is that the observed phase speed is slower than the phase speed of a simple dry Kelvin wave, as derived in Section 8.2. Another question concerns the mechanism which might excite the Kelvin wave. The evidence is that tropical moist convection may provide the necessary energy through some feedback between the large scale flow and the convective scale. 'Conditional instability of the second kind' (CISK) has been suggested as a model of how such a feedback might operate. The mean tropical atmosphere is conditionally unstable, but it is not saturated. Thus, for convection to begin in a given region, there must be convergence of a water vapour flux into that region. The release of latent heat within the convecting elements drives midlevel ascent and so serves to intensify the low level moisture convergence into the convecting region. The pattern of convergence and divergence driven in this way can also serve to generate large scale relative vorticity through the Rossby source term S (see Eqs. 8.7 and 8.8). Such a mechanism has also been suggested to account for the formation of hurricanes as well as other larger scale tropical weather

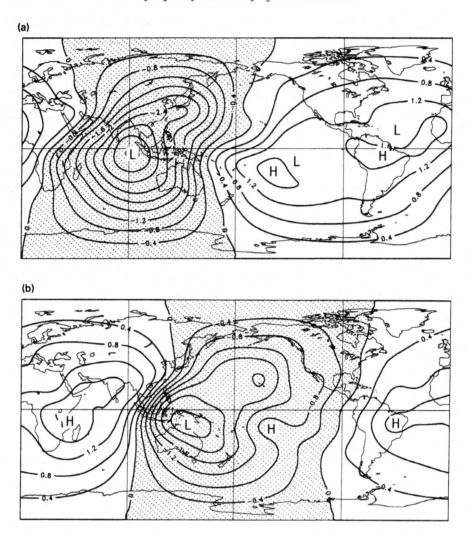

Fig. 8.12. EOFs 1 (a) and 2 (b) of the divergence field for eight DJF seasons of ECMWF data.

systems such as easterly waves. The problem with such mechanisms is that they depend upon some postulated relationship between cumulus scale motion and the large scale synoptic fields, that is, upon a parametrization of the cumulus convection. Given this, it is no surprise that different theoretical models of the intraseasonal oscillation give rather different predictions of the phase speed of the wave. The CISK mechanism can certainly act to slow a Kelvin wave down, by changing the coupling between the upper level and low level flow. Many current GCMs are capable of reproducing some

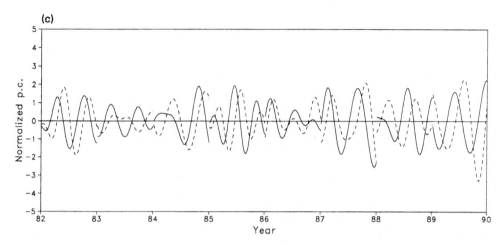

Fig. 8.12 (*cont.*). (c) The corresponding principal component time series; solid is PC1 while dashed is PC2. (Courtesy A. Matthews.)

kind of intraseasonal oscillation, but they overestimate the phase speed. The results are certainly sensitive to the particular convective parametrization used.

Since the intraseasonal oscillation was first identified, improved observations of the tropical atmosphere (especially the wider availability of satellite data) and careful analysis of earlier records have revealed its signature in many meteorological fields. For example, the 'active' and 'break' periods in the Indian monsoon, during which the monsoonal rainfall fluctuates, are revealed as a modulation of the monsoonal circulation by the intraseasonal oscillation. Cloudiness over South America and Africa also appears to vary according to the phase of the intraseasonal oscillation. Some global responses are also seen. For example, the length of the day, which directly reflects variations of the atmospheric angular momentum, also shows a 30 – 60 day oscillation.

8.5 The Southern Oscillation

Fluctuations with timescales of one year or longer contribute to the interannual variability, whereby one winter or summer season differs significantly from another. EOF or correlation analysis of seasonal or annual mean fields highlights whether any significant, spatially coherent patterns are associated with the interannual variability. One such pattern has been identified, and is recognized as the best-defined such structure. It accounts for a significant

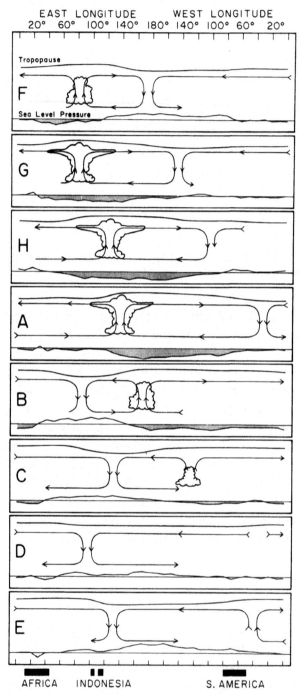

Fig. 8.13. Schematic summary of the relationship between anomalies of the surface pressure field, the azimuthal circulation, the tropopause height and the tropical cumulus convection through a typical cycle of the intraseasonal oscillation. (From Madden and Julian, 1972.)

fraction of the total interannual variance over much of the globe. It has two aspects: one is an oscillation of the atmospheric circulation and the other is an oscillation of the tropical ocean circulation. The former has been known for many years in the form of a negative correlation between the surface pressure at Darwin, Australia, and Easter Island, in the mid-Pacific. This was called the 'Southern Oscillation'. The oceanic part of the fluctuation was revealed by the sea surface temperatures off the coast of Peru. Normally, upwelling brings cold, nutrient-rich water to the surface here. But every few years, the upwelling ceases and the sea surface temperatures rise several degrees, with devastating consequences for the local fisheries. The phenomenon is called 'El Niño' by the Peruvian fishermen, for it tends to occur around Christmas.

Global monitoring reveals that both the Southern Oscillation and El Niño are large scale events which involve changes over most of the tropical Pacific. Teleconnections link the Southern Oscillation to circulation changes in the midlatitudes. Furthermore, the phenomena are intimately linked to each other, forming a coupled nonlinear oscillation of the tropical ocean and global atmospheric circulations. The coupled oscillation is called 'El Niño–Southern Oscillation' or, regrettably, 'ENSO'. Like other nonlinear oscillations, ENSO does not have a precisely defined period, but is a broad band feature. El Niño events differ in their intensity: some events are minor, while others can be very large indeed. The largest ENSO event in the last 40 years was recorded during 1982–3, and has been studied extensively. The interval between ENSO events can be as short as two years, or as long as seven years; the average period is about 40 months.

Figure 8.14 illustrates the two aspects of the 1982–3 ENSO. In Fig. 8.14(a) are shown anomalies of the 85 kPa wind for DJF 1982–3, compared to the six years 1983–9. A strong westerly anomaly extends over much of the central tropical Pacific, with weaker easterly anomalies over Indonesia. This pattern would be expected to result from the movement of the centre of tropical convection from Indonesia to the central or eastern Pacific. At the same time, strong cyclonic anomalies are seen at higher latitudes. These form wavetrains linking the tropics and midlatitudes. The northern hemisphere train is reminiscent of the PNA pattern shown in Fig. 8.1(b). Figure 8.14(b) shows anomalies of the sea surface temperatures during the El Niño in January 1983. The main features are cooling over the Indonesian region and warming towards the South American coast.

Attempts to model the ENSO oscillation are mainly based around a coupled oscillation of the combined atmospheric and oceanic circulation. Figure 8.15 is a schematic illustration of the sort of interaction which is

5.0

(a)

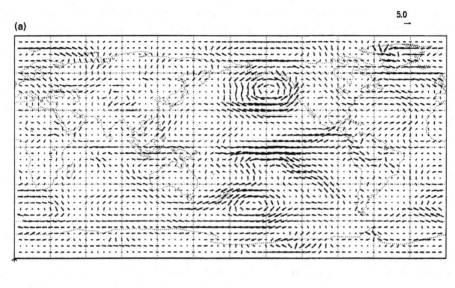

(b)

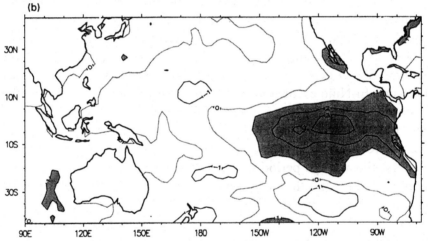

Fig. 8.14. The structure of ENSO during 1982–3, the largest ENSO event in the last 40 years. (a) Anomaly of the 85 kPa winds (ECMWF data); the sample vector represents 5 m s⁻¹. (b) Anomalies of sea surface temperature over the Pacific during 1982–3. Contour interval 1 K, values in excess of +1 K shaded.

probably involved. The usual state of affairs is shown in Fig. 8.15(a). The warmest sea surface temperatures are in the vicinity of Indonesia. These drive intense convection in this region, which releases latent heat and drives a Walker circulation, as discussed in Section 7.1. The low level winds associated with the Walker cell exert an eastward stress on the sea surface, feeding warm water into the western Pacific, and maintaining the upwelling along the

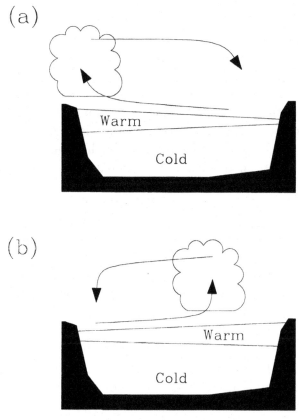

Fig. 8.15. A schematic illustration of the coupling between the ocean and atmospheric circulation during El Niño events. The diagrams show longitude–height sections of the atmosphere and the ocean along the equator over the Pacific Ocean: (a) Normal circulation, with convection over Indonesia; (b) anomalous or El Niño circulation, with the convection moving to the mid or western Pacific.

coast of South America. But now suppose that the sea surface temperature maximum moves towards the mid-Pacific. The convective maximum and the Walker cell will move further east. The low level easterly winds are replaced by westerlies over much of the tropical Pacific, moving the warm water further east and suppressing the South American upwelling. It can be seen that the changes will feed back on themselves and help to establish and maintain the anomalous El Niño state. Idealized models which represent such a coupling between the low level tropical winds and the flow in the upper layers of the ocean suggest that the process is an instability; an initial condition resembling that shown in Fig. 8.15(a) begins to oscillate between the El Niño state and the normal state (sometimes called the 'La Niña' state). The amplitude of the oscillation saturates at some maximum value. In a more complex situation, the influence of such irregularities as baroclinic instability in the midlatitudes will mean that the El Niño oscillations will be somewhat irregular. It has been suggested that their timescale may be determined essentially by the time taken for equatorial Kelvin waves in the ocean to propagate across the Pacific basin, and that changes in the ocean circulation could be used to forecast the evolution of such events.

The global implications can be understood in terms of Rossby wavetrains propagating away from the anomalous convective heating. In this way, the El Niño creates circulation anomalies over midlatitude North and South America, Australia and possibly even further away. Indeed, the El Niño is a major, if not the largest, component of interannual variability over much of the globe. Improved understanding of the El Niño holds out the possibility of imparting some sort of skill to weather predictions on the seasonal timescale or even longer. As a result of this hope, and because of the impact of the highly anomalous weather regimes associated with the massive 1982–3 event, studies of El Niño have become very widespread and fashionable.

8.6 Blocking of the midlatitude flow

The examples of low frequency behaviour that we have discussed so far have their seat in the tropical latitudes. The midlatitudes have much larger low frequency variance. But the midlatitude low frequency variability has a less well-defined range of frequency, and generally a much less clear spatial structure than tropical variability. In this section, we will describe one very characteristic low frequency fluctuation of the midlatitude circulation which is still imperfectly understood and which attracts considerable attention. This feature is called 'blocking'; it is often thought of as an alternation between a relatively zonal midlatitude flow with superimposed fast moving disturbances, and a more meridional flow with stationary or near stationary disturbances.

This alternation of flow regime is traditionally described by a 'zonal index'. This might, for example, be based on the mean midlatitude zonal geostrophic wind. A typical index is derived from the zonal mean geopotential heights at 55 °N and 35 °N. Zonal flow is characterized by large values of the zonal index, while blocked flow has a smaller zonal index. Such a zonal index is observed to oscillate erratically between low and high values, with typical periods in the range of several weeks. The low index flow is often marked by intense, nearly steady anticyclones. These have preferred locations; the eastern Atlantic and western European sector is one and the other is in the eastern Pacific. Both lie downstream of the major storm tracks, and in a region where the time mean wave pattern tends to be characterized by ridging (see Chapters 6 and 7). A blocking anticyclone may last for several weeks; in some major European droughts, it seems that a succession of such anticyclones keep forming in much the same location.

Blocking is a much less prominent feature of the southern hemisphere midlatitudes. Some regions where intense anticyclones are more prevalent have been identified, but they are less less long lived than their northern hemisphere counterparts. One such region lies to the south of New Zealand, where the tropospheric jet tends to split (see Fig. 7.11); another lies near South America.

Once a major block is established, it prevents the usual eastward motion of cyclonic weather systems. Instead, they are steered around the block, mainly around the poleward side. In so doing, they are distorted and tilted. This is illustrated for a characteristic case in Fig. 8.16(a), which shows the high pass filtered geopotential height field superimposed on the low pass filtered field at the same time during a blocking event over the UK. The discussion in Section 7.4 immediately suggests how a characteristic pattern of E-vectors will be associated with such distorted high frequency eddies. The E-vectors are shown in Fig. 8.16(b). They tend to converge into the block, suggesting that the interaction between the high frequency transients and the block will tend to reduce westerly flow in the region of the block. That is, the transients will tend to reinforce the block. This effect is one important way in which blocks tend to be maintained. On the other hand, one can imagine that a particularly intense cyclone may be strong enough to disrupt the block, explaining the apparently random collapse of the system. But these ideas do not explain the geographical location of the blocks especially well. If they were formed entirely by interactions with high frequency transients, one might expect to see them more frequently in the downstream part of a storm track, but one might also expect to see them form fairly frequently at other locations also. One might also expect to see a particular block drift away from its original location. It seems that interactions with features of the Earth's surface must also be involved in the formation of blocks. This might be directly, or indirectly, via dynamical interactions involving the steady waves forced by mountains or land–sea contrasts.

A particularly elegant theory of how the forcing of steady waves could lead to both blocked and zonal flow regimes was proposed by Charney and Devore (1981). It is a straightforward extension of the theory introduced in Section 6.2 of barotropic flow over a mountain. For simplicity, consider flow along a β-channel; the topography of the lower surface is simply wavelike

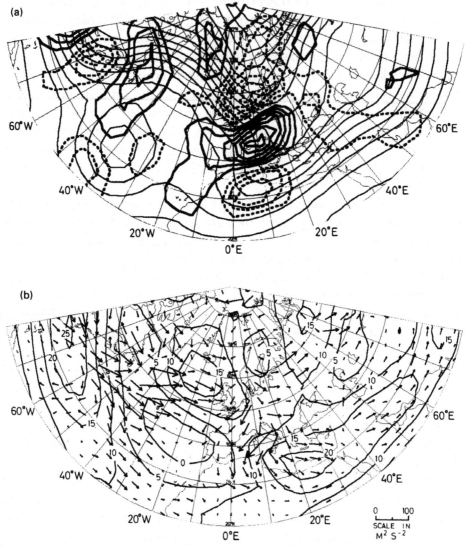

Fig. 8.16. The interaction of high frequency transient eddies with a blocking anticyclone over north-west Europe: (a) high pass eddies and low pass geopotential height; (b) **E**-vectors. (From Hoskins *et. al.*, 1983.)

in the x- and y-directions. This will force a steady wave whose phase and amplitude were given by Eqs. (6.9) and (6.10), and were illustrated in Fig. 6.8. The steady wave response is resonant, with a maximum amplitude when the zonal wind is:

$$U = \frac{\beta}{(k^2 + l^2)}. \tag{8.13}$$

The extra ingredient introduced by Charney and Devore was to allow inter-action between the zonal mean flow and the wave forced by the mountain. Suppose that the zonal mean flow obeys such an equation as:

$$\frac{\partial U}{\partial t} = \frac{U_E - U}{\tau_D} - \mathscr{D}.$$
(8.14)

Here, the first term simply represents relaxation towards some equilibrium velocity U_E on a timescale τ_D. It can be thought of as a parametrization of the total effect of the global circulation, including the Hadley cell and the midlatitude eddies, in driving the midlatitude westerlies. The second term $\mathscr{D}$ represents the 'form drag' exerted by the mountain on a unit mass of the flow, and can be written:

$$\mathscr{D} = -\frac{g}{p_0} \left[p_s \frac{\partial h}{\partial x} \right],$$
(8.15)

where h is the height of the orography. The surface pressure is related to the wind via the geostrophic relationship, and hence to the vorticity. That part of p_s which is in phase with the orography (and this includes the zonal flow) makes no contribution to the form drag, since the pressure force on the upstream side of the mountain exactly balances an opposite force on the downstream side. Thus, we may write Eq. (8.15) in the form:

$$\mathscr{D} = -\frac{g}{p_0} P h_0 k \sin(\delta),$$
(8.16)

where δ is the phase difference between the pressure wave and the orography wave (given by Eq. (6.10)), and P is the amplitude of the pressure wave, related to the vorticity amplitude by:

$$P = \frac{f p_0}{g H K^2} Z.$$
(8.17)

Here p_0 is the mean surface pressure and H is the pressure scale height. The form drag $\mathscr{D}$ reaches its maximum at the resonant flow speed β/K^2, and, if the drag is not too large, falls off sharply at higher or lower flow speeds.

In the steady state, the forcing of the mean flow must balance the form drag, that is,

$$\mathscr{D} = \frac{U_E - U}{\tau_D}.$$
(8.18)

This condition can be inverted to give values of U for zonal flows in equilibrium with the orography and the forcing. Such solutions are most easily appreciated by a graphical solution of Eq. (8.18), shown schematically in Fig. 8.17. Provided the resonance is sufficiently sharp, that is, that the

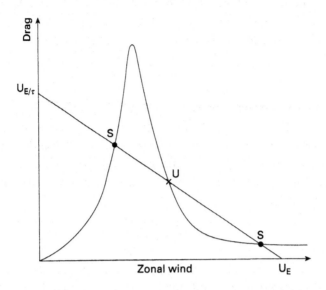

Fig. 8.17. Schematic illustration of the solutions of Eq. (8.18). The outer two solutions, marked S, are stable, while the central solution, marked U, is unstable.

drag τ_D is not too large, there are generally three different values of U which satisfy Eq. (8.18). In fact the central one turns out to be an unstable equilibrium, while the solutions for low U and high U are stable. They represent a 'low zonal index' solution with intense steady waves, and a 'high zonal index' solution with weak steady waves respectively. The system has two stable equilibrium configurations. If some sufficiently strong random forcing were introduced into Eq. (8.14), representing the effects of midlatitude transients, one could expect the flow to remain close to, say, the zonal state for some time, before being perturbed into the blocked state. The transitions between the blocked and unblocked states would occur more or less at random, just as is observed in the atmosphere. If more complex orography were introduced, there would be a possibility of multiple equilibria, not just two. The analysis can also be extended to baroclinic flows over orography.

The major drawback of such multiple equilibrium theories of blocking and the zonal index cycle is their reliance on resonance. In the case that we have just discussed, resonance is an artificial consequence of restricting the flow to a channel of fixed width by boundaries where $v^* = 0$. Such boundaries reflect Rossby waves, which are therefore confined to the channel and propagate zonally along it. Resonance occurs in a periodic channel when the Rossby wavetrain fills the channel and interferes constructively with the disturbance induced over the mountain itself. In a more realistic atmosphere, as we saw in Chapter 6, the Rossby wave activity would tend to be attracted to a critical line at low latitudes, where it would probably be absorbed if the drag were reasonably large. Only for the most contrived and artificial zonal flows would there be a zonal Rossby wave guide which could lead to resonance. Nevertheless, the theory of multiple equilibria is sufficiently attractive that a number of other ways of inducing a local resonant response which could lead to multiple equilibria have been suggested.

8.7 Chaos and ultra low frequency variability

Increasingly, as one considers lower frequency fluctuations of the atmospheric circulation, the effects of coupling between the atmosphere and more slowly varying components of the climate system become important. The typical dynamical timescale of the atmosphere is around five days, while the typical radiative timescale is about 30 days. In contrast, the timescales associated with ocean circulations range from weeks for the surface waters to as long as several thousand years for the deep ocean basins. The development and decay of major ice sheets requires some thousands of years, and may modify the atmospheric circulation profoundly, both locally and perhaps globally. Atmospheric composition, and hence the radiative forcing of the circulation, can fluctuate as atmospheric constituents are cycled through the biota. The relevant timescales cover a wide range, from a few years for trace constituents such as methane to around 10 million years for nitrogen. The orbital elements of the Earth vary periodically so that the amount and geographical distribution of solar radiation varies on timescales of as much as 10^5 years. These changes seem to trigger the major circulation and climate changes associated with the advance and retreat of ice sheets. Whether the solar output itself is subject to significant fluctuations, and if so, on what timescale, is not known.

With so many mechanisms which can modify the atmospheric circulation, explanations of very low frequency changes in circulation tend to centre around feedbacks between the atmosphere and these slower varying components of the climate system, or even on the direct forcing, without feedbacks, of circulation changes by external agencies. This must not blind us to the possibility that some low frequency behaviour can be generated internally, without any external excitation. The atmospheric circulation is such a nonlinear system that it is capable of all kinds of behaviour which are not revealed by the sort of linear calculations of instability and wave propagation which have underpinned much of the discussion in this book.

To illustrate, we will first consider an extremely simple analogue to the atmospheric circulation introduced by Lorenz. It consists of three ordinary differential equations governing the time evolution of three interacting quantities:

$$\frac{dU}{dt} = \frac{(8-U)}{4} - (A^2 + B^2), \tag{8.19a}$$

$$\frac{dA}{dt} = -4UB + (U-1)A + 1, \tag{8.19b}$$

$$\frac{dB}{dt} = 4UA + (U-1)B. \tag{8.19c}$$

Arguing loosely, U may be interpreted as a dimensionless measure of the midlatitude horizontal temperature gradient or, assuming thermal wind balance, the upper tropospheric zonal wind. The coefficients A and B represent the dimensionless sine and cosine coefficients of a wave. The individual terms on the right hand side of these equations are straightforward to understand. The first term on the right hand side of Eq. (8.19a) represents the radiative forcing of the zonal wind; in the absence of eddies, U would equilibrate to the value 8 on a timescale 4. This equilibrium value will be reduced by the second term on the right hand side when eddies of significant amplitude are present. Turning now to the equations governing the eddies, Eqs. (8.19b, c), the first terms in these equations indicate that the waves are progressive, with frequency $4U$. The second terms indicate that they are unstable, and that their amplitudes will amplify exponentially if U is held constant at some value greater than 1. The final, and crucial, ingredient is the last term on the right hand side of Eq. (8.19b). It represents a constant forcing of a stationary wave, for example by continent–ocean contrasts.

The set of Eqs. (8.19a–c) represents a deterministic dynamical system. That is, given initial values of U, A and B, their subsequent evolution would in principle be determined exactly. In practice, the nonlinearity of the equation set means that there is no general analytical solution, and so the integration has to proceed by numerical means. The approximations of the numerical calculation immediately introduce some uncertainty into the subsequent solution. In fact, for the particular choices of the various coefficients made in Eqs. (8.19a–c), any such errors tend to amplify exponentially so that arbitrarily close initial conditions will eventually lead to widely different solutions. An example of two such diverging solution trajectories is shown in Fig. 8.18. The separation between the two trajectories generally increases, though not monotonically. A mean of many such curves would initially produce a more or less exponential increase of the separation.

Clearly, this behaviour is analogous to that of the atmosphere. The exponential divergence of initially similar solutions is reminiscent of the 'forecasting problem', in which a deterministic set of equations is used to extrapolate an imperfectly defined initial state of the atmosphere to produce a forecast. It is well known that the atmosphere is not predictable in this sense; a numerical weather prediction generally loses any skill after about 6–10 days in most circumstances. Because of the exponential increase of errors, decreasing the errors of observation (principally by increasing the density of the observing network) would only lead to a limited extension of the useful period of a forecast.

The Lorenz system is an example of those equation sets which have been termed 'chaotic'. This is an unfortunate name, with its implications of randomness and lack of structure. In fact, the solutions to the equation set are highly structured, although that structure is exceedingly complex; a better term for this generic type of behaviour might be 'spontaneously complex'. In the interests of clarity, the trajectories shown in Fig. 8.18 are kept rather short. In Fig. 8.19, the complexity of the solutions is illustrated by a so-called 'Poincaré section' of a much longer integration of the same equation set. This is really a cross section of the solution trajectories, in which a point is plotted each time the trajectory crosses the $A = 0$ plane. It is noticeable that the solution trajectories tend to cluster in some regions, while other regions are scarcely entered at all. One such region is the vicinity of the point $\bar{U} = 1.031\,0$, $\bar{A} = 0.006\,4$, $\bar{B} = 0.317\,9$. The solutions oscillate around their mean value but spend little time close to the mean, which is an unstable point for the system.

Where the solution trajectories are clustered, the points on the Poincaré sections are arranged along folded sheets. Closer examination of longer integrations shows that this structure is repeated on a smaller scale. Indeed, the solution subspace is 'fractal', showing complex structure on any scale, no matter how small. This 'quasi-two-dimensional' character of the solutions can be expressed quantitatively by determining the effective dimension of the solution surface. Suppose that we have a very long numerical integration of an equation set such as Eqs. (8.19a–c), consisting of a large number of discrete points in (U, A, B) space. The dimensionality of the solution could be determined as follows. Take some arbitrary point within the attractor. Then count the number of solution points lying within the neighbourhood of this point, that is, inside a small volume of (U, A, B) space centred on the selected point. Denote the typical linear dimension of this neighbourhood

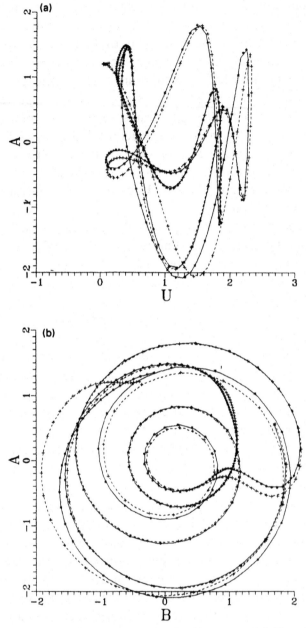

Fig. 8.18. The solutions of the Lorenz simple climate model, Eqs. (8.19a–c), for two nearby initial conditions: (a) projected on to the (U, A) plane; (b) projected on to the (A, B) plane.

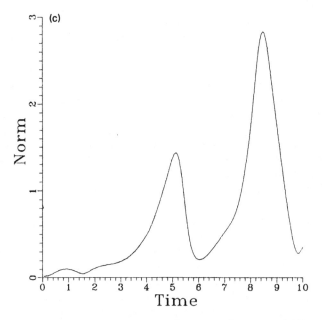

Fig. 8.18 (*cont.*). (c) The separation $(\Delta U^2 + \Delta A^2 + \Delta B^2)^{1/2}$ between the two solutions as a function of time.

as r. Now increase r and repeat the count. If the solutions lie along a line (possibly a curved line), then the number of points counted will increase as the first power of r. If they lie on a two-dimensional (possibly curved) surface, the number of points counted will increase as r^2, while if they fill (U, A, B) space uniformly, the count will increase as r^3. When this method is applied to the solutions of the Lorenz set, the resulting dimension is neither 2 nor 3, but somewhere in between, around 2.3. It is called a 'fractal dimension'; having a fractal dimension is a defining characteristic of truly chaotic, rather than merely rather complicated, solutions to a dynamical system such as Eqs. (8.19a–c).

A final way of summarizing the Lorenz system is in the form of a frequency spectrum of its fluctuations. Figure 8.20 shows such a spectrum for U. There are a number of peaks at high frequencies. These can be related to the various linear frequencies which emerge from linearizing the equation set about its mean state. But one effect of the nonlinearity of the system is to broaden these peaks across a range of frequencies. More significantly for our present purposes, there is a continuous background spectrum of variability which extends down to the lowest frequencies defined by the integration. This background spectrum is a manifestation of the infinite variety of solutions

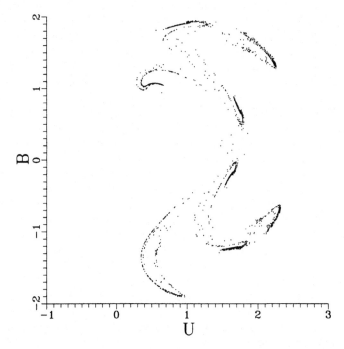

Fig. 8.19. A long integration of Eqs. (8.19a–c), showing the intersections of the solution trajectory with the $A = 0$ plane, forming a so-called 'Poincaré section'.

of Eq. (8.19a–c); they never repeat themselves exactly, although they are confined to the same neighbourhood of (U, A, B) space.

This simple system suggests the possibility that the observed low and very low frequency fluctuations of the global circulation could be due, at least in part, to such internally generated variability. In that case, it is not necessary to appeal exclusively to external sources of variability in terms of changing sea surface temperatures or radiative forcing in order to account for all the observed fluctuations. One way of testing this hypothesis is to carry out long integrations with a global circulation model with fixed boundary conditions and radiative inputs. An example, using the 'simple global circulation model' introduced in Section 2.5 with Eqs. (2.34) and (2.35), will be described here. In this model, the simple Newtonian cooling and Rayleigh friction eliminates any possibility of feedbacks between the circulation and its forcing. Any variability must be generated within the simulated atmosphere as a consequence of the nonlinearity of the governing dynamical equations.

In fact, long integrations of this model show that it is unsteady, even on

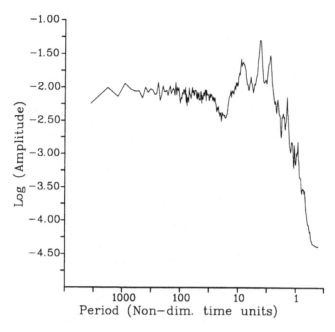

Fig. 8.20. An amplitude frequency spectrum of the fluctuations of U from a long integration of the Lorenz system, Eq. (8.19a–c).

the longest timescales. This 'ultra low frequency' variability is most marked when the zonal mean flow is considered. Figure 8.21 shows the frequency spectrum of the mean angular velocity of the modelled atmosphere. It rises to maximum values on the longest timescales encompassed by a 100 year run, with large amplitudes on timescales of 10 years and longer. This spectrum is a good example of a 'red noise' spectrum, with a large amplitude, random signal at low frequencies. It is reminiscent of the low frequency behaviour of the Lorenz system shown in Fig. 8.20. The main difference is that there is rather little evidence of the broadened peaks near the frequencies associated with the linear normal modes of the system. The nonlinearly generated low and ultra low frequencies are dominant.

EOF analysis shows that there is a distinct spatial structure to the ultra low frequency variability of our model atmosphere. The first EOF of [u] accounts for over 30% of the variability of the flow; its structure, shown in Fig. 8.22 consists of alternating easterly and westerly anomalies in the winter midlatitudes. In the 'negative phase' of the EOF, the anomaly consists of a midlatitude westerly jet with easterly jets to south and north. The picture is not unlike the changes to the zonal mean

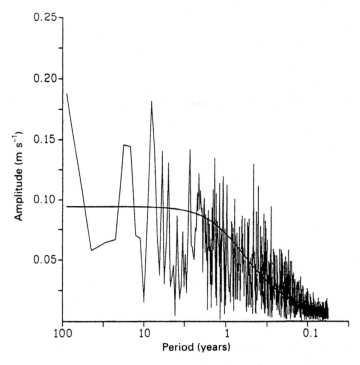

Fig. 8.21. Frequency spectrum of the fluctuations of the global average atmospheric angular velocity in a long integration of the simple global circulation model described in the text.

flow which result from a baroclinic lifecycle experiment, and strongly suggest that baroclinic instability of the winter midlatitudes plays a central role in driving the ultra low frequency variability of the circulation. The model has every appearance of being a truly 'chaotic' system. We cannot hope to predict its detailed behaviour very far into the future. But the hypothesis that some kind of feedback between the baroclinically unstable waves and the zonal mean flow might drive the ultra-low frequency variability would require that the behaviour of the system might be, at least qualitatively, describable by some low order set of nonlinear equations, rather like the Lorenz set. What we call a 'theory' of this system would indeed consist of such a low order model of it, in which the hundreds of differential equations, each governing the evolution of one spectral coefficient (or, equivalently, one grid point value) of each of the variables, were condensed into just a handful of separate equations.

The Lorenz set gives some clues about the sort of properties that this

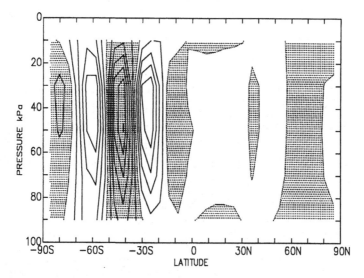

Fig. 8.22. Structure of the first EOF of [u] for the long integration of the simple global circulation model. This EOF accounted for 31% of the total variability of the time series, and an even larger fraction of its very low frequency behaviour.

low order equation set would have to possess. As a minimum, one variable would be needed to represent the baroclinicity of the zonal flow, and a further two variables would be needed to describe the (complex) amplitude of an unstable, propagating wave. These are the main ingredients of the Lorenz set. There is a feedback, such that strong developments of the waves weaken the baroclinicity which makes them unstable. The extra element which is needed in the Lorenz set to make its solutions chaotic is the interaction with a forced stationary wave, represented in the last term in Eq. (8.19b). If it is strong enough, this continuous forcing can disrupt any balance between the waves and the zonal flow, so that the system is always unsteady. There is no forced stationary wave in the simple global circulation model. Instead, we might appeal to interactions with a more slowly propagating long wave component to the flow, or to wave breaking, which would involve interactions with a harmonic of the unstable wave. Clearly, the possibilities rapidly multiply as extra degrees of freedom are introduced. One way of placing bounds on the sort of equation sets which would be needed to describe the long term behaviour of the global circulation model would be to determine the fractal dimension of its chaotic solutions. That, at least, would tell us how many variables would have to be retained in a low order approximation to it.

This turns out to be a very difficult task. The number of data points required to determine the fractal dimension of a solution becomes very large as the fractal dimension increases. Determinations of fractal dimensions greater than about 7 or 8 are probably unreliable. Attempts to determine a fractal dimension for the kind of integrations shown in Fig. 8.21 are not encouraging. Their dimension is probably well in excess of 8, unless some severe time smoothing is introduced. The most optimistic view is that the simple global circulation model embeds some simpler model with dimension perhaps 10 or 20; clearly, extracting such a model from integrations of the full model would be a very difficult task. The more pessimistic possibility is that the simple global circulation may not be reducible to simpler terms. In that case, it would be a waste of time seeking any 'explanations' of its ultralow frequency behaviour. More radically, we would be left with the concern that a higher resolution model would exhibit qualitatively different behaviour, since it is a set of many more equations. The use of global circulation models to study the behaviour of the atmosphere relies on the implicit assumption that, given enough resolution, the large scale, low frequency evolution of the flow will be independent of the model. At present, it is an open question whether this assumption is correct.

8.8 Problems

8.1 From a long numerical integration of the Lorenz set, Eq. (8.19a–c), the following data were obtained:

$$\overline{U} = 1.0606, \ \overline{A} = -0.0560, \ \overline{B} = 0.3326,$$

$$\overline{U^2} = 1.4129, \ \overline{A^2} = 0.7612, \ \overline{B^2} = 0.9319,$$

$$\overline{UA} = 0.0138, \ \overline{UB} = 0.2685, \ \overline{AB} = -0.0693.$$

Determine EOFs for the Lorenz system, and say how much variance each EOF accounts for. Plot your EOFs on the U–B and A–B planes, and relate them to Figs. 8.18 and 8.19.

8.2 Use Eqs. (6.9) and (6.10) to plot graphs of the mountain drag versus flow speed for a zonal wavenumber 2 variation of orographic height, amplitude 1000 m, drag timescale 5 days, channel width 5000 km. You may assume that $U_E = 50 \, \mathrm{m \, s^{-1}}$, and that the channel is centred at 45 °N. Hence determine whether multiple equilibria are possible according to the Charney–Devore theory. If they are, say what the mean flow is for each equilibrium state.

8.3 Suppose an isolated tropical heating anomaly were suddenly turned on, generating zonal wavenumber 3 Rossby waves at 25 °N. Given a typical DJF flow, estimate the poleward group velocity of these Rossby waves, and hence deduce how long it will be before the anomaly starts to perturb the flow at 55 °N. Repeat the calculation for zonal wavenumber 1.

9

The stratosphere

9.1 The seasonal cycle of the stratospheric circulation

Up until this point, we have concentrated almost exclusively upon the troposphere, which is characterized by a relatively weak stratification, with a temperature lapse rate of around 6–7 K km^{-1}. At the tropopause, the lapse rate becomes close to zero; the lower stratosphere is nearly isothermal. The corresponding change in stratification, as measured by the Brunt–Väisälä frequency, is by a factor of around two, from values of 10^{-2} s^{-1} in the troposphere to values of 2×10^{-2} s^{-1} in the lower stratosphere. In the upper stratosphere, from heights of 30 km to around 50 km, the temperature actually increases with height. The transition to stably stratified conditions is called the tropopause, which is extremely sharp in the tropics and midlatitudes. It is rather more gradual in polar latitudes, especially in winter when there is no incoming sunlight. The abrupt increase of stratification at the tropopause means that the stratosphere is dynamically very different from the underlying troposphere. Baroclinic instability is virtually suppressed and disturbances are mainly forced from below. The stratification acts as a filter, removing the smaller scale disturbances and allowing only the longest waves to propagate out of the troposphere to great heights in the stratosphere. Shorter wavelength disturbances are thereby trapped in the troposphere, which behaves as a waveguide, the upper boundary of which is the tropopause.

Figure 9.1 illustrates this filtering action. Fields have been taken from a set of ECMWF analyses to show the synoptic pattern of flow for the same winter analysis time but at different levels, ranging from the upper troposphere up to the middle stratosphere. At 30 kPa, a number of sharp troughs associated with surface depression systems can be seen, as well as an anticyclone to the west of Ireland and a number of other disturbances. At higher levels,

302

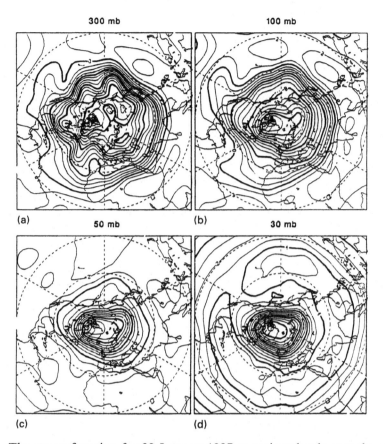

Fig. 9.1. The streamfunction for 22 January 1987 at various levels over the northern hemisphere. (a) 30 kPa (about 9 km) (b) 10 kPa (about 17 km) (c) 3 kPa (about 26 km) and (d) 1 kPa (about 34 km) . Contour interval $10^7 \, \mathrm{m}^2 \, \mathrm{s}^{-1}$.

a steady transition is noted; by the 3 kPa level (at about 24 km above the Earth's surface) the only features to be seen are the tight vortex displaced from the pole towards northern Europe and a weak anticyclone over the northern Pacific. A Fourier analysis of the streamfunction field at this upper level reveals that it is composed almost entirely of zonal wavenumbers 1–3.

In the summer season, a similar, but even more dramatic, filtering of wave like disturbances is seen. Plots of streamfunction for 22 July 1986 are shown in Fig. 9.2. At 30 kPa, the rather weak midlatitude westerlies are very disturbed, with large amplitude, transient systems particularly evident over the two ocean basins. By 10 kPa, the cyclonic vortex has nearly vanished, with just a vestigial feature over North America. Instead, the field at this level is dominated by the massive anticyclone over the Middle East

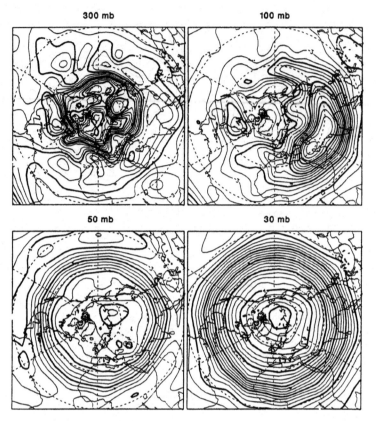

Fig. 9.2. Similar to Fig. 9.1, but showing the streamfunction for 22 July 1986. (a) 30 kPa, (b) 10 kPa, (c) 3 kPa and (d) 1 kPa. Contour interval $5 \times 10^6 \, \text{m}^2 \, \text{s}^{-1}$.

and central Asia associated with the Asian monsoon. This anticyclonic character becomes accentuated at 5 kPa, while at 3 kPa an almost precisely axisymmetric anticyclone is centred over the north pole and fills the entire summer hemisphere.

The theory of Section 6.4 gives a good account of the changing character of the flow with height shown in Figs. 9.1 and 9.2, and we will return to the application of this theory to the stratosphere in Section 9.2. But, first, it is necessary to give some account of the zonal mean flow in the stratosphere. This is primarily driven by radiative effects, but is strongly modified by dynamical heat transports. It has a characteristic seasonal cycle.

Approximately 1% of the sunlight incident on the atmosphere, mostly at ultraviolet wavelengths, is absorbed in the stratosphere. But because of the very low density of the air, substantial heating rates can result.

Figure 9.3 shows the daily mean flux of solar radiation incident on the top of the atmosphere as a function of latitude and time of year. The maximum insolation occurs at the summer pole, close to the time of the summer solstice. Around the winter solstice, high polar latitudes receive no sunlight at all. The most important absorber of ultraviolet radiation in the stratosphere is ozone, the triatomic form of oxygen. Ozone has a maximum partial pressure at around 25 km and a maximum mixing ratio at around 50 km. It is an extremely effective absorber of ultraviolet radiation with wavelengths of less than 300 nm. Thus the maximum heating rates, of up to 12 K day^{-1}, occur near the top of the ozone layer, at around 50 km. It is this pattern of heating which effectively determines the gross features of the vertical temperature profile of the atmosphere, with stable stratification from the tropopause to the stratopause at about 60 km, and rather less stable stratification in the mesosphere. The atmosphere develops strong temperature gradients in response to this heating, until long wave cooling is sufficient to balance the heating. Cooling is principally due to carbon dioxide, although both water vapour and ozone itself have important infrared emission bands which make large contributions to the cooling. Fig. 9.4 shows heating rates due to absorption of solar radiation by ozone, and the net heating rates due to all radiative processes. At the solstices, the heating follows the pattern we might expect from the distribution of the solar radiation, with heating throughout the summer hemisphere and intense cooling at high winter latitudes. At the equinox, the pattern of heating is roughly symmetric about the equator, with heating in the tropics and cooling at high latitudes in both hemispheres.

Although ozone has such a profound effect on the temperature structure of the stratosphere, its total concentration is extremely small, amounting to no more than 1 molecule in 10^5 at its most concentrated. The total volume of ozone in a column is sometimes stated in terms of the depth of a column of gas if the total ozone were isolated and compressed into a thin layer at 100 kPa and 273 K. Typical values range from 2.6 to 4.5 mm, depending upon latitude and season. In contrast, the rest of the atmosphere measured in this way would have a depth of 8 km. The situation is complicated by the fact that ozone is a highly variable constituent. The factors influencing its concentration will be discussed in more detail in Section 9.3.

In the winter, the heating rate is zero at high latitudes, since there is no incident sunlight. The heating rate builds up as tropical latitudes are approached. The result is an intense gradient of temperature near the poleward limit of sunlight at 66° of latitude, with very low temperatures near the pole. In thermal wind balance with this temperature gradient, the zonal winds increase with height at high latitudes, and are quite intense

Incident solar radiation

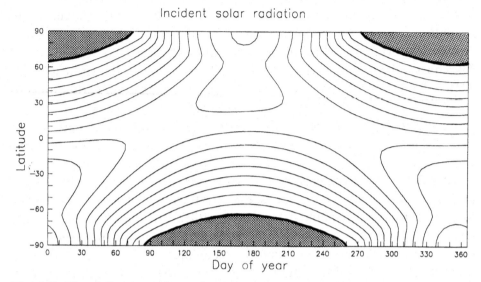

Fig. 9.3. The daily mean flux of solar radiation incident upon the top of the atmosphere as a function of latitude and time of year. Contour interval $50\,\mathrm{W\,m^{-2}}$. Shading denotes the polar winter regions of perpetual darkness.

above $5\,\mathrm{kPa}$. This high latitude jet stream or 'polar night jet' is generally far from axisymmetric, at least in the northern hemisphere; as seen in Fig. 9.1, it is distorted and displaced away from the pole.

Figure 9.5 shows latitude–height cross sections of the zonal mean zonal wind in the stratosphere. In the summer hemisphere, the tropospheric westerly jet dies away quickly in the lower stratosphere, and the wind regime becomes predominantly easterly. In the winter hemisphere, the tropospheric westerly jet also becomes weaker above the tropopause, while the polar night jet rapidly strengthens, attaining values of around $60\,\mathrm{m\,s^{-1}}$ at the stratopause. Winds in the tropics are more difficult to estimate. They cannot be inferred from thermal wind balance, and direct measurements of the stratospheric wind field are difficult to obtain above the usual ceiling for radiosonde ascents.

Although we have just discussed the gross features of the stratospheric zonal mean temperature and wind field in terms of the radiative equilibrium distributions of temperature, it must be appreciated that in some respects the stratosphere is very far from a state of radiative equilibrium. Figure 9.6 shows the results of a computation of the radiatively determined stratospheric temperature field, obtained by integrating a model which included the seasonal cycle of solar radiation but ignored dynamical heat transports, compared with that observed. The observed winter pole is considerably warmer than radiative equilibrium, implying that the atmospheric circulation is transporting heat into high latitudes. More strikingly, observations reveal a local maximum of temperature in the midlatitudes, with a cold equator and maximum temperature near 50 °N. The associated reversal of the usual temperature gradient is responsible for the easterly shear above the tropospheric jet and, consequently, for the characteristic jet maximum near the tropopause. The existence of this cold stratospheric equator is clear evidence for a very active circulation in the stratosphere which transports heat polewards.

The 'Charney–Drazin' theory introduced in Section 6.4 predicts the observed filtering effect of the stratosphere on the eddies. We saw there that in westerly flow, only long wavelength Rossby waves can propagate vertically. Shorter wavelengths are evanescent and will die away quickly with height. The stronger the westerly flow, the longer the wavelengths of vertically propagating Rossby waves. The rather strong westerlies associated with the polar night jet will prevent the upward propagation of smaller wavelength disturbances which are seen to become attenuated rapidly above the tropopause. At the highest levels, only wavenumbers 1, 2 and 3 are present with significant amplitude. If anything, the amplitude of these wavelengths seems to increase with height. This result is consistent with the density effect predicted in our Charney–Drazin theory.

In midsummer, the incident sunlight is a maximum at the pole. The relatively low angle of the sun is more than compensated by the long day at high latitudes. In the troposphere, the scattering and absorption of sunlight along the long atmospheric path of the beam means that insolation is still larger in the summer tropics. But in the stratosphere, scattering and absorption are less important. The temperature maximum is at the summer pole, and a poleward temperature gradient is characteristic of the summer stratosphere. Thermal wind balance suggests easterly shear within the stratosphere, and indeed by the 5 kPa level, easterlies are established throughout the summer hemisphere.

The transitions between the typical summer and winter flows are of considerable interest. The autumn transition is generally gentle and accomplished on radiative timescales (typically 5–20 days, depending upon height). As the solar heating declines at the pole around the equinox, cooling takes place and a small cyclonic vortex starts to form over the pole. The easterlies characteristic of the summer season persist in the midlatitudes. The polar vortex strengthens and expands until a highly disturbed polar night jet is established, while the easterlies retreat into the tropical latitudes. The transition is generally complete by the end of November. The spring transition

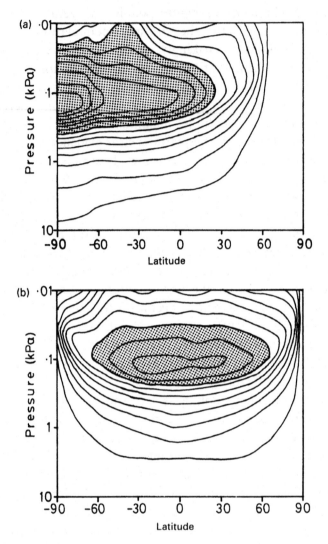

Fig. 9.4. Cross sections showing the heating rates in the stratosphere: (a) heating due to ozone, 21 December; (b) heating due to ozone, 21 March.

is much more dramatic. The westerlies break down rapidly, sometimes on timescales of only a few days. Anticyclonic circulations disrupt the polar vortex and displace it from the pole. At the same time, the temperature rises substantially over the polar regions. Such 'stratospheric sudden warmings' are among the most dramatic synoptic events in the neutral atmosphere. The sudden change of circulation establishes a summerlike easterly regime. When the warming occurs late in the winter, it can mark the final breakdown

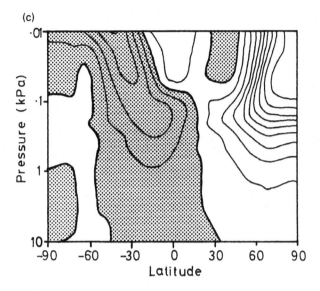

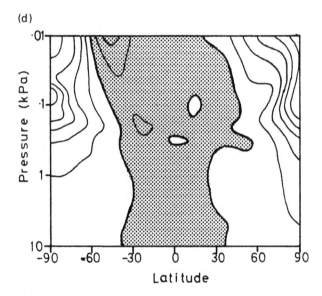

Fig. 9.4 (*cont.*). (c) Net heating, including long wave cooling, January; (d) net heating, March. Contour interval $1\,\mathrm{K\,day^{-1}}$. In (a) and (b), shading indicates values in excess of $8\,\mathrm{K\,day^{-1}}$, and in (c) and (d) shading indicates net heating. (After Gille and Lyjak 1986.)

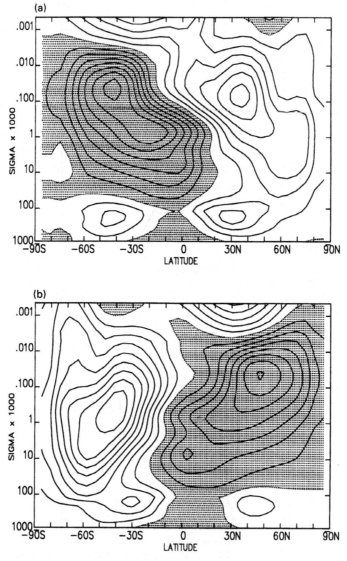

Fig. 9.5. Cross sections showing the zonal mean zonal wind $[\bar{u}]$ in the stratosphere, based on the climatology of Fleming *et. al.* (1990): (a) January; (b) July. The contour interval is $10\,\mathrm{m\,s^{-1}}$, with easterlies indicated by shading. In this, and succeeding similar plots, the vertical coordinate is $1000\,\log(p/p_R)$. It may be noted that the tick marks are therefore placed roughly every $14.7\,\mathrm{km}$.

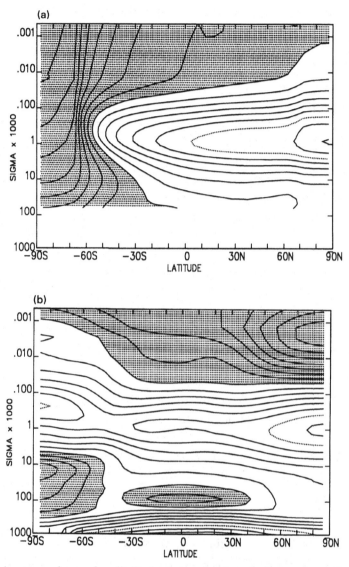

Fig. 9.6. A comparison of radiatively determined and observed zonal mean strato-spheric temperatures for July: (a) radiative equilibrium, kindly supplied by K. P. Shine, and based on an extension of the calculations in Shine (1987); (b) Observed, from the climatology of Fleming *et al.* (1990). Contour interval 10 K, with shading indicating temperatures less than −60 °C.

of the winter circulation. Easterlies persist until the following autumn. But when the warming comes earlier in the winter, or is less strong, a gradual relaxation towards the normal winter flow takes place, essentially on radiative timescales.

Figure 9.7 shows a typical sequence of fields during a stratospheric warming event. Although the actual reversal of the winds is indeed very abrupt, anticyclones build up outside the polar vortex for some time before the warming.

The dramatic changes during a warming, which take place on a timescale of only a few days, point clearly to a dynamic origin for the event. Propagation of disturbances from lower levels seems to be implicated, and there is evidence linking warmings to major anomalies (such as blocking) in the troposphere. The transition in the southern hemisphere stratosphere is less abrupt, and the polar night jet is less disturbed. This presumably reflects the somewhat weaker steady long waves in the southern hemisphere troposphere. The less disturbed state of the southern polar vortex is important in leading to the extremely large depletions of ozone observed in recent years above Antarctica. The lack of major disturbances means that air is trapped within the polar vortex for long periods and thus destructive catalytic reactions have time to occur.

The theory of upward propagating Rossby waves, outlined in Chapter 6, accounts in broad terms for many of the observations recounted in this section. At the same time, eddies in the winter stratosphere carry important fluxes of heat and momentum; these modify the zonal mean state. As our study of the Eliassen–Palm flux in Section 6.4 revealed, propagation of waves and their interaction with the mean flow are related matters. We return to the discussion of this in the next section.

9.2 Wave propagation and mean flow interactions

The requirements for the vertical propagation of Rossby waves rather neatly account for the absence of waves in the summer stratosphere and the long wave disturbances of the polar night jet observed in the winter. Figure 6.19 shows that steady Rossby waves will be evanescent in the vertical when the zonal wind is easterly. Thus, we would expect the mean flow of the summer stratosphere to be almost undisturbed by propagating waves. Indeed, it is highly axisymmetric. Incidentally, it is not greatly disturbed by transient eddies either. The theory of Section 6.4 is easily modified to the case of transient waves, with a phase speed c. The result is simply to replace U by $U - c$ in Eq. (6.43). The phase speed of tropospheric Rossby waves is generally fairly small compared to the magnitude of the summer stratospheric easterlies, so that there is little possibility for transients to propagate into the summer stratosphere. Vertical propagation is possible in westerly flow, such as would be expected in the winter stratosphere. The longest waves can propagate most readily, while shorter waves will be evanescent. The filtering effect of the stable stratosphere in westerly conditions, illustrated by Fig. 9.1, is accounted for in a straightforward way. Of course, the theory of Section 6.4 is highly idealized, and its assumption of constant U and N is hardly realistic. The Eliassen–Palm formalism provides a way of extending the theory to more general conditions. In this section, we will concentrate upon the winter situation in order to evaluate the transport properties of the stratospheric planetary waves.

The temperature flux carried by a vertically propagating Rossby wave was derived in Eq. (6.53) for the simplest situation of vertical propagation in a uniform zonal wind. In addition to the temperature flux carried directly by the eddies, there will be an indirect eddy transport of temperature by the eddy induced mean meridional circulation. Together, these effects determine the total heat transport resulting from the eddies. The same ideas will be required to discuss the transport of trace constituents, such as ozone, by the winter stratospheric circulation, an aspect to which we will return in the next section. The eddy and meridional mean components of the temperature flux tend to cancel out, a result which is qualitatively clear from the work of Section 4.4. There we showed that a maximum of the poleward eddy temperature flux in the midlatitudes would induce a thermally indirect mean meridional circulation. That is, the induced circulation will tend to transport heat downwards and equatorwards, while the eddies achieve the opposite. To make the argument more quantitative, we will restate the analysis of Section 4.4, here using pseudo-height z' as the vertical coordinate since we are concerned with the stratosphere.

The Eulerian mean equations in local Cartesian coordinates were given as Eqs. (4.27) and (4.28), together with the thermal wind equation, Eq. (4.29). In terms of the pseudo-height, they may be written:

$$\frac{\partial [u]}{\partial t} - f[v] = -\frac{\partial}{\partial y}[u^*v^*] + [\mathcal{F}_1], \tag{9.1}$$

$$\frac{\partial [\theta]}{\partial t} + \frac{\theta_R N^2}{g}[w] = -\frac{\partial}{\partial y}[v^*\theta^*] + [\mathcal{Q}], \tag{9.2}$$

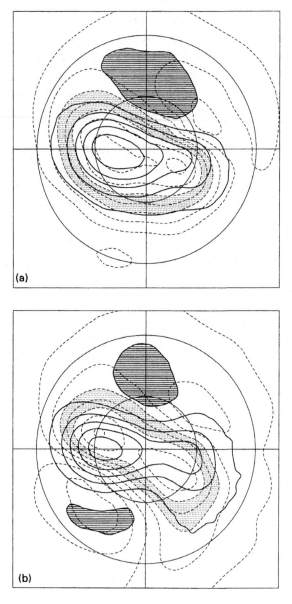

Fig. 9.7. Sequence of synoptic fields at 1 kPa during a stratospheric warming event. Solid contours show geopotential height, contour interval 500 m, with values greater than 31 km indicated by shading. Dashed contours show temperature, contour interval 5 K, with values between 210 K and 220 K shaded. The outer latitude circle is 30 °N. (a) 23 December 1981; (b) 26 December 1981.

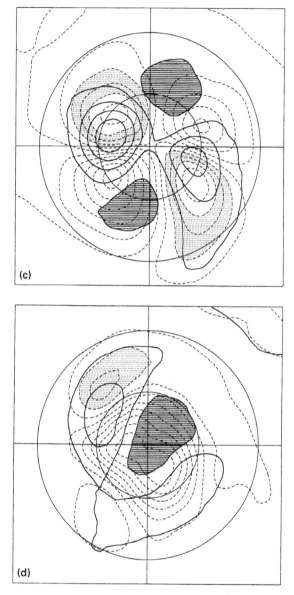

Fig. 9.7 (*cont.*). (c) 29 December 1981; and (d) 2 January 1981. (Plots kindly supplied by A. O'Neill.)

together with the thermal wind balance:

$$f\frac{\partial [u]}{\partial z'} = \frac{g}{\theta_R}\frac{\partial [\theta]}{\partial y},$$

(9.3)

and continuity:

$$\frac{\partial [v]}{\partial y} + \frac{1}{\rho_R} \frac{\partial}{\partial z'} (\rho_R [w]) = 0. \tag{9.4}$$

A useful transformation of these equations is obtained by defining a residual wind such that:

$$[v]_r = [v] - \frac{1}{\rho_R} \frac{\partial}{\partial z'} \left(\rho_R \frac{[v^*\theta^*]}{\theta_{Rz'}} \right), \tag{9.5a}$$

$$[w]_r = [w] + \frac{\partial}{\partial y} \left(\frac{[v^*\theta^*]}{\theta_{Rz'}} \right). \tag{9.5b}$$

The residual meridional wind still satisfies the continuity equation, and so it defines a residual meridional circulation. Substituting into the zonal mean momentum and thermodynamic equations, we have:

$$\frac{\partial [u]}{\partial t} - f[v]_r = \frac{1}{\rho_R} \nabla \cdot \mathbf{F} + [\mathscr{F}_1], \tag{9.6}$$

$$\frac{\partial [\theta]}{\partial t} + \frac{\theta_R N^2}{g} [w]_r = [\mathscr{Q}]. \tag{9.7}$$

These equations are called the 'transformed Eulerian mean' (TEM) equations. Note that the eddy term only appears explicitly in the momentum equation, where the combined effects of the temperature and momentum fluxes are contained in the $\nabla \cdot \mathbf{F}$ term. The thermodynamic equation reveals that in steady flows, the residual meridional circulation is purely a response to gradients of heating. The Eulerian mean circulation induced by the poleward temperature fluxes has been removed from the residual circulation by the transformations, Eqs. (9.5a, b).

In conditions of steady, frictionless and adiabatic flow, the TEM equations establish an important result, called the Eliassen–Palm theorem. Consider the special case of frictionless, adiabatic and steady flow. Then the TEM equations reduce to:

$$-f[v]_r = \frac{1}{\rho_R} \nabla \cdot \mathbf{F}, \tag{9.8a}$$

$$s^2 [w]_r = 0. \tag{9.8b}$$

From Eq. (9.8b) together with continuity, we conclude that $[w]_r = [v]_r = 0$, and consequently that

$$\nabla \cdot \mathbf{F} = 0. \tag{9.9}$$

This equation does not say that the eddy fluxes are zero. It does say that the variation of $[v^*\theta^*]$ in the vertical and of $[u^*v^*]$ in the horizontal must be such that the Eliassen–Palm flux has zero divergence. This is the case for the simple vertically propagating Rossby wave discussed in Chapter 6. We showed there that the poleward temperature flux associated with this wave is substantial. However, its vertical variation is such that:

$$\frac{\partial}{\partial z'}\left(\rho_R[v^*\theta^*]\right) = 0. \tag{9.10}$$

That is, Eq. (9.9) is satisfied if $[u^*v^*] = 0$. The physical interpretation of this result is that although there is a poleward temperature flux associated with the Rossby wave, it is exactly balanced by the equatorward temperature flux carried by the induced Eulerian mean circulation.

We have considered other situations where the Eliassen–Palm theorem does *not* apply. The temperature fluxes of a growing Eady mode, discussed in Section 5.4 do not satisfy $\nabla \cdot \mathbf{F} = 0$, even though the Eady model is frictionless and adiabatic. Strictly, there is infinite divergence of $\mathbf{F}$ in an infinitesimal layer near the lower boundary, and corresponding convergence near the upper boundary. The reason is that the Eady wave is exponentially amplifying, so we are not dealing with a steady flow. Although the induced meridional circulations offset the poleward temperature fluxes carried by the Eady mode, they do not exactly compensate for it, and so there is net poleward temperature transport. In the tropospheric global circulation itself, similar considerations apply. The Ferrel cell is thermally indirect and so tends to transport heat equatorward. But the flow is highly unsteady, and, furthermore, friction acts near the ground. Thus there is certainly net poleward temperature flux associated with midlatitude tropospheric eddies.

Our arguments so far have been largely based upon the linear theory of vertically propagating Rossby waves. The essence of a Rossby wave is a small meridional displacement of a fluid element. The element conserves its potential vorticity as it moves into a region where the ambient potential vorticity is different. The resulting circulations tend to return the Rossby wave to its 'home' latitude, providing the restoring tendency which results in wave motion. Note that this displacement is a totally *reversible* displacement.

However, as the Rossby wave propagates higher, its amplitude may vary, and can lead to large, irreversible distortions of potential vorticity contours or 'breaking'. At least two mechanisms for breaking are possible.

The first will operate even when $[u]$ is constant; according to Eq. (6.45), the amplitude will increase exponentially with height as the wave propagates to levels where the density is lower. The wave action density is conserved as the wave propagates to higher levels, but the mean particle displacement increases. This will mean that above some level, the linear assumptions of the theory will break down. Typically, this will happen when the meridional displacement of a fluid element becomes comparable with the wavelength of the waves. In this situation, contours of potential vorticity become highly distorted, and the potential vorticity is pulled out into long filaments. Such long, thin filaments are barotropically unstable and might be expected to break up rapidly into discrete vortices. The instability means that the distortion of the potential vorticity contours rapidly becomes irreversible. Various dissipative processes can act on these small scale fluctuations to smooth out the potential vorticity field. These processes are not simple to describe in analytical terms, and the rapid reduction of scales makes them difficult to represent in a numerical model. Consequently, the details of the saturation process are not fully understood. The overall effect, though, is to mix the potential vorticity throughout the region occupied by the saturated Rossby wave, giving a broad region in the midlatitudes where potential vorticity gradients are small.

The second breaking mechanism may be associated either with meridional or vertical propagation, and will occur when the zonal wind changes with height. In particular, if $[u]$ changes sign, a 'critical line', such as we discussed in Chapter 6, will be associated with the locus of points where $[u] = 0$. As a Rossby wave approaches a critical line, its amplitude will increase and its group velocity will decrease. The assumptions of linearity will break down, and irreversible shredding and mixing of the potential vorticity field will occur; the Rossby wave will break in much the same way as for the increase of amplitude with decreasing density.

There is an analogy between this saturation and breakdown of vertically propagating Rossby waves, and the well-known behaviour of surface gravity waves on deep water as they approach a shelving beach. As the waves propagate into regions of shallower water, they become steeper. Eventually, they become so distorted that they 'break', that is, the reversible displacements of water associated with wave motion become irreversible, and a rapid shrinking of the scales of motion takes place. This analogy has led various authors to speak of the 'breaking' of planetary scale Rossby waves in the middle stratosphere, and to refer to the region of small potential vorticity gradient as the 'stratospheric surf zone'.

Figure 9.8 shows a map of the Ertel potential vorticity (see Eq. (1.79)) on the 850 K potential temperature surface during the northern hemisphere winter. The vortex circumscribed by the polar night jet shows up as a strong maximum in the potential vorticity field, distorted by the upward

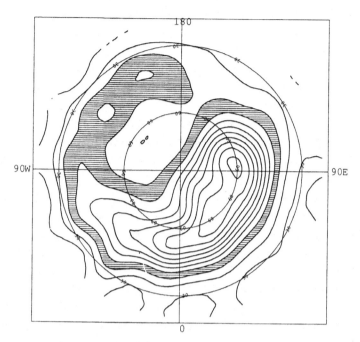

Fig. 9.8. Ertel's potential vorticity on the 850 K isentropic surface on 3 December 1981, showing an example of planetary wave 'breaking', Contour interval $10^{-4}\,\mathrm{K\,m^2\,kg^{-1}\,s^{-1}}$ (i.e., 100 PVU), with shading indicating values between 300 and 400 PVU. The outer latitude circle is 30 °N. (Data kindly supplied by A. O'Neill.)

propagation of mainly wavenumber 1 or 2 Rossby waves. Over the Pacific, a long streamer of high potential vorticity has apparently been pulled off the main vortex, and is being wrapped into a huge spiral. This large region of small and irregular potential vorticity gradient is the so-called 'surf zone'.

As well as mixing potential vorticity in midlatitudes, Rossby wave breaking tends to sharpen the gradients of potential vorticity around the polar vortex itself. This occurs as small distortions on the vortex are pulled out into long thin streamers, which are mixed into the surf zone, leaving the edge of the vortex smoother and sharper than previously. The effect is rather analogous to a plane removing the projections from a rough piece of wood, leaving the wood/air interface smoother and sharper. This process is particularly important in the southern hemisphere winter, where the vortex is less distorted. The sharpening of gradients around its periphery inhibits the exchange of matter and heat between the centre of the vortex and the surf zone. During the winter, the temperatures become very cold, and trace constituents can accumulate and take place in various chemical reactions, without mixing down to lower latitudes where they would be photolysed by sunlight. The formation of this cold 'containment vessel' is an important element in generating the infamous ozone hole.

Now let us return to a consideration of the vertical propagation of Rossby waves in the stratosphere. We can use the work of Section 6.5 which set up a framework for discussing the propagation of Rossby waves on a zonal flow $U = U(y, z')$. By making a 'slowly varying' approximation, we can follow the propagation of wave activity into regions where U is different. As we saw in Section 6.5, the Eliassen–Palm flux vector is parallel to the local group velocity. But in frictionless, steady and adiabatic conditions, the Eliassen–Palm theorem must hold, and so the Eliassen–Palm flux must have zero divergence. But now consider the situation depicted in Fig. 9.9, in which the zonal wind changes sign at some level in the stratosphere. Steady Rossby waves cannot propagate in easterly conditions. They are evanescent and, as we saw in Eq. (6.54), their poleward temperature flux is zero. In this region, both $\nabla \cdot \mathbf{F}$ and $\mathbf{F}$ itself will be zero. In the westerly flow, $\nabla \cdot \mathbf{F}$ must be zero, but $\mathbf{F}$ itself will generally be nonzero. In the simplest example of the Charney–Drazin model, $\mathbf{F}$ will be vertical and constant with height. In the vicinity of the 'critical level' where $U = 0$, there must be a large convergence of $\mathbf{F}$, and, consequently, the Eliassen–Palm theorem must break down there. Assuming that friction and heating remain small, the simplest solution is to predict that the flow will become unsteady at this level. Equation (9.6) suggests that this unsteadiness will be manifested as an easterly acceleration at the critical level. Thus, if there is a constant flux of wave activity into the base of the stratosphere, the region of easterly flow will descend until it fills the stratosphere.

This kind of process is the basis of current theories of sudden stratospheric warmings. Once easterly flow is established at some upper level (where the assumptions of frictionless, adiabatic flow may become invalid so that the atmosphere is not constrained by the Eliassen–Palm theorem), the region of easterlies can descend on a relatively fast dynamical timescale. Once the easterly regime is established, wave activity will virtually die away, and the usual westerly regime can only be re-established on a slower radiative timescale. The associated large temperature rises result from meridional circulations which maintain thermal wind balance as the zonal wind changes. It is easy to show that in the stably stratified stratosphere, quite small vertical displacements of air can result in large temperature changes.

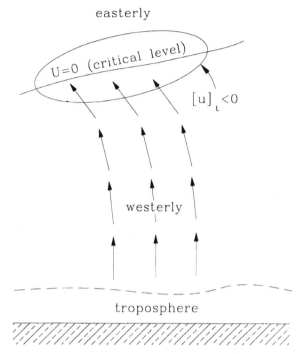

Fig. 9.9. Schematic illustration of steady Rossby waves propagating vertically into a region of easterly flow.

9.3 The production and transport of ozone

The distribution of stratospheric ozone was described in the preceding section. Figure 9.10 shows the global distribution of ozone as a function of height and latitude, while Fig. 9.11 shows how the total ozone in a column above the Earth's surface varies with latitude and time of year. The maximum concentration of ozone is generally at heights of 20–25 km and occurs in the tropics. However, the layer of ozone is generally rather thicker in the winter and spring polar regions, with the result that the column total ozone abundance reaches its maximum in the spring at high latitudes. The notorious 'ozone hole' which has recently developed in the southern hemisphere spring is a major depletion of this spring maximum at high southern latitudes. In this section, we will consider how the major features of the ozone distribution are to be understood, beginning with a consideration of the photochemical processes which produce ozone. But in the lower stratosphere, the meridional advection timescales are quite short compared with

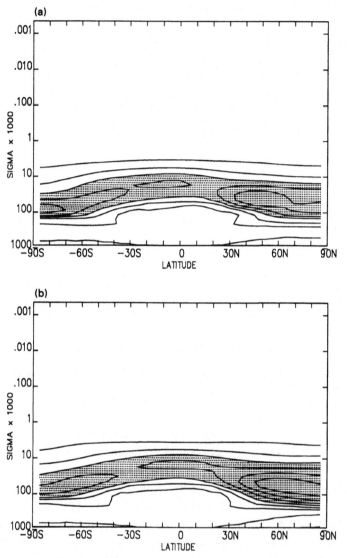

Fig. 9.10. Latitude–height cross sections of ozone concentration for (a) January; and (b) March, based on the CIRA climatology of Fleming *et. al.* (1990). Contour interval 10^{18} molecules m^{-3}, with values in excess of 3×10^{18} molecules m^{-3} indicated by shading.

the photochemical equilibrium timescales, and so the ozone distribution is strongly modified as a result of transport by atmospheric motions.

Ozone is formed as a result of photodissociation of ordinary diatomic oxygen molecules by ultraviolet light, and the subsequent recombination of free oxygen with diatomic oxygen. It is broken down by the action

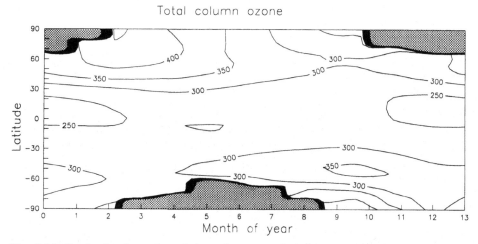

Fig. 9.11. Latitude–time plot of the column total abundance, in 'Dobson units'. The shaded region is unobserved since it receives no sunlight. Plot based on TOMS data for the period 1984–88, extracted from the SERC Geophysical Data Facility.

of sunlight and by the catalytic effect of certain trace constituents such as hydroxyl radicals, nitrogen oxides and anthropogenic chlorine atoms. Throughout much of the stratosphere, these creation and destruction rates are comparable to the dynamical timescales, so the advection of ozone is very important in generating concentration anomalies. This interaction between photochemical processes and transport processes has provided an important practical impetus for the study of stratospheric circulation in recent years. In this book, we confine ourselves to a very elementary account of the photochemistry of stratospheric ozone. The reader requiring greater detail is referred to more specialist texts on atmospheric chemistry.

The primary mechanism for the generation of ozone is called the 'Chapman scheme', being first suggested by Chapman in 1935. It involves a sequence of reactions which are consequent upon the splitting of an ordinary diatomic molecule of oxygen by a sufficiently energetic quantum of electromagnetic radiation:

$$O_2 + h\nu \rightarrow O + O. \tag{9.11}$$

The energy required to photodissociate the oxygen molecule means that this reaction can only take place for wavelengths less than 246 nm. The single oxygen radical produced by the reaction is highly reactive, and will readily bind on to suitable molecules, including that of diatomic oxygen. The prime reaction for producing ozone is:

$$O + O_2 + M \rightarrow O_3 + M. \tag{9.12}$$

where M is any third molecule, generally N_2 or O_2. This third molecule plays a catalytic role. It is needed to carry off excess momentum, enabling the O and O_2 molecules to bind during a collision while conserving both kinetic energy and momentum. Chapman also identified ozone destruction reactions; the balance between the production and destruction processes should define the observed ozone concentrations. The first process is the photolysis of ozone, effectively reversing reaction (9.12) by solar radiation with wavelengths of less than 1140 nm:

$$O_3 + h\nu \rightarrow O_2 + O. \tag{9.13}$$

Finally, a reaction involving both O and O_3 will destroy both forms of free oxygen:

$$O_3 + O \rightarrow 2O_2. \tag{9.14}$$

Because the concentrations of both O_3 and O are low, this last reaction is relatively slow.

These reactions can be used to estimate a photochemical equilibrium concentration of ozone. This is based on the law of mass action, which states that the rate of reaction of two constituents is proportional to the product of their concentrations. That is, for a simple reaction of the form:

$$A + B \rightarrow C \tag{9.15}$$

the rate of production of molecules of C is given by:

$$\frac{dn_C}{dt} = k n_A n_B. \tag{9.16}$$

Here, n_A, n_B and n_C are the so-called number densities, that is, the number of molecules per unit volume, of the reactants A, B and C respectively, and k is the reaction rate, which will generally be a function, possibly a rather sharply varying function, of temperature. Applied to the Chapman scheme, the law of mass action predicts the equilibrium number density of ozone molecules, n_3, to be:

$$n_3 = n_2 \left(\frac{J_2 k_2 n_M}{J_3 k_3} \right)^{1/2}, \tag{9.17}$$

where n_2 and n_M are the number density of the oxygen molecules and the catalyst molecules M respectively, k_2 and k_3 are the reaction rates of reactions (9.12) and (9.14), respectively, and J_2, J_3 are the photodissociation rates associated with reactions (9.11) and (9.13), respectively. These photo-dissociation rates themselves depend upon the ozone distribution, since they measure the amount of ultraviolet light of the appropriate wavelengths reaching the location under consideration. Some of this ultra-violet light will have been absorbed by O_3 and O_2 as it passes through higher layers of the atmosphere. The rate J_2 increases more rapidly with height than does J_3, with the result that Eq. (9.17) predicts a maximum of the number density of ozone molecules in the stratosphere. However, the scheme predicts rather too much ozone generally, and certainly cannot account for features of the distribution such as the spring maximum. Rather, photochemical arguments alone would suggest that the ozone distribution should correlate strongly with the distribution of short wave radiation.

Two important modifications to the simple photochemical equilibrium theory need to be introduced. First, further chemical reactions, often involving trace atmospheric constituents, can modify the chemistry of the Chapman scheme. Second, atmospheric transports can redistribute the ozone.

The most important modifications to the chemistry arise through catalytic destruction of ozone by trace constituents. Denote such a constituent by X. Then the ozone destruction process can be written:

$$X + O_3 \rightarrow XO + O_2, \tag{9.18a}$$

$$XO + O \rightarrow X + O_2. \tag{9.18b}$$

The net effect is to combine O and O_3 to form two molecules of O_2, leaving X unconsumed and ready to participate in further ozone destruction. The most important catalysts are OH, NO and Cl. The OH is formed naturally by the photodissociation of water vapour. The NO also has natural sources, such as lightning and photodissociation of ammonia and other biologically produced nitrogen compounds. It is also found in the exhaust gases of aircraft flying in the stratosphere, and there have been concerns that unrestricted growth in the number of commercial supersonic flights could have a significant impact on the ozone layer. A much more serious threat to ozone is now thought to come from Cl, which is almost entirely artificial in origin, being formed by a complex sequence of heterogeneous reactions, starting with the photodissociation of man-made chemicals such as chlorofluorocarbons. It is responsible for the large ozone depletions observed in the southern hemisphere spring and for the smaller depletion now becoming apparent in the northern hemisphere.

A full quantitative account of the photochemical equilibrium concentration

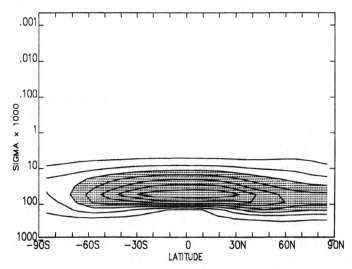

Fig. 9.12. The photochemical equilibrium distribution of ozone for March. Calculations kindly carried out by J. Haigh (see Haigh (1985) for details of the model used). Contouring as in Fig. 9.10.

of ozone, taking into account these catalytic destruction cycles, is rather complex. The various photochemical processes which generate the catalysts themselves need to be taken into account. Furthermore, the effects of the catalysts on ozone concentration are not simply additive, but can interact with one another. A large number of reactions, many with widely different rate constants, quickly becomes necessary. Figure 9.12 shows the zonal mean photochemical equilibrium distribution of ozone during spring. The concentration is a maximum at the equator, at altitudes of 40–50 km. But this is quite different from the actual distribution of stratospheric ozone at this time of year, shown in Figs. 9.10 and 9.11. The maximum ozone concentration is observed at high latitudes in the spring hemisphere, at levels below 20 km. This result strongly suggests that transport of ozone from production regions near the equator to the winter and spring polar regions is a very important process. Note that Fig. 9.11 may no longer be an accurate picture of the southern hemisphere spring, since it is based on data that are a few years old. Ozone levels in the spring stratosphere are now severely depleted over Antarctica, thanks to anthropogenic pollution by chlorine compounds.

In this book, our major concern is with how atmospheric motions will modify the large scale distribution of ozone. In the upper stratosphere, at heights between 30 km and 50 km, the answer is hardly at all. At these levels, the ozone concentration reaches its photochemical equilibrium in quite a short time, certainly short compared to the time for atmospheric motions to advect ozone to regions where the temperature, pressure or radiative regime is significantly different. In the lower stratosphere, the photochemical reaction rates become small, and so the time required to reach photochemical equilibrium becomes several weeks, much longer than the typical dynamical timescale. In the lower stratosphere, then, ozone is advected around almost as a conserved tracer. In fact, the largest abundances of ozone are found in the lower stratosphere, suggesting that the transport of ozone from production regions is a very important process. Furthermore, the column integrated ozone amount is a maximum at high latitudes in the spring, where photochemical production rates would be expected to be very low; there are simply not enough sufficiently energetic photons available to produce the observed abundances of ozone. Once more, there must be significant transport of ozone from lower latitudes.

The 'Brewer–Dobson' circulation was proposed in the 1940s to account for the observed distribution of ozone and other conserved trace constituents in the lower stratosphere. It is illustrated in Fig. 9.13 and consists of a meridional circulation in each hemisphere, with air rising into the stratosphere in the tropics, moving poleward, with descent and entrainment back into the troposphere at high latitudes. Such a mass circulation will transport ozone from the tropical production regions and accumulate it towards the poles, accounting for the spring polar maximum. Of course, such a circulation, deduced from the observed concentrations of trace constituents, is a Lagrangian circulation. Attempts to deduce a meridional circulation from observed heating rates and eddy fluxes of heat and momentum, as was described in Chapter 4, yield quite a different Eulerian mean circulation. Because of this misunderstanding, the 'Brewer–Dobson' circulation was regarded for a long time as incorrect. In more recent times, there has been renewed interest in the Lagrangian mean circulation and in approximations to it, such as the isentropic mean circulation or the residual circulation introduced in the last section.

The concentration of ozone in an air parcel can be written in the form:

$$\frac{\partial c_3}{\partial t} + \mathbf{v} \cdot \nabla c_3 + w \frac{\partial c_3}{\partial z} = C_3. \tag{9.19}$$

Here, c_3 denotes the mixing ratio of ozone, while C_3 denotes the photochemical production or destruction rate of ozone. Note that if c_3 represents a volume mixing ratio, it is proportional to the molecular number density n_3 given in Eq. (9.17). Conceptually, C_3 could be written in terms of an

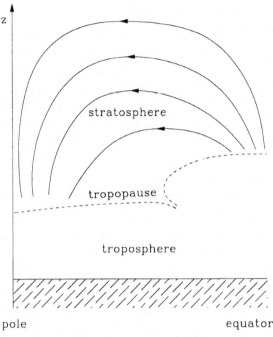

Fig. 9.13. Schematic illustration of the Brewer–Dobson circulation, proposed to account for the observed distribution of various conserved trace constituents in the lower stratosphere.

equilibrium concentration c_E and a photochemical timescale τ_E:

$$C_3 = \frac{c_E - c_3}{\tau_E}, \qquad (9.20)$$

which represents exponential decay of c_3 towards c_E on a timescale τ_E. This will be adequate if c_3 is not too different from c_E and could be a useful approximation for pedagogical purposes. Note that Eq. (9.19) is identical in form to the thermodynamic equation in terms of θ; the difference is that c_3 does not directly affect the atmospheric motions in the way in which θ does. Let us take the zonal mean of Eq. (9.19):

$$\frac{\partial [c_3]}{\partial t} + [v]\frac{\partial [c_3]}{\partial y} + [w]\frac{\partial [c_3]}{\partial z} + \frac{\partial}{\partial y}[v^* c_3^*] + \frac{1}{\rho_R}\frac{\partial}{\partial z}\left(\rho_R [w^* c_3^*]\right) = [C_3]. \quad (9.21)$$

It is difficult to see any way of simplifying this equation further. In particular, the ozone concentration may vary very sharply with height, so we are not justified in neglecting the vertical advection terms, as we did with the thermo-dynamic equation under a quasi-geostrophic scaling. The various terms in this transport equation are difficult to estimate from observations, since $[v]$

will be an $O(Ro)$ quantity, and vertical motions are suppressed even more in the stratosphere than in the troposphere because of the strongly stable stratification. When attempts are made to evaluate the terms, a strong cancellation between the mean transport by the mean circulation and the eddies is found. This problem is closely related to the transport of heat in the stratosphere, discussed in the last section; the mean and eddy fluxes of heat virtually cancel.

In certain limits, it can be shown that the residual meridional circulation, defined by Eqs. (9.5a, b), represents the effective transport velocity for chemical tracers. Then the transport equation, Eq. (9.21), becomes:

$$\frac{\partial [c_3]}{\partial t} + [v]_r \frac{\partial [c_3]}{\partial y} + [w]_r \frac{\partial [c_3]}{\partial z} = [C_3]. \tag{9.22}$$

The conditions for this to be valid are that the eddies be linear, steady and adiabatic. Thus tracers are advected by the circulations driven by the zonal mean heating and friction. In the idealized extreme when heating and friction are all zero, the nonacceleration theorem holds and the residual velocity will be zero. Equation (9.22) then shows that a corresponding non-transport theorem holds; the zonal mean concentrations of ozone will be simply related to the zonal mean sources or sinks:

$$\frac{\partial [c_3]}{\partial t} = [C_3]. \tag{9.23}$$

In reality, this extreme is not realized, and there is no doubt that the ozone distribution is strongly modified by the stratospheric circulation. The conditions that led to the nontransport theorem are violated in many ways. The eddies themselves are nonlinear and unsteady, and diabatic effects are significant on the long timescales implied by the residual circulation in the stratosphere. Furthermore, there is the challenging problem of feedbacks between the photochemical production and destruction processes and the circulation itself. A considerable research effort is ongoing in this area. The effort is centred on attempts to develop three-dimensional stratospheric circulation models which incorporate a reasonably complete account of the photochemistry of ozone. These are very large models which demand exclusive use of the largest computers currently available. That such an effort is proceeding is a measure of the rising concern about the deleterious effects of industrial activity on the ozone layer.

Models can be used to estimate the mass circulation in the stratosphere. Observations, including horizontal eddy heat transports, and the distribution of trace constituents, can also be used to generate a TEM meridional circulation, and so can approximate the Lagrangian mean mass circulation.

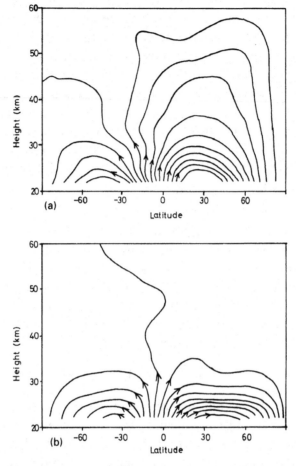

Fig. 9.14. Mass weighted meridional streamfunction in the stratosphere deduced from observed distribution of various advected chemical species for (a) January 1979; and (b) March 1979. (Redrawn from Solomon *et. al.* 1986.)

Figure 9.14 shows an estimate of the mean meridional circulation, based on one such observational study. Qualitatively, it is similar to the residual circulation, even though the conditions which formally allow an identification of the effective transport circulation with the residual circulation do not hold. It is also very similar to the Brewer–Dobson circulation. The main features are rising motion in the tropics and summer hemisphere, with descent over the winter polar regions. It is clear that the circulation is not confined to the stratosphere, and that an exchange of matter between the troposphere and the stratosphere is taking place. We will consider this mass exchange more fully in the next section.

9.4 Exchange of matter across the tropopause

The tropopause is an important, and unjustly neglected, structure in the global atmosphere. It is defined in terms of the lapse rate. This is typically 6–10 K km^{-1} in the upper troposphere but is small or even zero in the lower stratosphere. Correspondingly, the Brunt–Väisälä frequency increases by a factor of 2 or more across the tropopause. As a result of this change in static stability, the vertical velocity field (induced, for example, by gradients of heating) becomes considerably smaller above the tropopause. This can be verified by solving the circulation equation, Eq. (4.31), for a case in which the stability parameter $s(p)$ increases abruptly at the tropopause. In Chapter 6, we also saw that the change of stratification at the tropopause means that all but the longest wavelength Rossby waves are confined to the troposphere. Thus the tropopause acts rather as a lid to tropospheric motions. In some problems, such as the Eady model, it is simply represented by a rigid lid where $w = 0$. Nevertheless, there is some slow exchange of material across the tropopause, and understanding this exchange is of particular concern to stratospheric chemists. How do various chemically active constituents get into the stratosphere? And what is their likely residence time in the stratosphere? Answering these questions requires a knowledge of the location and magnitude of mass exchange across the tropopause.

As well as the discontinuity in vertical velocity and lapse rate at the tropopause, there is also a near discontinuity in the potential vorticity. Consider the Ertel potential vorticity q_E, given by Eq. (1.79) and approximated by Eq. (1.81). It is essentially the product of the absolute vorticity with static stability. As we would infer from Figs. 9.1 and 9.2, the vorticity itself is continuous across the tropopause, even though the vorticity tendency, dominated by the stretching term $f(\partial w / \partial z)$, will change substantially. But the large increase, by a factor of 4 or more, in $\partial \theta / \partial z$ at the tropopause means that the potential vorticity will be much larger in the lower stratosphere. Fig. 9.15 shows a schematic cross section of the surfaces of q_E and θ in the troposphere and lower stratosphere. Throughout the midlatitudes, the potential vorticity surfaces are more or less parallel to the tropopause. The tropics have a different character. The rapid change of sign of the absolute vorticity near the equator means that the potential vorticity surfaces must be nearly vertical in the deep tropics. Indeed, perhaps the most significant dynamical difference between the midlatitudes and the tropics is that potential vorticity surfaces are quasi-horizontal in the midlatitudes and quasi-vertical in the tropics. In the northern hemisphere midlatitudes, the tropopause lies close to the potential vorticity surface $q_E = 2 \times 10^{-6}$ K m^2 kg^{-1} s^{-1} (rather less

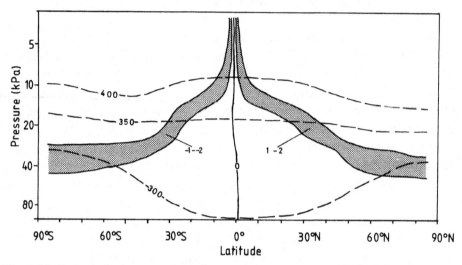

Fig. 9.15. Schematic cross section of potential temperature (dashed lines) and Ertel potential vorticity, q_E (solid lines). The shading indicates regions where $1 <| q_E |<$ 2 PVU, which roughly delineate the midlatitude tropopause.

clumsily referred to as 2 potential vorticity units or 2 PVU). This suggests a dynamical definition of the midlatitude tropopause as the $q_E = 2$ PVU surface which in many ways is more physically significant than the conventional definition in terms of lapse rate.

The evidence from both routine radiosonde ascents and from more detailed *in situ* aircraft observations is that the tropopause is a very sharp transition. It is clear that on radiative grounds we would expect low static stability in the lower atmosphere, but more stable layers in the stratosphere. In Chapter 3, we saw how the lower atmosphere is destabilized since much of the solar radiative flux reaches the ground, whereas long wave radiation is mostly emitted from higher levels. In the middle and upper stratosphere, the heating associated with ozone becomes large, as much as 15–20 K day^{-1} in the upper stratosphere. Hence, at these levels, we expect a temperature maximum simply on radiative equilibrium grounds. Thus there must be a transition between low static stability near the ground and high static stability towards the top of the ozone layer. What is less obvious is why this transition is sharp, rather than smooth and continuous. The explanation involves the dynamics of atmospheric circulation as well as radiative effects.

In the tropics, the atmosphere is generally conditionally unstable to moist convection. The heating at these latitudes is dominated by the release of latent heat in convective clouds. Saturated air parcels will rise, more or less

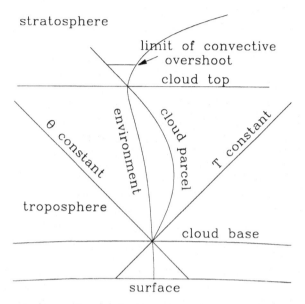

Fig. 9.16. Schematic view of a tropical atmospheric profile. The top of the clouds will be located where the environment curve crosses the saturated adiabat from the condensation level. Below the cloud tops, repeated convection will modify the environmental profile until it is close to moist convective neutrality.

along saturated adiabats, and will only stop rising when the environmental temperature once more exceeds that of the saturated parcel. The repeated development of convective cloud systems, with an exchange of heat and moisture between the ascending towers of cloud and the surrounding descending air, means that the mean tropical atmosphere is brought close to a state of saturated neutral stability. This is found to be a rather precise statement when the equation of state is modified to include the variable amount of water vapour in the ascending parcel, and account is taken of the effects of suspended cloud liquid water droplets on the density of an ascending parcel. Such a picture leads naturally to a discontinuity between the lapse rates in the troposphere and those in the lower stratosphere. A simple parcel theory of air in the cumulus clouds enables the top of the cloud layer to be determined. Below that top, the lapse rate will be close to the saturated adiabat. Above that level, radiative processes will determine the lapse rate, which will be more stable. Fig. 9.16 illustrates this.

In vigorous ascending cumulus updrafts, considerable kinetic energy may be associated with the rising air parcel. Vertical velocities of as much as $5\,\mathrm{m\,s^{-1}}$ or more are possible. This energy can go to lift the updraft air some way into the stable lower stratospheric air above the cloud tops, a process called 'convective overshoot'. Figure 9.16 illustrates a construction which places an upper bound on the distance by which this overshoot could penetrate into the stratosphere. The area between the environment curve and the cloud profile, where the latter is colder than the environment, represents the work needed to raise the overshooting parcel to a given level. This cannot exceed the kinetic energy per unit mass of the ascending air. Only a small proportion of the total vertical flux of mass in the cumulus cloud might be expected to overshoot in this way, but there is compelling evidence to suggest that this process provides the major route for the passage of tropospheric air into the stratosphere. For example, the humidity of the stratosphere is extremely low; it is measured in parts per million. Most of the stratosphere is very far from saturated; it is consequently virtually cloudless. However, the tropical tropopause, because of its height, is extremely cold and, indeed, apart from the winter polar regions, is one of the coldest parts of the stratosphere. Typical tropopause temperatures may be as low as −60 °C. At these temperatures, the stratosphere is not far from saturation. This observation led to the theory that most tropospheric air, and consequently water vapour, found its way into the stratosphere via convective overshoot in vigorous tropical cumulonimbus clouds. Such air is effectively 'freeze dried' by its passage through the very cold tropical tropopause. Estimating the flux of tropical air into the stratosphere is a difficult problem. It mainly takes place in intense but highly localized updraft regions, and so is difficult to observe. Convective overshoot is most definitely a subgridscale, parametrized process in global circulation models, and is therefore difficult to predict with any certainty from modelling studies.

The tropopause in midlatitudes is considerably lower than in the tropics. We turn now to a discussion of the processes which can form it. Moist convection is unlikely to be a major process. Although it is present and gives rise to important local weather events, convection has less impact on the large scale thermal structures in the midlatitudes than it does in the tropics. Most heat is carried by large scale motions in quasi-horizontal trajectories with typical gradients of 1 in 1000 (see Section 5.2), and deep convection occupies only a small fraction of the midlatitude troposphere. Instead, we will focus on the role of quasi-horizontal convection (i.e. baroclinic disturbances) in modifying the tropopause.

A helpful diagnostic is a map of potential vorticity on a θ surface which intersects the tropopause. Figure 9.15 shows that isentropic surfaces with θ between about 300 K and 340 K are in the lower stratosphere at high latitudes but in the troposphere at low latitudes. Since motion takes place on isentropic surfaces in frictionless, adiabatic conditions, such maps are very helpful in discussing the exchange of air parcels between the stratosphere and the troposphere on synoptic timescales. A sequence of maps of potential vorticity on the 350K isentropic surface is shown in Fig. 9.17. The shading indicates potential vorticity of around 2 PVU, and so can be taken as representing the tropopause. Deep baroclinic systems corrugate the tropopause, and bring about equatorward excursions of high potential vorticity air together with complementary poleward excursions of low potential vorticity air at neighbouring longitudes. As long as these excursions are reasonably small, they are reversible and do not result in any systematic long term change to the potential vorticity distribution. But a large amplitude, mature depression can cause patches of polar potential vorticity to be pulled out into long thin streamers and mixed irreversibly into the air of lower latitudes. The process is very similar to that discussed in Section 9.2, when the effects of breaking planetary waves on the polar vortex were discussed. The continual, irreversible peeling of high potential vorticity air off the edge of the stratospheric vortex by mature midlatitude cyclone systems sharpens the transition between the high potential vorticity vortex of the polar stratosphere and the low potential vorticity air of the subtropical upper troposphere. In other words, it sharpens the tropopause.

A run of the simple global circulation model introduced in Section 2.5 emphasizes the role of baroclinic eddies in generating a midlatitude tropopause. Some potential vorticity maps for this run are shown in Fig. 9.18. The radiative equilibrium state was defined with a gradual transition between a lapse rate of $7\,K\,km^{-1}$ in the lower troposphere and isothermal conditions in the middle stratosphere, and with horizontal temperature gradients in the lower stratosphere. The dynamics of the tropospheric flow were simplified by imposing a seven-fold symmetry, so that the circulation was dominated by developing, breaking and decaying wavenumber 7 baroclinic waves. The initial state was motionless, but with a temperature structure corresponding to the global mean radiative equilibrium state. Figure 9.18(a) shows an initial PV map. There is a gradual increase of PV towards the pole but no sharp transition and hence no sharp tropopause. As baroclinic instability breaks out, the PV contours at 310 K are deformed slightly. As the baroclinic waves deepen and mature, the distortion becomes large and irreversible. Figure 9.18(b, c) shows long streamers of high potential vorticity air being advected around large anticylonic gyres in the subtropics, where the gradients of PV are relatively gentle. The edge of the stratospheric vortex is sharpened, at least as far as the relatively coarse resolution of the model permits. There seems little doubt that it would continue to sharpen further in a

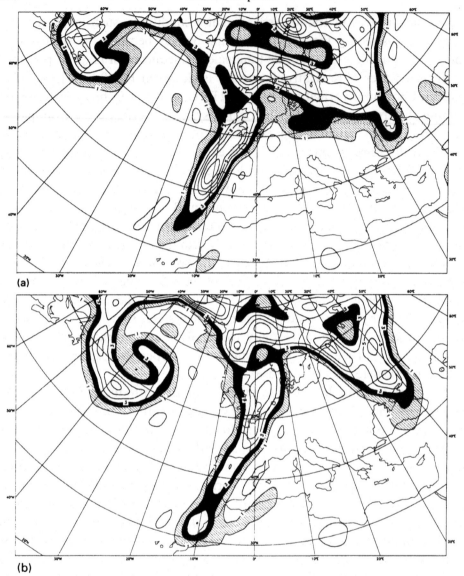

(a)

(b)

Fig. 9.17. A time sequence of maps of potential vorticity on the 310 K surface, at 12 hour intervals beginning at 00Z on 28 January 1980. Shading indicates values from 1 to 2 PVU and so locates the tropopause. (Based on ECMWF analyses; plots kindly supplied by C. Thorncroft.)

finer resolution model.

The midlatitude cyclone belt provides the recirculation of stratospheric air back into the troposphere. This is consistent with the Brewer–Dobson circulation within the stratosphere, which suggested a mass source at the tropical

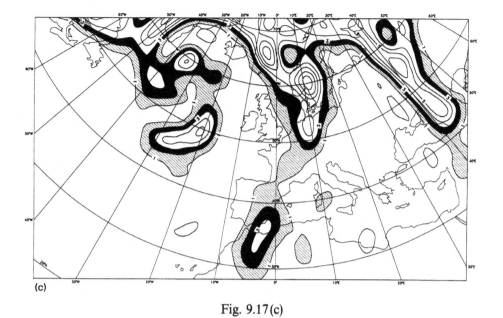

(c)

Fig. 9.17(c)

tropopause and a sink at mid- or high latitudes. One might suspect that a typical global circulation model coarsens this process, and observational studies suggest that this is indeed the case. Figure 9.19 shows a cross section through the upper part of an observed tropospheric front. Associated with the frontal circulations, an intrusion of stratospheric air is entrained into the troposphere, parallel to the frontal surface. Such an event is called a 'tropopause fold'. Theoretical studies of frontogenesis show that tropopause folds are a ubiquitous feature of frontal systems. The high potential vorticity air of stratospheric origin which intrudes along the frontal surface is eventually diluted by small scale mixing into the troposphere. The concentration of ozone and other chemical tracers brought into the troposphere in this way is quickly reduced.

Note that the exchange of mass in both directions across the tropopause appears to be highly localized, and so requires parametrization in models of the global circulation. This is a considerable practical difficulty in studying the spread of chemical pollutants into the stratosphere. It is also a fundamental problem, in that it suggests that the flux of mass associated with the Lagrangian circulation of the stratosphere is controlled by poorly understood processes at the tropopause. The rates of both convective overshoot in the tropics and entrainment of stratospheric material in tropopause

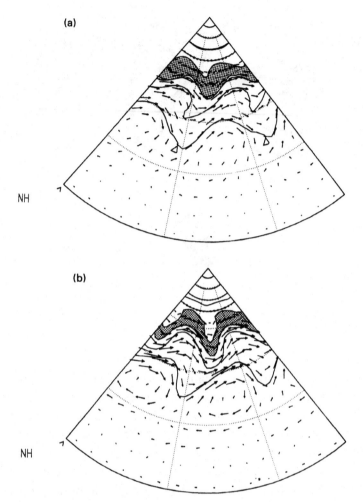

Fig. 9.18. Potential vorticity on the 310 K surface from a run of a simplified global circulation model. Three maps are given at intervals of five days: (a) prior to a major baroclinic wave event; (b) at the height of the wave event.

folds depend upon small scale turbulent mixing. Current theories on this are largely empirical in their foundation.

In concluding this section, it is worth remarking that the interface between the troposphere and stratosphere at high latitudes is rather different. Vigorous vertical heat transport, either by convection or large scale quasi-horizontal motion, is weak or absent at high latitudes. Observations suggest that there is weak descent in the meridional circulation in the troposphere. If anything, the deformation associated with this circulation would tend to weaken and spread vertical gradients of conserved quantities such as poten-

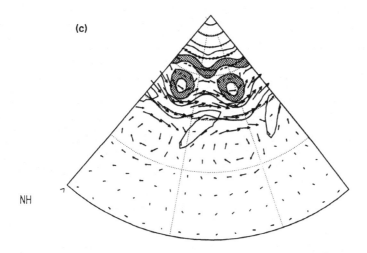

Fig. 9.18 (*cont.*). (c) As the waves begin to decay. Contour interval 0.5 PVU, shading for 1.5 – 2 PVU, roughly delineating the tropopause.

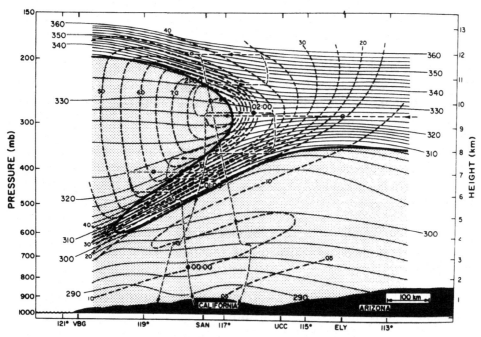

Fig. 9.19. Cross section through a tropopause fold. The heavy line delineates the tropopause, as defined by the 1 PVU surface. (From Shapiro, 1980.)

tial vorticity, rather than to sharpen them. It is no surprise that the observed polar tropopause is a rather weak affair. This is particularly true in the polar night, when at sufficiently high latitudes there is no direct solar heating at any level. Steady descent throughout the stratosphere and troposphere balances radiation of heat to space. While the stratification in the stratosphere is generally larger than in the troposphere, the transition between the two layers is vague in the polar winter. Large stratification, and hence very large values of potential vorticity, are found near the surface in the polar night, where a surface inversion forms as a result of the radiation of heat from the ice surface. There is a particularly strong surface inversion over Antarctica.

9.5 Problems

9.1 Use the hydrostatic relation to estimate the heights of (a) the 3 kPa, (b) the 1 kPa and (c) the 0.1 kPa pressure surfaces. You may assume that the stratospheric temperature is constant with height at 220 K.

9.2 Calculate a typical value for the Brunt–Väisälä frequency N in the lower stratosphere. Compare the period of oscillation of vertically disturbed air parcels with that in the troposphere.

9.3 Using the temperature cross section in Fig. 9.6(b), estimate the wind speeds in the southern hemisphere polar night jet at 30 km. What correction would you make if you replaced geostrophic balance with gradient wind balance at these latitudes?

9.4 If the stratosphere is characterized by an easterly wind of $25\,\mathrm{m\,s^{-2}}$, calculate at what height above the tropopause a wavenumber 5 disturbance at 50 °N will have decayed to half its amplitude. At what height will it have half the eddy energy?

9.5 The polar night jet is at a latitude of 60 °. Estimate the maximum wind speed which will still permit a wavenumber 2 disturbance to propagate vertically.

9.6 Show that an evanescent disturbance in the stratosphere has no poleward temperature flux associated with it.

9.7 Using Fig. 9.5(b), estimate the contribution to the potential vorticity gradient $[q]_y$ by the planetary vorticity, $[u]_{yy}$ and the term involving vertical derivatives in the Southern Hemisphere winter polar night jet.

9.8 Show that for steady Rossby waves, the vertical component of the group velocity can be written:

$$c_{gz} = \frac{2U^2(f^2/N^2)km}{\beta}.$$

Calculate c_{gz} for a wavenumber 1 disturbance at 60°N, and a zonal wind of $30\,\mathrm{m\,s^{-1}}$. Hence estimate the time taken for a wave packet starting at the tropopause to reach the stratopause at 50 km.

9.9 A typical meridional velocity associated with the Ferrel cell is $0.5\,\mathrm{m\,s^{-1}}$. What poleward temperature flux would be required to reduce the residual circulation to zero? Compare this to the observed $[v^*\theta^*]$ and comment on the result.

10

Planetary atmospheres and other fluid systems

10.1 Major influences on planetary circulations

Until recently, the study of global circulations has been confined to the circulation of a single system, namely that of the Earth. Throughout the earlier part of this book, we, too, have concentrated upon the Earth, showing how the poleward and upward transports of heat generate the kinetic energy associated with observed atmospheric circulations. We have described some of the forms which these heat fluxes can take, including the essentially axisymmetric circulations of the Hadley cells in low latitudes and the wave-like baroclinically unstable waves of the midlatitudes. These principles need not be restricted to the Earth's system alone. In this chapter, we will enquire how general are the particular heat transporting circulations observed in the Earth's atmosphere, and how they might be modified in different circumstances.

Such a discussion has become much more informed in the last 20 years or so, as the study of planetary atmospheres has advanced considerably. Spacecraft have now paid at least fleeting visits to every planet with a substantial atmosphere in the solar system, with the exception of Pluto, which may possess an atmosphere. In the case of Venus and Mars, direct *in situ* measurements of meteorological parameters have been made in addition to the more usual remotely sensed data. In the coming years, plans are under way for entry probes and direct measurements of other atmospheres, including those of Jupiter and Titan. Earth based observations and computer simulation studies have helped to interpret and to extend these observations. By contrasting and comparing the behaviour of a variety of planetary atmospheres, our subject has indeed matured and come of age scientifically.

Table 10.1 summarizes those physical elements of planets with dense atmospheres which are relevant to the dynamics of their atmospheric circula-

Table 10.1. *Physical properties of planets with substantial atmospheres*

Planet	a (10^6 m)	g (m s^{-2})	S (W m^{-2})	α	Ω (10^{-5} s^{-1})	τ_s (years)
Venus	6.05	8.60	2620	0.77	0.030	0.615
Earth	6.37	9.81	1370	0.29	7.29	1.00
Mars	3.40	3.72	590	0.15	7.09	1.88
Jupiter	71.4	22.88	50.6	0.45	17.57	11.86
Saturn	60.0	9.05	15.1	0.61	16.34	29.5
Titan	2.56	1.44	15.1	0.21	0.456	29.5
Uranus	26.15	7.77	3.72	0.37	11.26	84.0
Neptune	24.75	11.00	1.52	0.37	11.05	165
Pluto	(1.5)	(4.3)	0.88	(0.3)	1.14	248

tions. There is a wide range of radius and density, and hence of surface gravitational acceleration. A crucial parameter is the rotation rate Ω; this should be compared with the orbital period τ_s which determines the timescale of seasonal effects, if any. The amount of radiation received from the Sun is inversely proportional to the square of the distance of the planet from the Sun. Consequently, it varies more than do the other parameters shown in Table 10.1.

Table 10.2 gives some data on the atmospheric mass, temperature and composition. In this book, the complex and interesting questions of the chemical composition and genesis of the atmospheres will not be discussed. Similarly, the radiative transfer properties of the atmospheres will not be discussed in detail, but will be regarded as merely defining the heat sources and sinks which drive atmospheric motions. Only those constituents which contribute significantly to the mean molecular weight of the atmosphere are listed in Table 10.2. These are not necessarily the radiatively active gases which determine the thermal structure. For example, radiative transfer in the Earth's atmosphere is governed by carbon dioxide, water vapour and ozone and is significantly modified by trace constituents such as methane. These gases account for little over 1% of the mass of the atmosphere. Surface pressures and temperatures of the atmospheres are given where possible. In the case of the very deep atmospheres of Jupiter, Saturn, Uranus and Neptune, the temperatures and pressures at the visible cloud tops are given. The radiative time constant, τ_E, has been estimated for each atmosphere using Eq. (3.11). This has been based either on the surface pressure, when this is known, or on the pressure at the cloud tops otherwise. However, if the bulk of the sunlight is absorbed in the thin upper levels of the atmosphere, then the radiatively active layers of the atmosphere have a much smaller thermal capacity, and a smaller value of τ_E is appropriate.

Two distinctive groups of planets emerge from Tables 10.1 and 10.2. The first constitutes the so-called 'terrestrial planets', Venus, Earth and Mars. These are all relatively small, dense planets, with rocky crusts and mantles. Their atmospheres are largely composed of heavier gases such as carbon dioxide or nitrogen. The second group of 'gas giant' planets, Jupiter, Saturn, Uranus and Neptune, are much larger and less dense. They apparently possess no solid surface, at least not until very great depths within the interior of the planet. Their atmospheric composition is similar to that of the Sun, with large quantities of hydrogen and helium, and smaller traces of more complex compounds. With the exception of Uranus, the gas giants have substantial internal heat sources, and radiate considerably more heat than they receive from the Sun. Titan is in a class of its own, though, for our purposes, it is rather similar to a colder version of the terrestrial planets. It is a small body of intermediate density, probably largely composed of water ice and other frozen volatiles. Its atmosphere is dense and massive, and is predominantly composed of nitrogen. Very little is known of the atmosphere of Pluto, although a thin atmosphere has been inferred from spectroscopic data. Pluto is perhaps a rather similar type of body to Titan, but with a much thinner and less active atmosphere.

The circulations of the known planetary atmospheres show a considerable range of behaviour, though certain broadly defined regimes can be identified. These are illustrated in Fig. 10.1, which shows images of the cloud decks of Venus, Earth and Jupiter. Venus is entirely cloud covered; in ultraviolet light, the cloud tops exhibit dark markings which seem to originate in the tropics and spiral towards the poles. Earth is roughly 30% cloud covered. The image shows the line of convection associated with the ITCZ and the neighbouring clear areas under the descending branches of the Hadley circulation. The spiral cloud patterns associated with midlatitude depressions are prominent at higher latitudes. Jupiter is also entirely covered with cloud, with bright high layers of cloud and darker, more strongly coloured decks at lower levels. The clouds are organized into zonally orientated bands, with smaller scale, generally transient, disturbances embedded within them. These disturbances take the form either of oval spots or of more irregular turbulent eddies. Tracking of cloud features reveals some five or six westerly, and a similar number of easterly, jets associated with the zonal cloud bands.

The terrestrial regime is of course far and away the most closely studied and the best understood. Interpretation of the circulations observed on

Table 10.2. *Properties of planetary atmospheres*

Planet	Principal atmospheric constituents	$R^{(1)}$	$P_s^{(2)}$	$T_s^{(3)}$	$\tau_E^{(4)}$
Venus	CO_2 (96%), N_2 (3%)	193	90	730	10.8
Earth	N_2 (77%), O_2 (21%), H_2O (1%)	287	1	288	0.085
Mars	CO_2 (95%), N_2 (3%), Ar (2%)	192	0.007–0.010	218	0.002
Jupiter[5]	H_2 (90%), He (10%)	3779	0.42	125	3.15
Saturn[5]	H_2 (97%), He (3%)	4036	1.1	95	63.8
Titan	N_2 (82–99%), Ar (0–12%), CH_4 (1–6%)	280	1.8	92	7.87
Uranus[5]	H_2 (83%), He (15%), CH_4 (2%)	3200	5	57	680
Neptune[5]	H_2 (83%), He (15%), CH_4 (2%)	3200	1.5?	57	191
Pluto	CH_4, N_2?	16?	10^{-5}?	40	5×10^{-4}

[1] Units $J\,kg^{-1}\,K^{-1}$.
[2] Units 10^5 Pa.
[3] Units K.
[4] Units terrestrial years.
[5] Surface pressure refers to the main cloud deck and the temperature is the bolometric temperature.
Data are increasingly uncertain for the more remote bodies.

most of the other planets is controversial to some degree. The resolution of these controversies hinges upon more detailed observation, especially of the vertical structure of the atmospheres of the gas giants. As far as the circulation of an atmosphere is concerned, some features, such as chemical composition, are not directly relevant. Other properties are much more important. The principal features which determine the circulation regime are as follows:

(i) The strength and distribution of heat sources and sinks. In making this statement, we must recall that the circulation itself may help to organize and strengthen the heat sources and sinks. For example, in the terrestrial tropics, low level convergent flow imports moisture, initiating convection and the release of latent heat. This heating may reinforce the low level convergence.

(ii) The dynamical constraints imposed on the motions by the planet itself. The most important are the rotation rate and size of the planet.

(iii) The nature of the underlying surface. In particular, the amount of frictional drag exerted by the surface on the atmosphere is a crucial factor. The orography of the lower surface must also be considered under this heading; if the orography is sufficiently high, the circulation can be modified substantially.

We consider each of these factors in turn, sketching the range of variations observed in the planets of the solar system. More detail will be given in subsequent sections.

The atmospheres of the terrestrial planets gain virtually all their heating from intercepted solar radiation. The vertical distribution of the resulting heating is crucial to the character of the circulations. Atmospheric circulation requires an upward flux of heat, as we saw in Eq. (5.26). Thus, if all the incident solar energy were absorbed at some upper level in the atmosphere, no kinetic energy would be generated in the lower layers of the atmosphere. Radiative processes would establish an isothermal state in the motionless lower part of the atmosphere. This result, sometimes called Sandström's theorem, applies to the Earth's oceans; the currents observed in the ocean are not primarily thermally driven, but arise from mechanical forcing by the atmosphere. In the case of the deep, cloudy atmosphere of Venus, we might similarly expect the meteorologically active layers of the atmosphere to be above the cloud tops, around 60 km above the surface, with a deep isothermal layer beneath. Probes reveal that the Venus lapse rate is near adiabatic below the cloud tops, and that around 4% of the solar radiation absorbed by the planet reaches the surface. Thus, although the cloud top levels are the most meteorologically active, enough radiation reaches the lower atmosphere of Venus to drive weak motions near the surface. In the case of Earth and Mars, more radiation reaches the surface, so that the atmosphere is forced towards a state of static instability. The vertical heat transport by the planetary circulations helps to account for the observed state of static stability; at least in dust free conditions, such stabilization by large scale, dynamical heat fluxes is a dominant process on Mars. On Earth, the observed static stability is also governed by the moist adiabatic lapse rate in the tropics.

The giant planets Jupiter, Saturn and Neptune, on the other hand, all possess significant internal heat sources. This heat energy is derived from the slow gravitational collapse of the planets. In the case of Jupiter, it is estimated that a reduction of its radius by about 1 mm per century would account for the observed heat flux. Uranus differs from the other giant planets in apparently having no significant internal heat source, though in many respects its physical characteristics closely resemble those of Neptune. Apart from Uranus, the internal heat sources substantially exceed the amount of energy received from the Sun. The requirement for vertical heat transport is therefore strengthened, while the comparative weakness of solar heating means that the horizontal temperature gradients will be weak, and so poleward heat transports may be less important. The magnitude of these effects is not known accurately, with the result that the flow regime of these planets is still a matter of debate.

The strength of the heating is conveniently measured by the radiative relaxation time. The actual thermal structure is the result of two competing groups of processes. The radiative forcing generates temperature gradients, both in the vertical and the horizontal. Thermally excited fluid motions in the atmosphere will be such as to reduce these temperature gradients. If the radiative timescale is short compared to the dynamical timescales associated with the planetary circulations, then the thermal structure of the atmosphere will hardly be disturbed by the dynamics, and the mean thermal structure will be close to radiative equilibrium. On the other hand, in the limit of very long radiative relaxation times, fluid motions will virtually eliminate all temperature gradients.

The rate of rotation of the planet provides the strongest constraint on the regime of planetary flow. Rapid rotation inhibits vertical and poleward motion, and therefore changes the character of the circulation required to transport heat. At low rotation rates, Hadley circulations could transport virtually all the heat. At higher rotation rates, Hadley circulations become confined to a narrow tropical zone, while other processes such as baroclinic instability will dominate the heat transport in higher latitudes. The characteristic scale of the most unstable baroclinic scales, L_R, is given by

$$L_R = k_R^{-1} = \frac{NH}{f}. \tag{10.1}$$

This shows that the scale of the baroclinic waves decreases with the rotation rate, and so the efficiency of the heat transporting process will fall when L_R is significantly less than a, the planetary radius. A wide range of rotation rates is observed in the solar system, from the very slow rotation of Venus through to the extremely rapid rotation of Jupiter. The equatorial rotation velocity, Ωa, ranges from $1.8 \, \mathrm{m \, s^{-1}}$ for Venus to $12\,500 \, \mathrm{m \, s^{-1}}$ for Jupiter, a range of nearly 7000.

The character of the surface is also a critical factor. The surface of a terrestrial planet, being solid and rough, exerts a considerable drag on the the atmosphere. In the case of the Earth, the global mean 'spin up' time for tropospheric motions is around five days. The giant planets possess no solid surface. Theoretical models suggest that their mantles remain

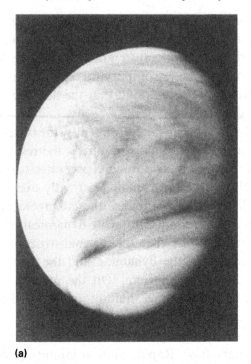

(a)

(b)

Fig. 10.1. Images of the cloud tops reveal some features of the atmospheric circulations of three planets. (a) Venus, taken in ultraviolet wavelengths, (b) Earth in visible light.

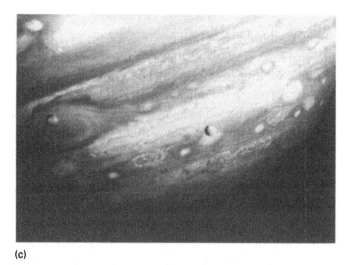

(c)

Fig. 10.1 (*cont.*). (c) Jupiter, also in visible light.

gaseous until very great depths. There may be a small rocky core near the centres of these bodies. But the exchange of momentum between the meteorologically active layers of the atmospheres and the deep interior is likely to be small, and therefore the spin down time for circulation will be much longer than for a terrestrial planet. Paradoxically, this does not imply that atmospheric circulation will necessarily be more vigorous. Consideration of energetics shows that the generation of kinetic energy by upward heat fluxes must ultimately balance the frictional dissipation of kinetic energy. If the friction were reduced, then the generation of kinetic energy, and hence the poleward and upward temperature fluxes, must also be reduced. The result would be a meteorologically inactive atmosphere, with the velocity vectors predominantly parallel to the temperature contours.

Significant orography on the solid surface of a planet will modify atmospheric circulation. Mars has considerably higher orography over wide areas than Earth, and so the contribution of steady, forced eddies to global heat transport might be larger. The most extreme example is provided by the Earth's oceans. The ubiquitous presence of longitudinal barriers dominates the circulation of most of the oceans.

In the following sections, we will consider groups of known planetary atmospheres, and attempt to relate these principles to what is known about their atmospheric circulations. We will also discuss briefly the circulations in related systems such as the oceans and rotating fluid systems in the laboratory.

10.2 Terrestrial circulations

Terrestrial global circulation is characterized by an essentially axially symmetric 'Hadley regime' at low latitudes, and a zonally disturbed 'quasi-geostrophic' regime at higher latitudes. The two regimes interact strongly, and forced steady waves, particularly in the northern hemisphere winter, introduce further complexity. These matters have all been dealt with in various degrees of detail in the earlier chapters of this book. The aim of the present section is to investigate the range of conditions for which this broad description remains helpful. In particular, the quasi-geostrophic regime will be discussed in this section, while the Hadley regimes will be considered further in the next section.

Heat transport in the terrestrial quasi-geostrophic regime occurs principally in baroclinic disturbances which exhibit a characteristic westward phase tilt with height. The Eady theory of Section 5.4 sets out a linear instability theory which gives a reasonable prediction of the scale and structure of disturbances observed in the Earth's midlatitudes. Recall that the zonal wavenumber of the most unstable wave is

$$k = 1.61 K_R, \tag{10.2}$$

where $K_R^{-1} = L_R$, the 'Rossby radius of deformation', given by

$$L_R = \frac{NH}{f}. \tag{10.3}$$

The Rossby radius is the characteristic length scale of disturbances in the midlatitudes. A first necessary condition for the existence of such disturbances is that the planet should be large enough to accommodate the unstable wavelengths, that is, that

$$L_R \leq a. \tag{10.4}$$

For the Earth, L_R/a is around 0.12, so that it falls comfortably within the range in which baroclinic instability is possible. For Mars, the same ratio is about 0.33, suggesting that baroclinic waves are still likely, but that they will have considerably larger scale relative to the planetary radius.

The growth rate of the most unstable wave was given in Eq. (5.52). It can be rewritten

$$\sigma = 0.31 Ri^{-1/2} f, \tag{10.5}$$

where Ri is the dimensionless Richardson number, defined as $N^2/(u_z)^2$, best thought of as a dimensionless measure of static stability. If thermal wind balance holds, *Ri* can be written entirely in terms of the potential temperature gradients:

$$Ri = \frac{f^2\theta_0}{g}(\theta_z)(\theta_y)^{-2}. \tag{10.6}$$

The quasi-geostrophic analysis which led to Eq (5.6) is valid only for large *Ri*, that is, for rapid rotation and strong stratification, but for relatively weak horizontal temperature gradients. The Earth's midlatitudes are characterized by $Ri \approx 50$, so that they fall within this range. But on the equatorward flank of the subtropical jet, the local values of *Ri* are much smaller, generally less than 10. The theory can be modified for smaller *Ri*. But as *Ri* approaches unity or less, the dynamical regime becomes entirely different, being dominated by 'symmetric instabilities', with zero zonal wavenumber, but a limited meridional scale. Thus, the static stability of an atmosphere is revealed as a crucial property in determining the character of its circulation.

The difficulty we have is that the horizontal and vertical gradients of potential temperature are not externally determined characteristics of an atmosphere. Radiative heat sources and sinks acting alone might serve to determine radiative equilibrium values of the temperature gradients, but the circulation of the atmosphere itself will modify, possibly profoundly, the temperature distribution. Apart from the Earth, it is generally not straightforward to determine the actual values of *Ri* for planetary atmospheres with any great degree of certainty. This is because it is difficult to measure the stratification sufficiently accurately. Indeed, for the gas giants, the lapse rate below the cloud tops is unknown, so that it remains an open question whether the circulations on these planets are quasi-geostrophic, baroclinic in character, or whether they belong to a totally different class.

To progress further, we need to parametrize the temperature transport of quasi-geostrophic circulations in terms of the imposed temperature gradients and other quantities. Despite many efforts, no very satisfactory way of doing this is possible. Indeed, this is one of the reasons why global circulation studies are now dominated by GCM modelling experiments, which endeavour to model the lifecycle of each individual baroclinic disturbance explicitly. The use of complex numerical models is also perhaps the best way forward in theoretical explorations of the dynamical regimes of planetary atmosphere flows. But for the purposes of this exposition, we will explore a limited attempt to parametrize baroclinic heat transports and hence to derive an analytical model of the structure of planetary atmospheres for a range of external parameters.

The analysis is based on the Eady linear model of baroclinic instability.

Expressions were derived in Section 5.4 for the horizontal and vertical temperature fluxes, given as Eqs. (5.54) and (5.56) respectively. For the purposes of the present order of magnitude estimates, the wavenumber of the most unstable wave will be substituted in these equations, and the fluxes will be evaluated at a midlevel and midlatitude point. For these particular choices, the temperature fluxes reduce to

$$[v^*\theta^*] = 3.9\frac{f^2\theta_0}{gNH^2}\Psi^2, \tag{10.7}$$

$$[w^*\theta^*] = -1.2\frac{f^2[\theta]_y}{N^3H^2}\Psi^2. \tag{10.8}$$

Here, Ψ is the streamfunction amplitude. A typical value of Ψ must be estimated so that temperature fluxes can be calculated from Eqs. (10.7) and (10.8). Some relationship between Ψ and the mean state of the atmosphere must be assumed; such a relationship is called a 'closure hypothesis'. A number of such closure hypotheses are possible, all equally defensible; the closure assumption made in this section is that the waves reach a typical amplitude such that

$$v^* = -\frac{\partial\psi}{\partial x} \approx \Delta U. \tag{10.9}$$

This implies that the streamlines make angles of around $45°$ with the undisturbed zonal background flow when growth ceases. Using Eq. (10.9), in conjunction with the form of the normal mode solution for the most unstable wave, given by Eqs. (5.44) and (5.46), one obtains estimates for the total temperature flux:

$$[v^*\theta^*] = 1.5\frac{g^{3/2}H^2}{f^2\theta_0^{3/2}}[\theta_y]^2[\theta_z]^{1/2}, \tag{10.10}$$

$$[w^*\theta^*] = 0.46\frac{g^{3/2}H^2}{f^2\theta_0^{3/2}}[\theta_y]^3[\theta_z]^{-1/2}. \tag{10.11}$$

These expressions relate the expected typical poleward and vertical temperature fluxes to the imposed horizontal and vertical temperature gradients. Were they accurate, much current effort in global circulation modelling would be wasted; we could simply parametrize the baroclinic systems of midlatitudes. As a result, climate and global circulation studies would be much more straightforward. In reality, these estimates are extremely crude; at best, they provide merely an approximate upper bound on the actual temperature fluxes. Our purpose in this chapter is not to obtain very accurate predictions of the actual magnitudes of the fluxes in particular circumstances, but, rather, to suggest the parametric dependence of temperature fluxes on planetary properties such as the rotation rate.

The temperature fluxes associated with the baroclinic waves will modify both the horizontal and vertical temperature gradients. These effects will be offset by radiative processes which, acting in isolation, will set up a state of radiative equilibrium. Let us write the thermodynamic equation as follows:

$$\frac{\partial [\theta]}{\partial t} = -\frac{\partial}{\partial y}[v^*\theta^*] - \frac{\partial}{\partial z}[w^*\theta^*] + \frac{[\theta_E] - [\theta]}{\tau_E}. \tag{10.12}$$

The smaller, opposing temperature fluxes carried by the induced zonal mean circulation have been ignored. Differentiating with respect to y or to z then yields equations for the poleward and vertical potential temperature gradients. According to Eady theory, $[v^*\theta^*]_{yz} = 0$; the cross derivative $[w^*\theta^*]_{zy}$ will also be neglected. The second derivatives of the temperature fluxes may be approximated by finite difference formulae such that

$$[v^*\theta^*]_{yy} \simeq [v^*\theta^*]/a^2, \quad [w^*\theta^*]_{zz} \simeq [w^*\theta^*]/H^2. \tag{10.13}$$

The first of these expressions assumes that there is a single zone of baroclinic waves between the equator and pole; in the case which we will consider in Section 10.4 when there may be several parallel storm tracks, a rather smaller scale than a would be needed in this formula. The equations for the temperature gradients may be written

$$B_t + 0.75\frac{g^{3/2}H^2}{\Omega^2\theta_0^{3/2}a^2}V^{1/2}B^2 = \frac{B_E - B}{\tau_E}, \tag{10.14}$$

$$V_t - 0.23\frac{g^{3/2}}{\Omega^2\theta_0^{3/2}}V^{-1/2}B^3 = \frac{V_E - V}{\tau_E}, \tag{10.15}$$

where, for convenience, B and V are defined as

$$B = -[\theta]_y, \; V = [\theta]_z, \tag{10.16}$$

so that B is positive for the usual situation when the poles are colder than the equator; B_E and V_E are the radiative equilibrium values of B and V.

The steady state climatology of the planet is found by setting the time derivatives to zero in Eqs. (10.14) and (10.15); the problem reduces to solving a pair of simultaneous algebraic equations in B and V. By way of example, suppose that the equilibrium atmosphere is neutrally stratified so that $V_E = 0$. Then the horizontal potential temperature gradient is given by

$$\gamma^2 \left(\frac{B}{B_E}\right)^3 + \left(\frac{B}{B_E}\right) - 1 = 0 \qquad (10.17)$$

and the vertical potential temperature gradient is

$$V = 0.55 \frac{\gamma B_E}{\delta} \left(\frac{B}{B_E}\right)^2. \qquad (10.18)$$

The structure of the atmosphere is determined by the value of a single dimensionless parameter γ where

$$\gamma = 0.68 \frac{g \tau_E^{2/3} \delta}{\Omega^{4/3}} \frac{B_E}{\theta_0} \qquad (10.19)$$

and δ is the geometrical aspect ratio, H/a. Figure 10.2 illustrates the solution as γ is varied. For very large γ, baroclinic instability is so effective that it destroys virtually all the horizontal potential temperature gradients and leaves the vertical profile very close to neutral. When γ tends towards zero, the atmosphere is close to radiative equilibrium. For intermediate γ, the horizontal temperature gradients are reduced and the atmosphere is stabilized. The maximum stability is achieved for $\gamma = 2$. For the Earth, γ has the value 5.0, while for Mars, γ is 1.3. Both are fairly close to the regime where the eddies achieve maximum stabilization of the atmosphere. This simple calculation suggests that horizontal temperature gradients will be closer to radiative equilibrium for Mars than for Earth, though the vertical profile for both atmospheres will be significantly stabilized by the eddies.

The baroclinic disturbances of the Earth's atmosphere are highly irregular. Individual systems have a wide range of intensities, phase speed and lifetime. The Earth's tropospheric flow is therefore inherently unpredictable. Small differences in a given initial state lead to very different evolutions of individual baroclinic systems, and hence rapidly to very different global flow fields. Surface observations suggest that the Martian atmosphere, while essentially similar to that of the Earth in that heat transport by baroclinic transients dominates the winter circulation, is much more predictable. Figure 10.3 shows the pressure record from the Viking landers. During the summer, the atmosphere is very inactive meteorologically, with only small fluctuations of pressure. A diurnal variation of the surface wind, related to the local slope of the ground, is almost the only significant weather feature. In autumn and winter, the pressure rises; this is due to the evaporation of large amounts of carbon dioxide which condensed out of the atmosphere on to the southern polar cap during the southern hemisphere winter. The flow

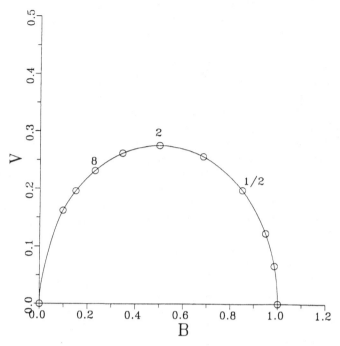

Fig. 10.2. Showing the horizontal and vertical temperature gradients predicted by an analytical model of a quasi-geostrophic planetary atmosphere for various values of the dimensionless parameter γ. Selected values of γ are indicated by circles. B has been scaled by B_E and V by B_E/δ.

associated with the condensation of carbon dioxide at the winter pole and its corresponding evaporation at the summer pole is an important element in Martian global circulation. Simple calculations suggest that it could intensify the winter jets by as much as 15–$20\,\mathrm{m\,s^{-1}}$. During the winter, fluctuations of the surface pressure as baroclinic cyclone systems passed over the lander were observed. These fluctuations were highly regular, with similar amplitudes and periods, suggesting the regular passage of wavenumber 4 disturbances of rather uniform amplitude over the landing site. A spectral analysis of the pressure record shows a single sharp peak centred at a period of about four Martian days.

The simple theory of this section is severely limited. It predicts the typical length scale of baroclinic eddies, and suggests the degree of modification of the radiative equilibrium temperature profile. It suggests how these quantities might depend upon such imposed parameters as rotation rate and radiative timescale. But it cannot be expected to give an accurate quantitative prediction of the lapse rate or horizontal temperature gradient. Neither can

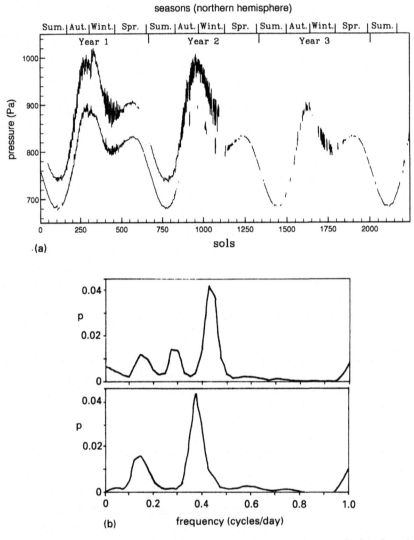

Fig. 10.3. (a) Pressure versus time (in Martian rotation periods) for the Viking landers on the Martian surface. (b) Spectral analysis of the winter pressure record, showing the concentration into a single sharp peak.

it give any indication of the momentum fluxes by the eddies, which are not predicted by Eady theory, or about the interaction between the midlatitude quasi-geostrophic regime and the tropical Hadley regime. Investigating these questions requires a global circulation model. Figure 10.4 shows the results of a sequence of runs with a global circulation model with different rotation rates. The parametrizations of boundary layer processes, radiative heating and cooling, and of convection and other moist processes were essentially terrestrial. The diagram simply shows the zonal mean wind, $[\bar{u}]$, for each run. The zero rotation rate run shows virtually no systematic zonal flow, as one would expect. The low rotation rate runs have a strong jet at the poleward edge of a Hadley cell which extends to very high latitudes. Around the terrestrial rotation rates, the familiar pattern of a subtropical jet and a weaker high latitude jet begins to appear. At higher rotation rates, the zonal flow becomes quite weak, consistent with thermal wind balance. But several midlatitude jets, each with its own associated maximum of eddy activity, are present at midlatitudes. At the highest rotation rates, a strong westerly equatorial jet dominates the flow. Such a pattern is more reminiscent of the gas giants than the terrestrial planets, and suggests that the variations of rotation rate might be the most important difference between terrestrial and Jovian regimes. This is a contentious view which will be discussed further in Section 10.4.

A sharp reminder of how misleading a parametrization of baroclinic eddies in terms of Eady (or Charney) type baroclinic instability might be is provided by a final example. Illustrated in Fig. 10.5 are two contrasting runs with the simple global circulation model introduced in Section 2.5. Both runs are for identical radiative equilibrium conditions, designed to simulate the terrestrial winter circulation moderately faithfully, and for identical planetary properties, with one exception. The difference is that the first run has a realistic frictional spin-up time of about five days, while the second has a very long spin up time of 1250 days. The effect on the eddies is dramatic. In the low surface friction case, there are virtually no eddies, and the temperature field is essentially the radiative equilibrium field. The high friction case has active eddies, and a thermal structure which is consistent with the simple model illustrated in Fig. 10.2. The surface winds in the high friction case are small, but in the low friction run they are very large, and the horizontal wind shears are large. It is this latter feature which causes the Eady parametrization to break down. The Eady model includes no horizontal variation of $[u]$; its inclusion leads to an intractable, nonseparable problem. Numerical solutions of the linear baroclinic instability problem with large horizontal shears show that such shears reduce the growth rate of unstable waves, primarily by restricting their structure so that they cannot tap the available potential energy effectively.

The last result emphasizes the very important role that the upper and lower boundaries play in the baroclinic instability process. The Eady and Charney models both depend upon the presence of rigid boundaries to support instability. Even for 'internal baroclinic instability' in which a change of sign of the potential vorticity gradient at some level makes instability possible,

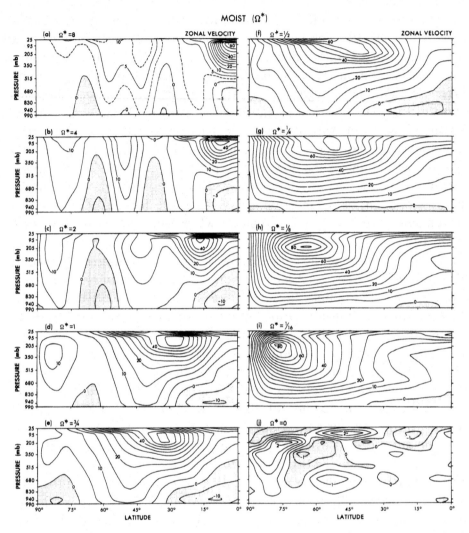

Fig. 10.4. A sequence of pressure-latitude cross sections of the mean zonal wind [$\overline{u}$] from numerical simulations of an Earth-like planetary atmosphere, in which the rotation rate of the planet has been varied. (From Williams, 1988a.)

without some layer where friction can occur, horizontal shears will build up during the growth of baroclinic waves to finite amplitude, and these shears will eventually stabilize the fluid. The implications for a planet with a very deep atmosphere, or for an atmosphere with no solid lower boundary, are that baroclinic instability might not be a very important mechanism. This is a consideration to which we will return in Section 10.4 when we consider the circulation of the giant planets.

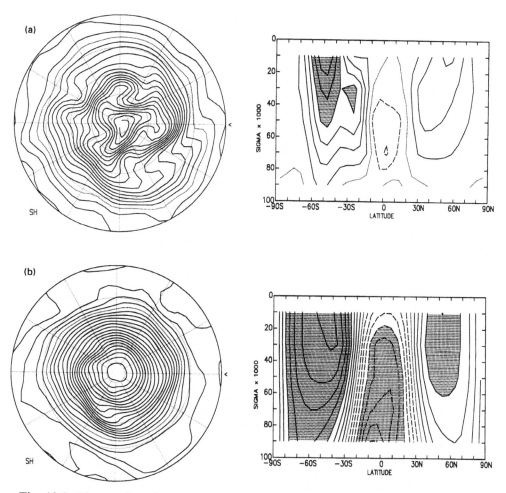

Fig. 10.5. The results of numerical experiments with the 'simple global circulation model' described by Eqs. (2.34) and (2.35) with different surface drag, showing (i) the temperature field at an arbitrary time at the lowest model level (contour interval 4 K) and (ii) cross sections of the zonal mean wind $[\bar{u}]$ (contour interval $5\,\mathrm{m\,s^{-1}}$). (a) $\tau_D = 1$ day, (b) $\tau_D = 250$ days.

10.3 Slowly rotating atmospheres

Figure 10.4 shows that as the rotation rate is decreased, other things being equal, the Hadley regime expands and eventually dominates the circulation throughout the tropics and midlatitudes. The jet at the poleward edge of the Hadley cell intensifies and moves poleward. This is qualitatively consistent with the Held–Hou model of the Hadley cell of Section 4.2. Equation (4.12) suggests that the width of the Hadley cell should be inversely proportional to the rotation rate, other factors being held fixed. When the width of the

Hadley cell becomes comparable to the planetary radius, the Hadley cell will dominate the circulation and more or less squeeze out the quasi-geostrophic wave regime. The condition for a dominant Hadley cell is therefore

$$Ro_T = \frac{gH\Delta\theta}{\Omega^2 a^2 \theta_0} \geq 1. \tag{10.20}$$

The dimensionless number Ro_T is simply a Rossby number, in which the characteristic velocity scale has been estimated by assuming thermal wind balance between the temperature and zonal wind field, and in which the characteristic length scale is the planetary radius.

Pure angular momentum conservation, as assumed in Section 4.2, would lead to very strong zonal winds at high latitudes, and a state of zero absolute angular momentum throughout most of the midlatitudes and tropics. Friction with the surface will moderate these strong winds, and will have an increasing effect as the rotation rate decreases, so that the overturning timescale of the Hadley cell becomes very long in comparison with the drag timescale. Accordingly, in Fig. 10.4, the zonal winds do not exceed $80\,\mathrm{m\,s^{-1}}$ when Ω is one quarter that of the Earth, and they decrease for smaller Ω.

Venus provides the best-studied example of a slowly rotating atmosphere in the solar system. The atmosphere of Venus is dense and deep, but we will largely concern ourselves with motions near the cloud top levels, about $60\,\mathrm{km}$ above the surface, where the pressure is around $30\,\mathrm{kPa}$ and where the bulk of the incoming solar radiation is absorbed. A small fraction of the incoming solar radiation reaches the surface, and it serves to maintain the temperature gradients in the lower troposphere close to the adiabatic lapse rate. But measurements from descending probes suggest that the winds in the lower troposphere are very small, and that the meteorologically active layers are in the vicinity of the cloud tops, where very strong winds are observed. The zonal component dominates, and is such that the cloud tops superrotate, with the atmosphere circulating the planet once every 4 Earth days or so. Relative to the solid planet, the equatorial winds are around $100\,\mathrm{m\,s^{-1}}$. (They are in fact easterlies, but since Venus has a slow retrograde rotation, they represent a superrotation.) The winds decline at lower and higher levels; at latitudes higher than about $40°$, the cloud tops are roughly in solid body rotation about the poles. Figure 10.6 summarizes the observed wind fields.

When the zonal winds are as strong as they are on Venus, the 'centrifugal force' acting on air parcels is appreciable. In the simplest case, the horizontal component of the centrifugal force would be expected to balance the remaining large force, namely the pressure gradient force. For steady,

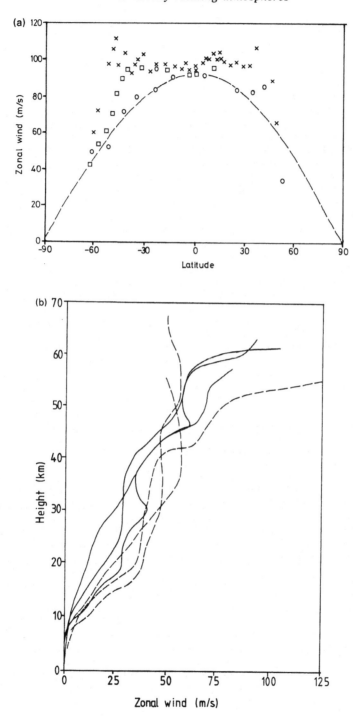

Fig. 10.6. Illustrating the zonal winds of the Venus atmosphere. (a) Winds near the cloud tops as a function of latitude, deduced from the motions of cloud features observed by the Pioneer spacecraft. (b) vertical profile of winds, from the Pioneer sounders and various Russian probes.

frictionless motion, the zonal mean of Eq. (1.33b) reduces to

$$\frac{[u]^2}{a} \tan \phi + 2\Omega[u] \sin \phi = -g \frac{\partial [Z]}{\partial y},$$ (10.21)

where pressure has been used as a vertical coordinate. When $\mid [u] \mid \gg \Omega a$, the centrifugal term dominates over the Coriolis term, and so the meridional momentum equation can be simplified yet further to a balance between centrifugal force and pressure gradient force, called 'cyclostrophic balance', described by:

$$\frac{[u]^2}{a} \tan \phi = -g \frac{\partial [Z]}{\partial y}.$$ (10.22)

A cyclostrophic thermal wind equation can be obtained from this relationship by using the hydrostatic equation to eliminate the geopotential height. This can be written:

$$\frac{\partial}{\partial p}[u]^2 = -\frac{ha}{\tan \phi} \frac{\partial [\theta]}{\partial y},$$ (10.23)

where $h(p)$ was defined in Eq. (1.54). Note that cyclostrophic balance is not possible for arbitrary temperature fields. Assume that $[u]$ is close to zero at the surface. Then at any arbitrary lower pressure

$$[u]^2 = -\frac{a}{\tan \phi} \int_p^{p_s} h(p) \frac{\partial [\theta]}{\partial y} dp.$$ (10.24)

Clearly, if the integral on the right hand side of this equation were positive, cyclostrophic balance would imply imaginary $[u]$, and so a more complex balance of forces must apply. A minimum condition for cyclostrophic balance is that the poles should generally be colder than the equator; even when this condition is fulfilled, it is possible that there could be regions where cyclostrophic balance is not consistent with the temperature field.

At least above the cloud tops, it is easier to monitor the temperature field of Venus than the wind speeds, using remote sensing techniques. The implied cyclostrophic wind is in good agreement with direct measurements of the zonal wind near the cloud tops, but gives an imaginary wind at higher levels. Figure 10.7a shows a cross section of the temperature field. However, the details of the section are a good deal more complex than simple axisymmetric overturning transporting heat polewards and upwards would suggest. In particular, there is a minimum temperature around a circumpolar collar, and higher temperatures over the pole.

The general prograde circulation of the atmosphere of Venus is very difficult to explain, and certainly cannot be accounted for by axially symmetric Hadley circulations. The difficulty in obtaining prograde equatorial winds

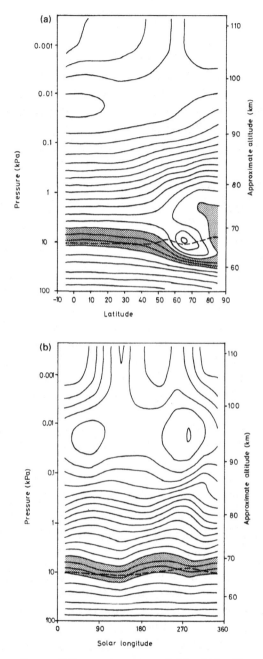

Fig. 10.7. Cross sections of the temperature field of the Venus atmosphere, based on remotely sensed data from the Pioneer Venus orbiter in 1978 (a) Latitude–height section, (b) Longitude–height section. Contour interval 5 K for temperatures below 250 K, 10 K thereafter. Shading indicates temperatures between 235 K and 250 K, and the dotted line indicates the mean cloud top levels. (Redrawn from Schofield & Taylor, 1983.)

from purely axially symmetric circulations was hinted at in Section 8.3. Consider such an axially symmetric circulation, in which there may be friction at the ground, but in which the flow is otherwise inviscid. Such a flow will conserve the specific angular momentum of air parcels, unless a torque about the rotation axis acts. The only such torque is provided by surface friction, and the most this can do is to reduce the air parcel to rest relative to the solid planet. Thus, the maximum angular momentum which any air parcel can have is Ωa^2; conserving this, its zonal wind at any latitude ϕ cannot exceed $\Omega a \sin^2 \phi / \cos \phi$ (see Eq. (4.1)). The observed superrotation indicates that departures from axial symmetry are crucial to the dynamics of the atmosphere of Venus. Similar arguments must also apply to the strong equatorial jets observed on Jupiter and Saturn. These must be maintained by eddies carrying a flux of westerly momentum into the equatorial regions.

The question is: what mechanisms can generate the eddies needed to drive the Venus superrotation? With such a slow rotation rate, baroclinic instability may be discounted. Other forms of instability, such as barotropic instability, might generate Rossby waves which could propagate equatorwards. But these would have the wrong sign of momentum flux, tending to give a westerly acceleration to their source latitudes, and an easterly acceleration to the tropics. Vertically propagating waves, especially trapped equatorial waves of the kind discussed in Section 7.1, can give a westerly acceleration to upper levels, as we saw in our discussions of the QBO (Section 8.3). We still have to explain how such waves can be excited, and we still have to estimate how strong a prograde acceleration they are capable of imparting to the atmosphere.

One important candidate for the forcing of waves is the thermal tide raised by the contrast between the heating of the sunlit side of the planet and the cooling of the night side. According to Table 10.2, the radiative timescale is very much longer than the rotation period, and so the thermal tide would be very small. However, this long radiative timescale was based on the mass of the entire atmosphere, with its huge thermal inertia. Most of the incoming solar radiation is absorbed near the cloud top levels where the pressure is around 30 kPa. The radiative timescale at these levels is a great deal shorter than at the surface, and an appreciable thermal tide could be raised. Tidal theory suggests that the largest response will be in the zonal wavenumber 2 tide. It is possible for this tidal forcing to excite planetary scale waves which will propagate vertically away from the source levels. Figure 10.7(b) shows a longitude height section based on remotely sensed temperature data from the Pioneer Venus spacecraft. The wavenumber 2 thermal tide is very clear near the cloud tops. The phase tilt indicates that the tide is forcing westerly acceleration near the cloud top levels.

The role of thermal tides in modifying the zonal flow is illustrated by Fig. 10.8. This is a second sequence of integrations of a terrestrial global circulation model with different rotation rates due to G.P. Williams, similar to the sequence in Fig. 10.4. On the left are shown $[\bar{u}]$ for a sequence of low rotation rate calculations in which the radiative heating is zonally symmetric. The jet is strongest for a rotation rate 1/64th that of the Earth, where the strongest winds are at 75 °N. When a realistic diurnal variation of the incoming solar radiation is introduced, the westerly momentum is transported to lower latitudes, and, indeed, westerlies are seen at upper levels on the equator. Although these experiments reveal that a thermal tide can drive a superrotation, it is much less strong than that observed on Venus.

So while the thermal tide hypothesis is plausible, it is still quantitatively inadequate. Indeed, no theoretical model capable of predicting the observed strength of the Venus superrotation has yet been proposed. The most successful models tend to involve a number of crucial but arbitrary parameters in the form of anisotropic 'eddy viscosities' or other parameters. Most of these models can support a prograde flow, but fail to achieve velocities as large as those observed. So, while thermal tides are very likely to be involved in generating the superrotation, other effects almost certainly play an important role. Further progress awaits a more detailed monitoring of the circulation of Venus for a longer period than the Pioneer probe was able to achieve. For instance, there is some evidence that the zonal winds may change substantially over long periods; quasi-periodic phenomena analogous to the QBO in the Earth's tropics may possibly be present.

The atmosphere of Titan is another example of an atmosphere on a slowly rotating planet. No direct observations of winds have been made, since a virtually featureless cloud or haze layer fills the atmosphere. But infrared measurements indicate fairly large horizontal temperature gradients, especially at upper levels of the atmosphere. Near the surface, the temperature is nearly uniform, presumably because the extremely long radiative timescale means that even very weak motions can smooth out any temperature variations. But above the haze layer, cyclostrophic balance with the observed temperature fields indicates strong prograde zonal winds. Once again, the thermal tide provides one obvious mechanism for generating such winds. Perhaps strong superrotation is a consequence of any thermally driven flow on a slowly rotating planet; if so, we scarcely understand the relevant mechanisms.

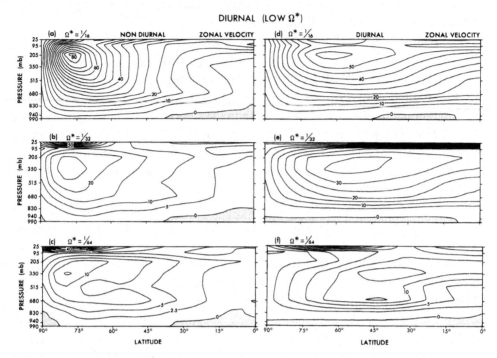

Fig. 10.8. Similar to Fig. 10.4, but comparing two sequences of numerical experiments at low rotation rate. The left hand column shows the zonal winds when the thermal forcing is axisymmetric, while the right-hand column shows the corresponding experiment when a diurnal variation of the incoming solar radiation is included. (From Williams, 1988b.)

10.4 The atmospheric circulation of the giant planets

The gas giant planets are characterized by their large size and rapid rotation. Because of their distance from the Sun, the solar heat flux they intercept is small. But, with the exception of Uranus, all possess an internal heat source which is comparable in magnitude to the solar energy flux. Thermal energy is released by their slow gravitational collapse. The other major factor which must be included in any discussion of their atmospheric circulations is the absence of any solid surface.

Jupiter is the best studied of the gas giant planets, and the observed features of its atmospheric circulation will be outlined here. It is covered by cloud decks, the highest of which are near the 30 kPa level. Temperature measurements suggest that the brighter white clouds consist of ammonia ice and are at higher levels, while the darker coloured clouds are warmer and therefore lower in the atmosphere. The alternating pattern of light and dark cloud bands suggests a sequence of meridional cells, with ascending motion dominating at the latitudes of the light bands and descent dominating at the latitudes of the darker bands. Earth based telescopic observations show that the major features of these bands are long lived, although the individual bands are interrupted by transient spots and eddies.

Tracking of individual cloud features leads to estimates of the winds at the cloud top levels, although the interpretation of these measurements is complicated by the difficulty of assigning an accurate pressure to them. The winds are referred to a frame of reference rotating with the radio period of the planet, assumed to reflect the rotation rate of the deep interior. Figure 10.9 shows the pattern of zonal winds observed. A broad region of strong westerlies, around $100 \, \mathrm{m \, s^{-1}}$, blows around the tropics, roughly within $10°$ of latitude of the equator. At higher latitudes, alternating easterly and westerly jets are seen. The westerly jets tend to coincide with the higher cloud layers and the easterly jets with the lower, warmer layers. The bright elliptical spots which are a prominent feature of the images, some of them very long lived, all exhibit anticyclonic circulations. Some dark cyclonic spots, highly elongated in the zonal direction, were observed around $18°\mathrm{N}$ by Voyager.

Similar observations exist for Saturn and Neptune. Uranus is a special case which will be discussed separately. The zonal jets on Saturn are stronger than on Jupiter, with the equatorial jet being broader and with a maximum wind speed of some $400 \, \mathrm{m \, s^{-1}}$. Although more difficult to observe because of the lack of contrast at the cloud tops, similar features by way of a banded cloud deck and anticyclonic spots are also characteristic of Saturn.

The presence of equatorial jets on the gas giants is a pointer to the importance of eddies in maintaining the observed circulation. Purely axisymmetric motions can merely redistribute angular momentum in the meridional plane and so cannot produce any maxima where the specific angular momentum exceeds Ωa^2. Eddies, sheared and distorted by the ambient flow, can provide the necessary up gradient fluxes of angular momentum to drive equatorial jets. They can also drive jets at other latitudes. One group of theories of the circulation of the major planets would suggest that the observed circulation is essentially driven by small scale eddies. These are illustrated in their purest form by the concepts of two-dimensional turbulence on a β-plane. Rapid rotation of a barotropic fluid serves to suppress vertical motions and leads to an essentially two-dimensional flow field.

Consider a random set of strong eddies in a layer of rapidly rotating fluid. We will suppose that the domain is such that boundary fluxes are either zero or periodic, and for the moment we ignore the β-effect. We have

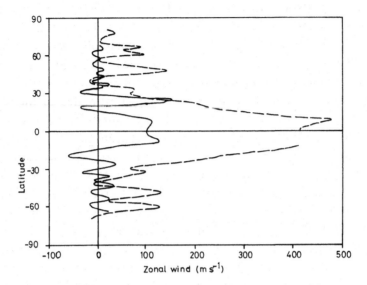

Fig. 10.9. Zonal winds at the cloud top levels on the gas giants, inferred from tracking cloud features in the Voyager images. Solid curve is for Jupiter, dashed for Saturn. (After Ingersoll, 1990.)

already seen examples of the way in which such eddies evolve, in the finite amplitude evolution of baroclinic waves in Chapter 5 or in the breaking of Rossby waves in the stratosphere in Chapter 9. The vorticity of the eddies becomes concentrated in thin shear regions separating large regions of smaller vorticity. That is, the mean square vorticity (the 'enstrophy') cascades to smaller and smaller scales until dissipation (due, for example, to molecular viscosity) can remove it from the flow. At the same time it is easy to see from the momentum equations and vorticity equation for such inviscid two-dimensional flow that both the total kinetic energy $< \frac{1}{2}u^2 >$ and the total enstrophy $< \frac{1}{2}\xi^2 >$ must be conserved. Scaling arguments can be used to show that if some forcing with wavenumber k_f excites eddies, then a steady state is achieved in which there are two distinct spectral regions:

(i)   For $k > k_f$, the kinetic energy varies as k^{-3}. There is a flux of enstrophy from large to small scales, and no flux of energy.

(ii) For $k < k_f$, the kinetic energy varies as $k^{-5/3}$. There is a flux of energy from small to large scales and from high to low frequencies, while the flux of enstrophy is zero.

For applications to the gas giants, we will suppose that $k_f^{-1} \ll a$, and we will concentrate on the eddies whose scales are larger than k_f^{-1}. As the eddies become larger, the β-term becomes progressively more important. At the wavenumber k_β where:

$$k_\beta = \left(\frac{\beta}{U}\right)^{1/2}, \tag{10.25}$$

U being a characteristic eddy velocity, the nonlinear and β terms become equal; eddies begin to propagate as coherent Rossby wave packets. Now the frequency of Rossby waves can be written (see Eq. (6.13)) in the form:

$$\omega = \frac{\beta}{k}\cos\alpha, \tag{10.26}$$

where α is the angle between the wavenumber vector and the x-axis, while the frequency of waves in the turbulence regime is simply

$$\omega = Uk. \tag{10.27}$$

Once energy reaches the wavenumber k_β, further cascade to low frequencies is blocked by the Rossby wave propagation. Hence k_β is sometimes called the 'Rhines blocking wavenumber'. The cascade can proceed only if the total wavenumber remains fixed, while the wavenumber vector rotates towards the meridional direction. That is, eddies become elongated in the zonal direction and contract in the meridional direction. The end result is a set of alternating easterly and westerly currents, whose characteristic width is $L_\beta = \pi/k_\beta$. Figure 10.10 illustrates this process schematically.

Numerical experiments confirm these rather heuristic arguments. We conclude that the end result of arbitrary forcing of eddies at some large wavenumber k_f on a rapidly rotating planet will be a set of zonal jets whose width L_β depends upon the rotation rate and the level of eddy kinetic energy. Figure 10.11 shows the result for a Jovian parameter setting. Friction induced meridional circulations could then account for the banded clouds associated with the jets.

In the model used to produce Fig. 10.11, the forcing was simply a pre-scribed mechanical forcing. The model serves to demonstrate that if isotropic kinetic energy is generated at some large wavenumber k_f, two-dimensional turbulence with a β-effect will eventually generate a number of zonally ori-entated jets. The question for the gas giants then concerns the nature of the eddies that force the large scale flow. Since some hydrodynamical instability of the flow must be present to generate such eddies, one would anticipate some interaction between the jets and the eddies. For example, baroclinic instability might be concentrated in the jet regions, where the vertical shears are large, and would be weaker in the horizontally sheared, but more weakly baroclinic, regions either side of the jets.

Two major theories exist for the eddies on Jupiter and similar planets. The

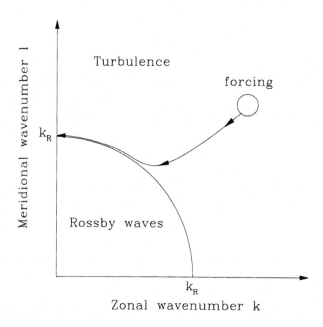

Fig. 10.10. Schematic illustration of the evolution of geostrophic turbulence on a β-plane towards a pattern of parallel zonal jets.

first is that baroclinic instability is responsible. The temperature contrast between the poles and the equator on Jupiter is very small. Presumably, variations of the solar input across the planet are compensated by variations in the heat flux from the interior of the planet. Larger horizontal temperature gradients might be present within the individual jets. Even then, it is not clear that baroclinic instability will be present, other than in a very weak form. The Eady instability relies on the presence of rigid boundaries at top and bottom where $w = 0$. The sudden change of static stability in the Earth's tropopause approximates this boundary condition for deep baroclinic waves. The Charney model requires a boundary with $w = 0$ at the bottom, and a critical level above. In a planet with no solid boundary, and with an adiabatic lapse rate in the deep interior, these situations are not encountered. There is a weak tropopause on Jupiter, and this might sustain Charney-type modes of instability in regions of easterly vertical shear, but it is difficult to see how baroclinic instability could be strong in regions of strong westerly shear.

The other possibility for baroclinic eddies is 'internal' baroclinic instability, the necessary condition for which is that the potential vorticity gradient, $[q]_y$, should change sign at some level in the atmosphere. While such a condition can be envisaged, it is difficult to see how it could be sustained over large

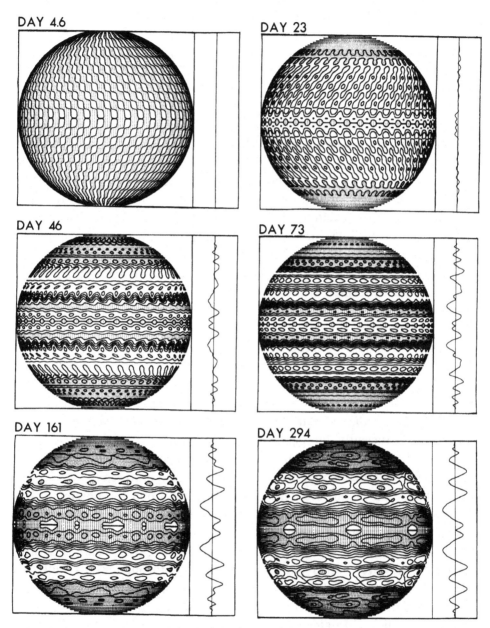

Fig. 10.11. Evolution of the streamfunction in a rapidly rotating thin shell of barotropic fluid in which small scale eddies are randomly excited. (From Williams, 1978.)

parts of the planet. Such internal instability is unusual on the Earth or in laboratory systems; its finite amplitude evolution has not been studied very thoroughly. Such studies as are documented suggest that the eddies would generally be weaker than those in the Charney or Eady idealizations.

Vertical convection, driven as a result of the destabilizing effect of the internal heat flux, might be an alternative source of eddy kinetic energy. On Earth, small scale convection is capable of organizing larger scale flows, in cloud cluster systems, or tropical cyclones. Similar processes are thought to be important in generating the shallow mesoscale systems known as 'polar lows' at high latitudes. The predominantly anticyclonic circulation and high cloud associated with the spots observed on Jupiter and Saturn suggest an analogy with the anticyclonic upper tropospheric outflow from convectively driven tropical cyclones in the Earth's atmosphere. However, terrestrial tropical cyclones depend crucially upon large convergence of latent heat fluxes in the planetary boundary layer. It is not clear whether analogous systems could be formed in the absence of a rigid boundary.

The crucial test to distinguish between these possibilities is to estimate the Richardson number, Eq. (10.6), for Jupiter or Saturn. If baroclinic instability dominates, Ri should be large. If the flow is driven by vertical convection, Ri will be very small. However, since both θ_y and θ_z are quite small, rather accurate determinations of θ_z and θ_y (or, equivalently, u_z) are needed to establish Ri accurately. Present estimates of θ, which are based on infrared and microwave radiance data, have too coarse a vertical resolution and do not extend sufficiently far beneath the upper cloud layers to give useful estimates of Ri. A major goal of future space missions will be to obtain more accurate vertical soundings of winds and temperatures.

The discussion in this section has concentrated upon the possibility that the observed circulation of Jupiter is confined to the optically thin, meteorologically active layers of the planet. It is seen as an extreme development of the sort of circulation which we have studied on Earth. There is another possibility, however. In a planet which is fluid throughout its bulk, the motions we observe might simply be the surface expression of circulations extending throughout the deep bulk of the planet, driven by the internal heat flux. Studies of Jupiter's magnetic field indicate that at deep levels, the hydrogen which dominates its composition becomes electrically conducting, so that dynamo effects generate strong magnetic fields. Determining the circulation then becomes a complicated problem in thermally driven magnetohydrodynamics. A description of the circulation in these terms would be highly speculative and grounded on poorly understood physics.

Before concluding this section, it is worth making a few remarks about

Uranus. This is a very similar planet to Neptune, except that its internal heat flux is not more than 10% of the intercepted solar heat flux, and indeed is probably almost totally absent. Its other peculiarity is its very large axial tilt, which means that, averaged through its year, maximum insolation is received at the poles and minimum insolation at the equator. The depth of the atmosphere and the weak solar constant mean that the radiative timescale is very long, and so seasonal effects will be negligible, at least until very high reaches of the atmosphere are considered. Thus, the meridional temperature gradients will be reversed from those of the terrestrial planets. Images from the Voyager probe indicate a very inactive atmosphere, with few clouds and therefore weak vertical motions. This suggests that the destabilizing effect of the internal heat fluxes is a most important feature of the other gas giants. Very large zonal winds were determined from the few clouds that could be observed; a profile is shown in Fig. 10.12. The zonal winds are not consistent with a radiative equilibrium temperature variation. This would have a warm pole and cold equator, implying that the winds would become increasingly retrograde with height. Instead, strongly prograde winds were observed at the cloud top levels in the summer hemisphere, especially at high latitudes. The temperature distribution, too, is very far from radiative equilibrium. Voyager 2 recorded the largest temperatures at the equator. Despite its bland appearance, it seems that the atmospheric structure of Uranus is profoundly modified by dynamical transports of heat and momentum.

10.5 Large scale ocean circulation

The oceans form a planetary scale fluid system whose dynamics has much in common with the dynamics of the atmosphere. Away from the equator, the small flow speeds mean that the Rossby number is very small, so that quasi-geostrophic flow dynamics are appropriate. In the tropics, various equatorially trapped waves such as those discussed in Section 7.1 are important, though with different characteristic space and timescales than those of their atmospheric counterparts. However, although the dynamics of individual oceanic systems are essentially the same as in the corresponding atmospheric systems, the global scale circulation of the ocean is very different. This is due to two major factors:

(i) The distribution of heating is totally unlike that of the atmosphere.
(ii) The flow is mainly confined to nearly closed ocean basins, with boundaries on their eastern and western sides.

We consider each of these factors in turn.

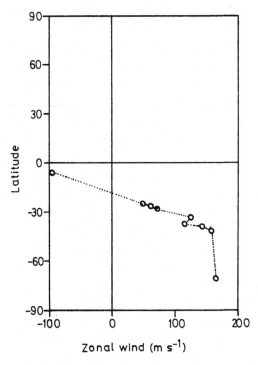

Fig. 10.12. As Fig. 10.9, but for Uranus.

The sunlight which enters the ocean is entirely absorbed in its uppermost layers. The typical vertical temperature structure which results consists of a thin layer of warm, highly stratified surface layers overlying the much colder, nearly isothermal, abyssal ocean. The highly stratified layer is called the thermocline; it is generally not more than 1.5 km deep. The abyssal layers of the ocean have temperatures between 2 and 4 °C, and may be 3–5 km deep. Thus, at least for the bulk of the oceans, there is no possibility of releasing potential energy through upward heat fluxes, and so the ocean circulation cannot be driven thermally. The oceans act as a heat pump, rather than as a heat engine. The source of mechanical energy is provided by the surface winds of the atmosphere, which apply a surface stress to the ocean. At the same time, the fluxes of heat and water vapour from the ocean surface provide a most important source of thermal energy for atmospheric circulation, especially in the tropics.

The exception to this is at high latitudes, where strong winter cooling produces surface water which may be denser than the cold abyssal waters. If this occurs, thermally driven convection can result, and new bottom water can be

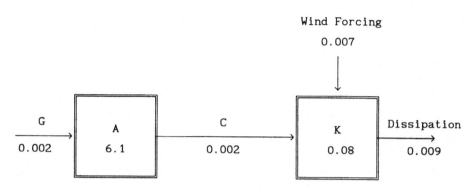

Fig. 10.13. Energetics of the ocean circulation. Wind forcing by the atmospheric circulation provides an additional input to the kinetic energy. Units as Fig. 5.15. (Based on data in Peixóto & Oort, 1992.)

formed. Two regions have been identified where such generation of bottom water occurs. One is in the north Atlantic, and the other is in the Weddell Sea. The downward convection of cold water in these regions must be compensated by slow upwelling elsewhere, a circulation called the 'thermo-haline circulation'. The thermohaline circulation is important in exchanging water between the thermocline and the abyssal ocean. These exchanges are important in considering the global budgets of dissolved constituents such as salt and carbon dioxide. The thermohaline circulation is thermally direct, and so converts available potential energy into kinetic energy. However, measurements reveal that it makes a rather small contribution to the kinetic energy budget of the oceans. The energetics of the oceanic circulation are illustrated in Fig. 10.13. Available potential energy and kinetic energy are given in the same units as were used for the atmospheric energy diagram Fig. 5.15. The kinetic energy of the oceans is much smaller than that of the atmosphere. A small amount of mechanical energy is generated thermally, but over 75% of the observed kinetic energy comes from wind forcing.

The confinement of the flow by continental boundaries gives rise to impor-tant asymmetries in the circulation in each individual ocean basin. Consider the flow in a barotropic ocean which is set into motion by wind stresses at its surface. The flow is described by means of the barotropic vorticity equation

$$\frac{\partial \xi}{\partial t} + \mathbf{v} \cdot \nabla \xi + \beta v = \frac{\mathbf{k} \cdot \nabla \times \tau_w}{\rho_w H}, \tag{10.28}$$

where τ_w represents the stress resulting from the winds, ρ_w is the density of seawater and H is the depth of the ocean. If, as we expect, the currents are weak, the advection terms will be small and so our equation will be dominated over most of the basin by a balance between βv and the curl of the wind stress, $\mathbf{k} \cdot \nabla \times \tau_w / \rho_w H$. Thus, where there is forcing of anticyclonic vorticity there will be an equatorward drift of water, while there will be a poleward drift where cyclonic forcing operates. This so-called 'Sverdrup balance' is illustrated in Fig. 10.14, which shows the circulation induced in a square ocean basin by a wind field which has a midlatitude maximum of westerly flow. That is, the vorticity forcing is cyclonic in the north of the basin and anticyclonic in the south. In a narrow region near the western boundary, however, Sverdrup balance cannot operate. The boundary condition of $u = 0$ at meridional boundaries means that there must be a meridional flow which balances the Sverdrup drift at the western boundary. This generates vorticity which reinforces the wind stress forcing, and so the western boundary current becomes narrower and stronger.

In the example of Fig. 10.14, an artificial diffusion term was added to Eq. (10.28). This prevented the western boundary current becoming infinitely narrow. In the real ocean, the western boundary current becomes hydrodynamically unstable, and so eddy transports prevent it from collapsing entirely. Considerable current oceanographic research is presently devoted to exploring the nature and role of ocean eddies. The narrow western boundary currents are clear in any map of mean surface currents in the oceans, and are very important in transporting warm water poleward in the subtropics, and cold water equatorward in the high latitude parts of the ocean basins. It is perhaps no coincidence that the northern hemisphere storm tracks start just where these tropical and polar waters meet, forming strong meridional gradients of sea surface temperature. But cause and effect are difficult to separate; the location of the atmospheric jet itself determines the ocean circulation and hence the latitude of the largest sea surface temperature gradients.

Steep orography can lead to analogues to the western boundary currents in an atmospheric context. The low level jet over east Africa, which is an important component of the Asian monsoon circulation described in Section 7.1 is one such example. Mars is characterized by very high orography, with plateaux and depressions more than a Rossby radius across, and with a vertical scale in excess of a pressure scale height. It has been suggested that boundary currents might be an important component of the circulations here also.

10.6 Laboratory systems

Fluid can be forced into motion in the laboratory in a way which is closely analogous to the development of global circulation. Such systems are amen-

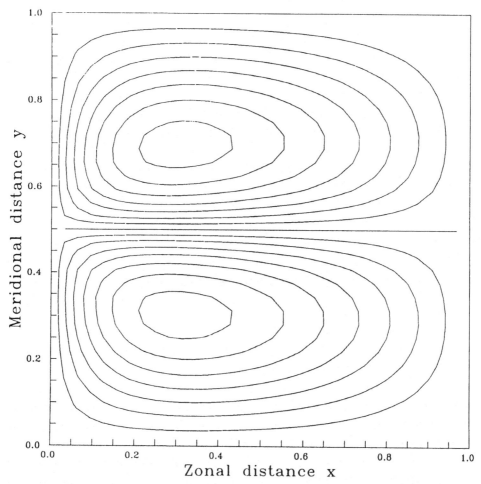

Fig. 10.14. An example of the circulation of a barotropic ocean confined to a square ocean basin, based upon a numerical integration of Eq. (10.28).

able to experiment and suggest ways in which circulation might be modified under changes of imposed parameter. They enable certain primary dynamical mechanisms to be isolated. On the other hand, there are some respects in which they introduce elements which have no analogue in a planetary atmosphere.

The basic apparatus consists of fluid in an annular container. This is mounted on a rotating platform, and subjected to a horizontal temperature gradient. A typical configuration is shown in Fig. 10.15. Different upper boundary conditions can be imposed. The fluid may be in contact with a solid lid, so that the top and bottom boundary conditions are symmetric, or

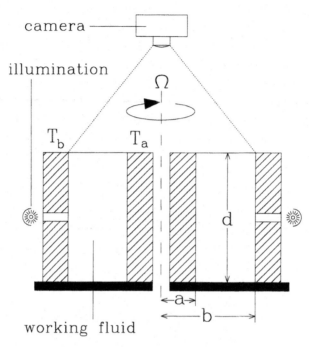

Fig. 10.15. The configuration of the basic 'annulus' experiment. The notation used in the text is defined.

else it may be in contact with air. This latter stress-free boundary condition is more closely analogous to atmospheric configurations and will be assumed in what follows. The fluid is commonly water, but other substances can be used in order to change such parameters as the ratio of thermal diffusion to viscosity (the 'Prandtl' number). A variety of methods of diagnosing the resulting flow can be used. These include mounting temperature sensors in the working fluid, or following the motion of neutrally buoyant plastic beads.

The equations governing the flow in the annulus system are much the same as those which govern the atmospheric circulation. The equation of state is written:

$$\rho = \rho_0 \left(1 - \alpha(T - T_0)\right), \tag{10.29}$$

where α is the expansion coefficient. In fact, for water, a quadratic term in $(T - T_0)$ is also significant. Since the arguments of this section would not be altered by including such a complication, it will be ignored. The fluctuations of density are quite small, and so the Boussinesq approximation is valid. The other difference from the atmosphere is that the Coriolis parameter is a constant, 2Ω. The thermal wind relationship for the zonal flow in the system is

$$\frac{\partial[u]}{\partial z} = \frac{g}{2\Omega\rho_0}\frac{\partial[\rho]}{\partial r}, \tag{10.30}$$

where the horizontal gradient of density is related to the horizontal temperature gradient by Eq. (10.29). Denoting the temperature difference across the system by ΔT, the thermal wind relationship can be integrated to estimate the zonal wind shear ΔU between the top and bottom of the annulus:

$$\Delta U = \frac{dg\alpha\Delta T}{2\Omega(b-a)}. \tag{10.31}$$

Thus, the zonal flow will increase with height when the outer wall is warmer than the inner. This zonal flow can be thought of as the result of Coriolis forces deflecting meridional overturning motions driven by the temperature gradients.

Such a vertically sheared flow will be baroclinically unstable. Assuming that the curvature of the system is not too strong, the Eady model of Section 5.4 can be applied directly to the system. The Eady model is characterized by a minimum wavelength for which there is instability, that is,

$$\lambda \geq 1.31\frac{Nd}{\Omega}. \tag{10.32}$$

The Brunt-Väisälä frequency N is estimated by assuming that the temperature difference between the top and bottom of the system is comparable to the imposed horizontal temperature difference. That is, warm fluid rises against the warm wall and spreads across the top of the annulus, while cold fluid sinks against the cold wall and spreads across the base. Hence

$$N^2 = -\frac{g}{\rho_0}\frac{\partial\rho}{\partial z} = \frac{g\alpha\Delta T}{d}. \tag{10.33}$$

The longest wave which can be accommodated by the apparatus has wavelength $\pi(b+a)$, and so the condition for baroclinically unstable waves to be present can be written

$$\frac{g\alpha\Delta T d}{4\Omega^2(b+a)^2} \leq 1.44. \tag{10.34}$$

At slow rotation rates, for a given apparatus and ΔT, instability will not be possible, and the flow will be purely axisymmetric. At rapid rotation rates, condition (10.34) can be satisfied and unstable baroclinic waves will be expected to develop. The precise form of the flow will depend upon nonlinear effects as the unstable waves reach finite amplitude. Condition (10.34) can be rewritten in terms of a Rossby number based on ΔU (Eq. (10.31)) and the gap width, $(b - a)$. This is simply the thermal Rossby number Ro_T (Eq. (10.20)) with the gap width $(b - a)$ replacing the planetary radius:

$$Ro_T = \frac{\Delta U}{2\Omega(b - a)} \leq 1.44 \frac{(b + a)^2}{(b - a)^2}. \tag{10.35}$$

The appearance of waves requires that the thermal Rossby number be less than some critical value.

Figure 10.16 shows the flow for a given annulus with fixed ΔT but increasing Ω. At the lowest rotation rates, the flow is axisymmetric, in agreement with Eq. (10.35). At larger rotation rates, the axisymmetric flow is unstable and waves appear. As long as the wavelength of the most unstable disturbance is long compared to $(b - a)$, a 'regular regime' is observed, in which the wave equilibrates at a finite amplitude. The zonal flow is concentrated into a tight jet stream which meanders between the inner and outer walls of the apparatus, and the entire pattern drifts around the apparatus. The jet stream transports heat across the apparatus, even though the motion is quasi-horizontal and geostrophic, so that $[v]$ tends to zero as Ω increases. The narrow, meandering jet stream is rather similar to the midlatitude jet observed in the Earth's troposphere. This regular flow, characterized by a single dominant wavenumber with a constant drift speed, is highly predictable. The Martian midlatitude flow is apparently somewhat similar. At the largest rotation rates accessible to the apparatus, the most unstable wavelength becomes smaller than $(b - a)$. Then the waves become irregular both in space and time. The irregular flow resembles that of the Earth's atmosphere in that it is inherently chaotic and unpredictable.

The transition from axisymmetric flow to regular waves depends upon factors other than just Ro_T. Viscosity and thermal diffusion can stabilize baroclinic waves, and so affect the value of Ro_T at which the transition takes place. Figure 10.17 shows a regime diagram, in which the wavenumber and character of the flow has been plotted against Ro_T, and a dimensionless measure of the viscosity, the Taylor number Ta. The Taylor number is defined as

$$Ta = \frac{4\Omega^2(b - a)^5}{v^2 d} \tag{10.36}$$

and is closely related to the Ekman number by $Ta \propto Ek^{-2}$. Inviscid theory, as outlined above, suggests that the transition between axisymmetric and wavy flow should simply be a horizontal line; this is indeed so at large Ta. In contrast, at sufficiently large viscosity, waves are absent. At intermediate viscosity, as well as the upper symmetric transition of inviscid theory, there

Fig. 10.16. Flow just beneath the upper surface in an annulus system at different rotation rates. The flow was visualized by time exposure photographs of neutrally buoyant beads. The working fluid was a water-glycerol solution with a viscosity of $1.56 \times 10^{-6} \, \mathrm{m^2 \, s^{-1}}$, the temperature difference was 9 K, and the dimensions of the annulus were $a = 55 \, \mathrm{mm}$, $b = 101 \, \mathrm{mm}$, $d = 135 \, \mathrm{mm}$. The rotation rates were: (a) $0.41 \, \mathrm{rad \, s^{-1}}$; (b) $1.07 \, \mathrm{rad \, s^{-1}}$; (c) $1.21 \, \mathrm{rad \, s^{-1}}$; (d) $3.22 \, \mathrm{rad \, s^{-1}}$; (e) $3.91 \, \mathrm{rad \, s^{-1}}$; and (f) $6.4 \, \mathrm{rad \, s^{-1}}$. (From Hide & Mason, 1975.)

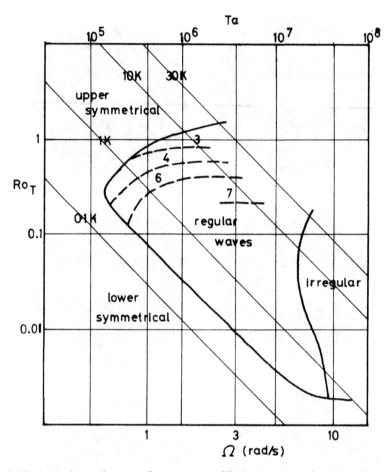

Fig. 10.17. A regime diagram for a water filled annulus, showing the regions of symmetric, regular wavy flow and irregular flow on a (Bu, Ta) plane. (After Hide & Mason, 1975.)

is a lower symmetric transition at very small ΔT which results from viscous diffusion. A modified linear theory which includes the effect of viscosity in the Eady model is also able to account for this transition.

Within the regular regime, a number of interesting 'vacillation' phenomena have been observed. While the waves are highly regular in their spatial structure, their properties may change periodically over time. Two kinds of vacillation have been identified: amplitude vacillation and shape vacillation. In amplitude vacillation, the waves periodically grow and collapse, with various eddy quantities such as the heat flux fluctuating in sympathy. It has been suggested that amplitude vacillation is analogous to the much more irregular index cycles observed in the Earth's midlatitudes. Shape vacillation is characterized by a more constant amplitude and hence heat flux; instead the direction of the trough line fluctuates, leading to a large fluctuation in the momentum flux carried by the eddies.

One of the most interesting aspects of the annulus system is that it suggests several ways in which nonlinearity can be manifested in a baroclinic system. All the wavy flows we have discussed have finite amplitude waves whose mean intensity has equilibrated to the zonal flow; the linear processes generating eddy kinetic energy balance the nonlinear cascade of kinetic energy down to smaller scales where diffusive processes can remove it. This nonlinear limiting may lead steady, regular waves, to vacillating waves with periodic large amplitude, low frequency fluctuations of the flow, or to highly irregular, 'chaotic' flows of the kind discussed in Section 8.7.

We have to be careful not to stretch the analogy between the annulus system and planetary atmospheres too far. One difference is the absence of a β-effect in the laboratory systems. To some extent, this can be simulated by using top or bottom boundaries such that the depth d increases with radius. This is exactly equivalent to a β-effect for a barotropic fluid, but only approximately so for a baroclinic system. But perhaps the most important difference is the controlling effect of boundary layers on the cylindrical inner and outer walls in the annulus. The inner cylinder can be reduced to small radius (such a system is sometimes called a 'dishpan'), but the outer boundary condition is really very different from the interface with a tropical Hadley circulation in a terrestrial type atmosphere, or perhaps with parallel jets in the case of a Jovian-type atmosphere. Weightless environments in spacecraft have provided an opportunity to experiment with a truly spherical laboratory system. Electrostatic forces have been used to provide a radially symmetric body force to replace the gravitational force in a planetary atmosphere. But such experiments are necessarily very expensive, and perhaps numerical experimentation offers a more convenient way of exploring regimes of planetary circulation now that large computers are reasonably readily available to atmospheric scientists.

The scientific study of global circulations presents peculiar difficulties since the number of natural systems which can be observed to check our theoretical predictions is limited. Numerical models offer the chance to carry out controlled and detailed experiments. But laboratory experiments provide another independent way of testing hypotheses. This is always going to be a stern and challenging task which is likely to lead to those unexpected results which make any science a continuous adventure into the unknown.

10.7 Problems

10.1 Show that the closure hypothesis, Eq. (10.9), for the simple parametrization of baroclinic eddies in a planetary atmosphere is equivalent to specifying a 'mixing length' L, that is, a distance over which heat is transported by individual eddies, where

$$L = 3.2 L_R.$$

10.2 Turbulent eddies are observed in the vicinity of Jupiter's Great Red Spot, at 28 °S. Their typical horizontal scale is 1200 km. Assuming they are baroclinic waves, estimate the Brunt–Väisälä frequency for Jupiter's atmosphere. Estimate the difference between the implied and adiabatic lapse rates.

10.3 A pole–equator temperature difference of around 35 K is observed in the Titan atmosphere, between heights of 50 km and 70 km above the surface. Assuming cyclostrophic balance, estimate the zonal wind speed at these levels.

10.4 Images of Neptune's cloud features suggest the presence of four jets between the equator and the pole. Assuming these are generated by the Rhines mechanism, estimate a typical eddy velocity.

10.5 Use the simple analytical model of the global circulation driven by baroclinic waves, with $V_E = 0$ (Eqs. (10.17) and (10.18)), in the limit of large α to show that the slope of the meridional heat flux vector is about $1/3$ of the slope of the isentropes.

10.6 Estimate Ro_T and Ta for the Earth's midlatitudes. Ta can be estimated if the thickness of the midlatitude boundary layer, which is around 1 km, is used to estimate an effective eddy viscosity. Use your values of Ro_T and Ta to compare the regime of flow of the Earth's midlatitudes with that observed in the annulus for similar parameters.

10.7 Assuming a 'Sverdrup balance' such that

$$\beta v = \mathbf{k} \cdot \nabla \times \tau_w / (\rho_s H),$$

estimate a typical poleward current in the surface layers of the Southern Ocean at 60 °S. Assume that the surface wind stress is given by $\rho C_D \mid \mathbf{v} \mid^2$, where ρ is the density of the air and the drag coefficient C_D may be taken as 10^{-3} over the ocean.

10.8 In the midlatitude ocean, away from the continents, the sea surface temperature is found to be 15 °C, decreasing to 4 °C at the base of the thermocline at a depth of 1.5 km. Assuming a constant salinity and a coefficient of thermal expansion of $10^{-4}\mathrm{K}^{-1}$, estimate the Brunt–Väisälä frequency for the upper ocean. Determine the Rossby radius and hence deduce the typical scale of eddies.

Appendix
Solutions to Problems

A.1 Solutions to Chapter 1

1.1 Use Dalton's law of partial pressures, and denote the gas constant of dry air R_d and the gas constant of any constituent R_c. The volume occupied by unit mass of a consituent with partial pressure e_c is simply the specific volume $\alpha = R_c T / e_c = R_m T / p$. But if we compress the sample to pressure p at the same temperature T, the volume it will occupy is $\alpha_c = R_c T / p$. The volume mixing ratio $r_v = \alpha_c / \alpha = e_c / p$. Now consider a unit volume of dry air: its mass is $\rho = p / R_d T$. The mass of the constituent is $\rho_c = e_c / R_c T$. Therefore the mass mixing ratio r_m is given by $\rho_c / \rho = R_d e_c / (R_c p)$, i.e., $r_m = (R_d / R_c) r_v$. Since the gas constant $R = R^* / m$, the final result is:

$$r_m = \frac{m_c}{m_d} r_v.$$

Using this formula and the data in Table 1.1, the mass mixing ratios in dry air are 0.755 22 for N_2, 0.231 38 for O_2, 0.012 56 for Ar and 0.000 50 for CO_2.

1.2 Since $dQ = c_p dT - \alpha dp$, and using the hydrostatic relationship $\partial p / \partial z = -\rho g$, we have:

$$Q = \int_0^z c_p \frac{\partial T}{\partial z} dz + \int_0^z g dz.$$

Again, using the hydrostatic relation, the Brunt–Väisälä frequency N^2 can be written as $(g/T)(\Gamma_a + \partial T / \partial z)$, where the adiabatic lapse rate $\Gamma_a = g/c_p$. Hence

$$\frac{\partial T}{\partial z} = \frac{N^2 T}{g} - \Gamma_a.$$

386

If N^2 is constant this can be integrated to give:

$$T = \left(T_s - \frac{g\Gamma_a}{N^2}\right)\exp\left(\frac{N^2 z}{g}\right) + \frac{g\Gamma_a}{N^2}.$$

The integral can now be evaluated, and we obtain:

$$Q = gz - c_p\left(T_s - \frac{g\Gamma_a}{N^2}\right)\left(1 - \exp\left(\frac{N^2 z}{g}\right)\right).$$

Substituting values, we find $Q = +8369\,\text{J}\,\text{kg}^{-1}$. This is *positive*, i.e. heat must be supplied to the parcel. This is obvious, for if it were lifted adiabatically, its temperature would follow the adiabatic lapse rate, Γ_a and it would arrive at 3000 m considerably colder than its surroundings.

1.3 Since the Coriolis force is always at right angles to the velocity vector, the Coriolis force cannot change the kinetic energy of the parcel, only its direction of motion. Equating the magnitude of the Coriolis force to the magnitude of the centripetal force required to maintain circular motion, radius r, we have

$$2\Omega U \sin\phi = \frac{U^2}{r} \text{ or } r = \frac{U}{2\Omega \sin\phi}.$$

The time taken to execute a complete circuit is simply $2\pi r/U$ or $2\pi/(2\Omega \sin\phi)$. Since the length of the day is $2\pi/\Omega$, the time taken for each orbit is conveniently written $(2\sin\phi)^{-1}$ days. Such motion is termed 'inertial motion'.

1.4 The temperature can be written

$$T(z) = T_g + \Gamma(z_s - z),$$

where T_s is the surface temperature, Γ is the lapse rate and z_s is the elevation of the surface above mean sea level. The hydrostatic relation gives:

$$\frac{\partial p}{\partial z} = -\frac{pg}{R(T_g + \Gamma(z_s - z))},$$

Integrating from $z = 0$ where $p = p_0$, to $z = p_s$ where $p = p_s$, gives the required relationship:

$$p_0 = p_s\left(1 + \frac{\Gamma z_s}{T_s}\right)^{\frac{g}{\Gamma R}}.$$

If Γ were set to zero (isothermal case), the integration would yield:

$$p_0 = p_s\exp\left(\frac{gz_s}{RT_g}\right).$$

In fact, the two formulae give nearly identical results. Using the given numbers, p_0 is 100.2693 kPa using the full formula and 100.2665 kPa using the isothermal formula.

1.5 Let the length of the jet exit be L and its width W. Assuming that the jet is orientated parallel to the x-axis, Eq. (1.65a) reduces to:

$$u\frac{\partial u}{\partial x} = \frac{\partial}{\partial x}\left(\frac{u^2}{2}\right) = fv_a,$$

or, in finite difference terms:

$$v_a = \frac{1}{2f}\frac{u_1^2 - u_2^2}{L}.$$

From continuity, $\partial v_a/\partial y = -\partial\omega/\partial p$, and so the typical vertical velocity is

$$\omega = -\Delta p\frac{u_1^2 - u_2^2}{2fLW}.$$

Choose Δp to be 37.5 kPa; then a typical ω works out to be $0.25\,\mathrm{Pa\,s^{-1}}$, upward on the poleward flank of the jet and downward on the equatorward flank.

1.6 Start with Eq (1.81) and break up the absolute vorticity, density and potential temperature into a 'reference' part and an anomaly:

$$\zeta = f_0 + \zeta_A, \theta = \theta_R(z) + \theta_A, \rho = \rho_R(z) + \rho_A.$$

Substitute in and linearize by neglecting the products of anomaly quantities to obtain:

$$\frac{\rho_R q_E}{\theta_{Rz}} = (f_0 + \zeta_A)\left(1 + \frac{\theta_{Az}}{\theta_{Rz}}\right)\left(1 + \frac{\rho_A}{\rho_R}\right)^{-1}$$

$$\simeq f_0 + \zeta_A + f_0\frac{\theta_{Az}}{\theta_{Rz}} - f_0\frac{\rho_A}{\rho_R}.$$

Note from the equation of state and the ideal gas equation that $\rho = h\theta$, from which it follows that

$$q = f_0 + \zeta_A + f_0\frac{\theta_{Az}}{\theta_{Rz}} - f_0\frac{\rho_A}{\rho_R}, \text{ i.e. } q \simeq \frac{\rho_R q_E}{\theta_{Rz}}.$$

A.2 Solutions to Chapter 2

2.1 Let $Q = \bar{Q} + Q'$, $R = \bar{R} + R'$. Then:

$$QR = (\bar{Q} + Q')(\bar{R} + R') = \bar{Q}\bar{R} + Q'\bar{R} + R'\bar{Q} + Q'R'.$$

Take the time average, noting that $\overline{\overline{Q}} = \overline{Q}$ while $\overline{Q'} = 0$, and the required result follows:

$$\overline{QR} = \overline{Q}\,\overline{R} + \overline{Q'R'} \text{ or } \overline{Q'R'} = \overline{QR} - \overline{Q}\,\overline{R}.$$

This expression is useful when handling a long time series of data, since it only requires one pass through the data to compute the covariances.

2.2 If Q is conserved, then the value of Q for a parcel of fluid displaced a distance η in the y-direction represents its value at its initial position y. It follows that:

$$Q' = Q(y) - Q(y + \eta).$$

But, using Taylor's expansion:

$$Q(y + \eta) = Q(y) + \eta \frac{\partial Q}{\partial y} + O(\eta^2).$$

Then assuming that the *mean* displacement is zero, it follows that:

$$Q' = -\eta \overline{Q}_y \text{ and so } \overline{\eta^2}^{1/2} = \frac{\overline{Q'^2}^{1/2}}{\overline{Q}_y}.$$

2.3 Using Taylor series expansions, write:

$$Q_{n+1}^m = Q_n^m + \Delta x Q' + \frac{\Delta x^2}{2} Q'' + \frac{\Delta x^3}{6} Q''' + O(\Delta x^4),$$

$$Q_{n-1}^m = Q_n^m - \Delta x Q' + \frac{\Delta x^2}{2} Q'' - \frac{\Delta x^3}{6} Q''' + O(\Delta x^4),$$

where $Q' \equiv \partial Q/\partial x$, etc. Then adding and re-arranging:

$$\frac{\partial^2 Q}{\partial x^2} = \frac{Q_{n+1}^m - 2Q_n^m + Q_{n-1}^m}{\Delta x^2} + O(\Delta x^2).$$

Using this, a finite difference substitution for the diffusion equation may be written:

$$Q_n^{m+1} = Q_n^{m-1} + \frac{2K\Delta t}{\Delta x^2} (Q_{n+1}^m - 2Q_n^m + Q_{n-1}^m).$$

Since this is a linear equation, the error obeys the same finite difference equation as the analytical solution; seek solutions to the error of the form $\delta \propto e^{m\Delta t} e^{i2\pi n/N}$ or $\epsilon^m e^{i2\pi n/N}$. The condition for stability is that $|\epsilon| \le 1$. Substituting in, we have:

$$\epsilon^2 + \frac{4K\Delta t}{\Delta x^2} (1 - \cos(2\pi/N)) \epsilon - 1 = 0.$$

By inspection, this quadratic equation has distinct real roots for ϵ whose product is 1. It follows that for one root, $|\epsilon| > 1$ and so there is instability, no matter what Δt, Δx are chosen.

Using the same method, it is easy to verify that the substitution:

$$Q_n^{m+1} = Q_n^{m-1} + \frac{2K}{\Delta x^2}(Q_{n+1}^{m-1} - 2Q_n^{m-1} + Q_{n-1}^{m-1}),$$

while only first-order accurate, is stable for sufficiently small Δt. The slightly more elaborate 'Dufort–Frankel' substitution:

$$Q_n^{m+1} = Q_n^{m-1} + \frac{2K}{\Delta x^2}(Q_{n+1}^m - Q_n^{m+1} - Q_n^{m-1}Q_{n-1}^m)$$

is second-order accurate and stable for all Δt.

A.3 Solutions to Chapter 3

3.1 This is a straightforward application of Eqs. (3.3), (3.8) and (3.11). The results are: Venus–1234 Earth days, Earth–38 Earth days, Mars–0.23 Earth days. Using data in Table 10.1, these work out at 5.1, 38, and 0.23 planetary rotation periods or 5.4, 0.105 and 3.3×10^{-4} planetary years respectively. We conclude that diurnal and seasonal effects will be large on Mars, only seasonal effects will be large on Earth, and neither diurnal nor seasonal effects will be significant on Venus. The Venus result is open to question, since most of the dynamical activity is in the layers above the cloud tops, at 30 kPa. If this pressure is used in place of p_s, then a value of 4.3 Earth days is obtained for τ_E, suggesting strong diurnal effects. See Chapter 10 for more details.

3.2 The linearized, perturbed thermodynamic equations for the atmosphere and the surface may be written:

$$C_a \frac{\partial \Delta T_a}{\partial t} = 4\epsilon\sigma T_g^3 - 8\sigma T_a^3 \Delta T_a,$$

$$C_g \frac{\partial \Delta T_g}{\partial t} = -4\sigma T_g^3 \Delta T_g + 4\sigma T_a^3 \Delta T_a,$$

respectively. Assuming that ΔT_a, ΔT_b vary as $\exp(qt)$, we obtain two simultaneous homogeneous equations for q, whose determinant must be zero, i.e.,

$$C_a C_g q^2 + \left(\frac{8\sigma T_a^3}{C_a} + \frac{4\sigma T_g^3}{C_g}\right)q + \frac{16(2-\epsilon)\sigma^2 T_a^3 T_g^3}{C_a C_g} = 0.$$

It is convenient to write this in terms of $q_a = 8\sigma T_a^3/C_a$ and $q_g = 4\sigma T_g^3/C_g$; the roots of this quadratic equation are:

$$q = \frac{-(q_a + q_g) \pm \left(q_a^2 - 2(1 - \epsilon)q_a q_g + q_g^2\right)^{1/2}}{2}.$$

Assuming that $q_a \gg q_g$, the two roots can be simplified further, and we find:

$$q = -q_a - \epsilon q_g/2 \text{ or } q = -(2 - \epsilon)q_g.$$

The first root corresponds to a 'fast' relaxation time, more or less equal to the radiative equilibrium timescale introduced with Eq. (3.11), while the second represents a 'slow' timescale characteristic of the surface.

3.3 The thermodynamic equation can be written:

$$\frac{\partial T}{\partial t} + \frac{T}{\tau_E} = \frac{T_E + \Delta T_E e^{i\omega t}}{\tau_E}.$$

Here, $\omega = 2\pi/\tau_s$. This first-order ordinary differential equation is straight-forward to solve using standard methods, giving

$$T = T_E + A\exp(-t/\tau_E) + \frac{\Delta T_E}{(1 + i\omega\tau_E)}\exp(i\omega t).$$

A is a constant of integration which depends upon the initial conditions; the term proportional to A tends to zero at long times, in which case, the solution is a sinusiodal oscillation of T about T_E, with amplitude $\Delta T_E/(1 + 4\pi^2\tau_E^2/\tau_s^2)$ and phase lag $\tan^{-1}(2\pi\tau_E/\tau_s)$. If τ_E is small compared to τ_s, it follows that the time between the maximum equilibrium temperature and the maximum actual temperature is approximately τ_E. Temperatures calculated in this way, with a time-dependent forcing but no dynamical effects, are sometimes called 'radiatively determined' as distinct from 'radiative equilibrium' temperatures.

3.4 The circuit we consider consists of two adiabatic processes and two isobaric processes. The heat added during an isobaric process is $c_p\Delta T$. If the pole–equator temperature difference at the surface is ΔT_s, then the upper level temperature difference must be $\Delta T_s(p/p_s)^\kappa$. Thus, the total heat added during the circuit, $\oint T\,dS$, is $c_p\Delta T_s\{1 - (p/p_s)^\kappa\}$, and the mean rate at which it is added is $c_p\Delta T_s\{1 - (p/p_s)^\kappa\}/\tau_c$. This must balance the rate of dissipation of kinetic energy, U^2/τ_D. Finally, we have the result that $\Delta T_s = U^2\tau_c/\{c_p(1 - (p/p_s)^\kappa)\tau_D\} = 18\,\text{K}$. This is rather less than observed, although, given the crudity of the calculation, better agreement is perhaps not to be expected. The main shortcoming is the assumption that all air parcels circulate between pole and equator: the *average* air parcel will undoubtedly undergo a more limited circuit.

A.4 Solutions to Chapter 4

4.1 Take Fig. 4.1(a), showing DJF. The typical poleward wind V in the upper branch of the Hadley cell is $2.5\,\mathrm{m\,s^{-1}}$ and the horizontal extent L of the cell is $4100\,\mathrm{km}$. Hence the time taken for an air parcel to traverse the cell, L/V, is around 19 days. Rather than trying to measure the vertical velocities directly, use continuity to show that the typical vertical velocity $W \simeq (\Delta p/L)V$; the time taken to sink from the upper to the lower level is $\Delta p/W$, i.e. L/V. The time taken for a complete circuit in the Hadley cell is therefore around $4L/V$, i.e. 76 days.

4.2 This is a similar calculation, except that it is easier to estimate the vertical velocities, rather than the horizontal velocities, from Fig. 4.1(a). A typical vertical velocity in the Ferrel cell is $10^{-2}\,\mathrm{Pa\,s^{-1}}$ and so the time to go from $30\,\mathrm{kPa}$ to $100\,\mathrm{kPa}$ is 81 days. Hence the time for a complete circuit would be $4\Delta p/W$, i.e. 324 days.

4.3 The Brunt–Väisälä frequency is written in finite differences:

$$N^2 = \frac{g}{\theta_{65}} \frac{\theta_{30} - \theta_{100}}{\Delta z}.$$

Since the diagrams are plotted with pressure as the vertical coordinate, use the hydrostatic relationship to write:

$$\Delta z \simeq \frac{h\theta}{g}\Delta p,$$

so that

$$N^2 = \frac{g^2 \Delta\theta}{h(p)\theta^2 \Delta p}.$$

Take Fig. 4.2(a) for DJF, and read off some values of θ, as follows:

p kPa	θ at 30°N	θ at 60°N
100	290 K	260 K
65	310 K	285 K
30	330 K	310 K

Noting $h(p) = 3.90 \times 10^{-3}\,\mathrm{J\,K^{-1}\,Pa^{-1}\,kg^{-1}}$, then $N^2 = 1.5 \times 10^{-4}\,\mathrm{s^{-2}}$ at 30°N and $2.2 \times 10^{-4}\,\mathrm{s^{-2}}$ at 60°N.

4.4 Take Fig. 4.2(a) for DJF at 45°N by way of example: $[u]$ at $30\,\mathrm{kPa}$ is $23\,\mathrm{m\,s^{-1}}$ and $[u]$ at $100\,\mathrm{kPa}$ is about $3\,\mathrm{m\,s^{-1}}$. Thus $\partial[u]/\partial p$ is $2.86 \times 10^{-4}\,\mathrm{s^{-1}}$. The right hand side of the thermal wind equation, Eq. (1.53) is $(h/f)\partial[\theta]/\partial y$.

At 45 °N f is $1.03 \times 10^{-4}\,\mathrm{s}^{-1}$; other quantities required at the midlevel (65 kPa) were calculated in problem 4.3. We find that $(h\Delta[\theta])/(f\Delta y) = 2.85 \times 10^{-4}\,\mathrm{s}^{-1}$, which is in far better agreement with $\Delta[u]/\Delta p$ than we have any right to expect!

4.5 From $T_E = ((1-\alpha)S/\sigma)^{1/4}$, it is straightforward to calculate T_E as 256 K (globe), 173 K (pole) or 274 K (equator), so that $\Delta T_E/T_E = \Delta\theta_E/\theta_E = 0.39$. Hence, choosing $H = 10\,\mathrm{km}$, we find from Eq. (4.12) that $Y = 3460\,\mathrm{km}$. It is not quite clear whether it would be preferable to use the radiative equilibrium $\Delta\theta/\theta$ or the observed $\Delta\theta/\theta$ to estimate Y; the use of the equilibrium value compensates for the heat extracted from the poleward edge of the Hadley cell by the midlatitude eddies, ignored in the Held–Hou model.

4.6 The precipitation rate P is 3000 mm per year, i.e. $3000\,\mathrm{kg}\,\mathrm{m}^{-2}\,\mathrm{year}^{-1}$. Thus the rate of heating is $LP/3.2 \times 10^7\,\mathrm{W}\,\mathrm{m}^{-2}$, ie $243\,\mathrm{W}\,\mathrm{m}^{-2}$. According to the Held–Hou model, the heating rate for the symmetric Hadley cell will be $c_p p_s(\theta_{E0} - \theta_{M0})/(g\tau_E)$. Using the values in the text, this works out at around $6\,\mathrm{W}\,\mathrm{m}^{-2}$. To estimate the vertical velocity in the moist Hadley cell, convert the heating rate into units of $\mathrm{K}\,\mathrm{s}^{-1}$ by multiplying by $g/(c_p p_s)$. This gives $2.3 \times 10^{-5}\,\mathrm{K}\,\mathrm{s}^{-1}$. Then $w \simeq 2g/(T_E N^2) \simeq 6.8 \times 10^{-3}\,\mathrm{m}\,\mathrm{s}^{-1}$. From continuity, a typical $v \simeq (L/H)w$, ie $1.5\,\mathrm{m}\,\mathrm{s}^{-1}$.

4.7 Assuming angular momentum conservation for a flow which emerges with $[u] = 0$ at $y = y_0$, the zonal wind at any other latitude will be:

$$[u] = \frac{\Omega}{a}(y^2 - y_0^2)$$

within the Hadley cell, falling discontinuously to smaller values beyond $y = Y$ according to Held–Hou theory. Thus reading off values of ϕ_w from Fig. 4.7(a) and converting latitude into distance from the equator, y, we find for $\phi_0 = 2°$, $y_w = 2.73 \times 10^6\,\mathrm{m}$ and $[u]_w$ is therefore $84.8\,\mathrm{m}\,\mathrm{s}^{-1}$ while for $\phi_0 = 4°$, $y_w = 4.00 \times 10^6\,\mathrm{m}$ and $[u]_w$ is $173.7\,\mathrm{m}\,\mathrm{s}^{-1}$.

4.8 First deal with the units of s^2. The function $h(p)$ has units $\mathrm{J}\,\mathrm{K}^{-1}\,\mathrm{kg}^{-1}\,\mathrm{Pa}^{-1}$, and $\partial\theta/\partial p$ has units of $\mathrm{K}\,\mathrm{Pa}^{-1}$. So the units of $s^2 = h\partial\theta_R/\partial p$ may be written as $\mathrm{m}\,\mathrm{s}^{-2}\,\mathrm{Pa}^{-2}$, although there are a number of alternative ways of expressing this combination! Now consider the section shown in Fig. 4.2(a) for DJF, and take 45 °N as a representative midlatitude point. Between 100 kPa and 30 kPa, $\Delta\theta$ is around 43 K. At 65 kPa, $h(p)$ is $3.9 \times 10^{-3}\,\mathrm{J}\,\mathrm{K}^{-1}\,\mathrm{kg}^{-1}\,\mathrm{Pa}^{-1}$. Hence $s^2 \simeq h\Delta\theta/\Delta p$ works out as $2.4 \times 10^{-6}\,\mathrm{m}\,\mathrm{s}^{-2}\,\mathrm{Pa}^{-2}$.

4.9 From Eq. (4.31), we wish to estimate typical midlatitude values of the terms $(f/s^2)[\overline{u'v'}]_{yp}$ and $(h/s^2)[\overline{v'\theta'}]_{yy}$. A suitable value of s^2 was estimated in the last problem. Now, for the first term, simple finite differences enable $[\overline{u'v'}]_{yp}$ to be written as $[\overline{u'v'}]_{max}/(\Delta p \Delta y)$. The maximum momentum flux is at 25 kPa, with a value of $40\, m^2\, s^{-2}$; take $\Delta p = 75\, kPa$ and $\Delta y = 2 \times 10^6\, m$. We conclude that $(f/s^2)[\overline{u'v'}]_{yp}$ is around $1.4 \times 10^{-8}\, Pa\, m^{-1}\, s^{-1}$. In estimating the thermal term, note that the section shows $[\overline{v'T'}]$; convert to $[\overline{v'\theta'}]$ by multiplying selected values by $(p/p_R)^{-\kappa}$ if you wish, but it really does not make a significant difference. The maximum of $[\overline{v'T'}]$ is at 85 kPa, with smaller values above in the troposphere. Hence a mean $[\overline{v'\theta'}]$ is estimated to be $11.4\, K\, m\, s^{-1}$ (see problem 5.2 for a quick way of estimating a vertical mean) and $[\overline{v'\theta'}]_{yy} \simeq [\overline{v'\theta'}]_{mean}/\Delta y^2$; h/s^2 is simply $\Delta p/\Delta\theta$ (see solution 4.8) and so $(h/s^2)[\overline{v'\theta'}]_{yy}$ is estimated to be $4.6 \times 10^{-9}\, Pa\, m^{-1}\, s^{-1}$. Given the crudity of these estimates, we can only conclude that both forcings are roughly comparable. Note that both forcings have the same sign. Write the total typical forcing term as $2 \times 10^{-8}\, Pa\, m^{-1}\, s^{-1}$. To estimate the typical magnitude of the induced meridional wind $\psi_p \simeq \Delta\psi/\Delta p$, suppose the two terms on the left hand side of Eq. (4.31) have comparable magnitude; then

$$2\frac{f^2}{s^2}\psi_{pp} \simeq \frac{2f^2\Delta\psi}{s^2\Delta p^2} \simeq \text{forcing.}$$

Hence $[v]$ may estimated as roughly $s^2\Delta p/(2f^2) \times$ (forcing) which, using values of f and s for midlatitudes, works out at around $0.2\, m\, s^{-1}$. This is reasonably consistent with the values shown in Fig. 4.1 for the winter Ferrel cell.

4.10 Your picture should show a dipole of meridional circulation, with poleward flow at the level of wave breaking and equatorward flow above and below. The poleward flow creates acceleration $f[v]$ which offsets the decceleration due to wave drag. In the steady climatological state, $f[v]$ must exactly balance $\partial[u]/\partial t$ in the absence of other processes, and so we can estimate that the midlatitude $\partial[u]/\partial t$ required to balance a $1\, m\, s^{-1}$ meridional flow would be about $10\, m\, s^{-1}\, day^{-1}$.

A.5 Solutions to Chapter 5

5.1 $K = \langle u^2 \rangle/2 = 113\, J\, kg^{-1}$. The rate of destruction of K is K/τ_D which is $2.6 \times 10^{-4}\, W\, kg^{-1}$. Multiply by p_R/g to convert this to $W\, m^{-2}$, which comes out to be $2.65\, W\, m^{-2}$. This is only about 1% of the global mean insolation of $243\, W\, m^{-2}$, which is why the heating due to friction at the base of the atmosphere is not usually included in global energy budgets.

5.2 For DJF in the NH, $[\overline{v'T'}]$ has a maximum at 85 kPa, 45°N of

$14 \,\mathrm{K\,m\,s^{-1}}$. The average through the troposphere is less than this; one way of estimating this is to read off the temperature fluxes at three equispaced levels and then to apply Simpson's rule: the average of any quantity Q is then given by $(Q_0 + 4Q_1 + Q_2)/6$. Applying this formula, the depth averaged $[\overline{v'T'}]$ at $45\,°\mathrm{N}$ is probably closer to $10 \,\mathrm{K\,m\,s^{-1}}$. It has dropped to $2 \,\mathrm{K\,m\,s^{-1}}$ by $20\,°\mathrm{N}$. The distance between these latitude circles is $1.11 \times 10^5\,\mathrm{m} \times 25\,° = 2.8 \times 10^6\,\mathrm{m}$. Hence, $\partial[\overline{v'T'}]/\partial y = 8/2.8 \times 10^6 = 2.9 \times 10^{-6}\,\mathrm{K\,s^{-1}}$. Multiply by c_p to convert this to $2.9 \times 10^{-3}\,\mathrm{W\,kg^{-1}}$, i.e. $29\,\mathrm{W\,m^{-2}}$. Similar calculations can be performed for other latitudes.

5.3 From Fig. 5.5(a), the maximum $[\overline{u'v'}]$ is $40\,\mathrm{m^2\,s^{-2}}$ at $30\,°\mathrm{N}$; estimating the average as in problem 5.2 yields $20\,\mathrm{m^2\,s^{-2}}$. This falls to close to zero by $10\,°\mathrm{N}$, and so a typical $[\overline{u'v'}]$ in the subtropics is $20/(2 \times 10^6) = 10^{-5}\,\mathrm{m\,s^{-2}}$, often conveniently expressed as $1.2\,\mathrm{m\,s^{-1}\,day^{-1}}$. The depth averaged $[\overline{u}]$ is $9\,\mathrm{m\,s^{-1}}$ (see problem 5.6), and so the spin up time is $[\overline{u}]/[\overline{u'v'}]_y$, which is $9 \times 10^5\,\mathrm{s}$, i.e. about 10 days.

5.4 From Eqs. (5.6), the scales in the x- and y-directions (taken to be half a wavelength) are $(\pi g/f)(\overline{Z'^2}/\overline{v'^2})^{1/2}$ and $(\pi g/f)(\overline{Z'^2}/\overline{u'^2})^{1/2}$ respectively. For the given eddy at $45\,°\mathrm{N}$, these work out as $1850\,\mathrm{km}$ and $2620\,\mathrm{km}$ repectively, which have a ratio of 0.71. From Eq. (5.10), the phase tilt is $22.5\,°$.

5.5 Assuming that the amplitude and temperature flux do not vary with height, this is a straightforward application of Eq. (5.17), from which we deduce that δ is $42\,°$, that is, the phase difference between the $90\,\mathrm{kPa}$ wave and $50\,\mathrm{kPa}$ wave is $84\,°$. The zonal wavelength is $(2\pi g/f)(\overline{Z'^2}/\overline{v'^2})^{1/2}$, that is, $3590\,\mathrm{km}$. Therefore the horizontal separation between the upper and lower troughs is $3590 \times (84/360) = 837\,\mathrm{km}$.

5.6 $AZ = \langle h^2[\theta_A^2]/(2s^2)\rangle = \langle h[\theta_A^2]/(-2\theta_{Rp})\rangle$. From the definition of N^2, it follows that $-\theta_{Rp} = RT\theta N^2/(pg^2) = 4.7 \times 10^{-4}\,\mathrm{K\,Pa^{-1}}$, while for typical midtropospheric pressures (say $62.5\,\mathrm{kPa}$), $h(p) = (R/p)(p/p_R)^\kappa = 4.7 \times 10^{-3}\,\mathrm{J\,K^{-1}\,Pa^{-1}\,kg^{-1}}$. From Fig. 5.15, $AZ = 3.7 \times 10^6\,\mathrm{J\,m^{-2}}$, i.e., $370\,\mathrm{J\,kg^{-1}}$. Hence, a global mean value of $[\theta_A^2]^{1/2}$ is $8.6\,\mathrm{K}$. If we assume that the temperature varies with latitude according to Eq. (4.4), then $[\theta_A^2]^{1/2} = \Delta\theta/3$ and so $\Delta\theta = 26\,\mathrm{K}$. A typical θ^* is related to AE in much the same way. Since AE is observed to be $1.2 \times 10^6\,\mathrm{J\,m^{-2}}$, i.e., $118\,\mathrm{J\,kg^{-1}}$, we have $\langle\theta^{*2}\rangle \simeq 24\,\mathrm{K^2}$, i.e., a typical temperature fluctuation is $4.9\,\mathrm{K}$.

The kinetic energies are more straightforward. From $KZ = 4.7\times 10^5\,\mathrm{J\,m^{-2}}$, that is, $46\,\mathrm{J\,kg^{-1}}$, the typical zonal wind must be $9.6\,\mathrm{m\,s^{-1}}$, and from $KE = 7.5 \times 10^5\,\mathrm{J\,m^{-2}} = 74\,\mathrm{J\,kg^{-1}}$, the typical eddy wind is $12.2\,\mathrm{m\,s^{-1}}$.

5.7 This problem requires estimates of the terms $[\overline{u}]_y[\overline{u^*v^*}]$ and

$(h/s)^2[\bar{\theta}]_y[\overline{v^*\theta^*}]$. Take values for DJF at 45°N: $[\bar{u}]_y$ is $15/2 \times 10^6 = 7.5 \times 10^{-6}\,\mathrm{s}^{-1}$ and the average $[\overline{u^*v^*}]$ is $20\,\mathrm{m}^2\,\mathrm{s}^{-2}$. Hence the magnitude of the integrand of the first term is $1.5 \times 10^{-4}\,\mathrm{W\,kg}^{-1}$, i.e. $1.53\,\mathrm{W\,m}^{-2}$. The factor $(h/s)^2$ can be estimated from values given in problem 5.6 and comes to $10\,\mathrm{J\,K}^{-2}\mathrm{kg}^{-1}$. A typical midlatitude $[\bar{\theta}]_y$ is $-30/(3 \times 10^6) = -10^{-5}\,\mathrm{K\,m}^{-1}$, while for $[\overline{v^*\theta^*}]$ take a value of $10\,\mathrm{K\,m\,s}^{-1}$. Hence the thermal term is $-10^{-3}\,\mathrm{W\,kg}^{-1}$ or $-10.2\,\mathrm{W\,m}^{-2}$. We conclude that the thermal (or 'baroclinic') conversion dominates. Both estimates are likely to be overestimates since they are based on midlatitude values rather than global means, but the ratio of the two terms is unlikely to change much with a better global mean.

A.6 Solutions to Chapter 6

6.1 Figure 6.4(a) shows two maxima and two minima around 60°N, suggesting wavenumber 2 is the dominant wavenumber at 60°N. Now the dimensionless wavenumber n is related to the dimensional wavenumber k by:

$$k = n/(a\cos\phi),$$

ϕ being latitude. So we have $k = 6.28 \times 10^{-7}\mathrm{m}^{-1}$. From the same diagram, we note that the phase tilt between 100 kPa and 30 kPa is around 40° of longitude, i.e. $\delta = 10°$ and $\Delta p = 35\,\mathrm{kPa}$ in Eq. (5.17). In order to invert Eq. (5.17) and estimate $[\bar{v}^*\overline{T}^*]$, we must estimate the amplitude A of the steady waves; inspection of the diagram suggests that 150 m might be a good value. In that case, $[\bar{v}^*\overline{T}^*]$ is around $6\,\mathrm{K\,m\,s}^{-1}$. This is entirely comparable with the tropospheric average for 60°N which may be read off Fig. 6.5(a).

6.2 From Eq. (6.8),

$$Z = \frac{fU}{(\beta/K^2 - U)}\frac{h}{H_0}.$$

To estimate the wavenumber, assume the ridge represents half a wavelength in each direction, so that $k = \pi/L_x$, $l = \pi/L_y$, i.e. $k = 3.14 \times 10^{-6}\,\mathrm{m}^{-1}$, $l = 6.28 \times 10^{-6}\,\mathrm{m}^{-1}$ and $K = \sqrt{k^2 + l^2} = 3.20 \times 10^{-6}\,\mathrm{m}^{-1}$. At 45°N, $\beta = 2\Omega\cos\phi/a = 1.62 \times 10^{-11}\,\mathrm{m}^{-1}\mathrm{s}^{-1}$. Hence, $Z = -1.15 \times 10^{-4}h/H_0\,\mathrm{s}^{-1}$. Assume H_0 is 7 km; then $Z = 2.47 \times 10^{-5}\,\mathrm{s}^{-1}$. The negative sign indicates that there is anticyclonic vorticity above the summit of the mountain. To estimate the typical meridional wind, write $v^* \simeq k\Psi$, where the streamfunction amplitude $\Psi = -Z/K^2 = 2.41 \times 10^6\,\mathrm{m}^2\,\mathrm{s}^{-1}$. Then $v^* \simeq 7.6\,\mathrm{m\,s}^{-1}$.

6.3 When drag is included, the vorticity amplitude is

$$Z_D = \frac{fU(h/H_0)}{((U - \beta/K^2)^2 + (1/\tau_D k)^2)^{1/2}}.$$

Substituting values, we note that $(\tau_D k)^{-1} = 0.72\,\mathrm{m\,s}^{-1}$ and $(U - \beta/K^2) = 13.42\,\mathrm{m\,s}^{-1}$; we anticipate that the effect of friction is not very large in this case. In fact, comparing $|Z|$ and $|Z_D|$ and using the binomial theorem, we find that the difference between the vorticity amplitude with and without friction is only around 0.3%.

6.4 On the β-plane, $K_s^2 = \beta/U = (2\Omega\cos\phi/a)/(U_0\cos\phi) = 2\Omega/(aU_0)$. This is independent of latitude ϕ or y. Since $l = \sqrt{K^2 - k^2}$, and k is conserved following the ray, it follows that the direction of the ray does not vary with y, i.e. it is a straight line. If this argument is recast in full spherical geometry, it turns out that the rays follow great circle paths. For the given flow, $K_s = 1.07 \times 10^{-6}\,\mathrm{m}^{-1}$. It follows that the longest propagating wave has zonal wavenumber $k = K_s = 1.07 \times 10^{-6}\,\mathrm{m}^{-1}$, i.e. $n = 4.82$. Zonal wavenumber 4 is therefore the largest propagating wavenumber at 45°N, and meridional wavenumber $l = 5.9 \times 10^{-7}\,\mathrm{m}^{-1}$. A packet with these wavenumbers will propagate at an angle $\tan^{-1}(l/k) = 34°$ to the zonal direction.

6.5 The meridional component of the group velocity of a steady Rossby wave is $c_{gy} = 2U\cos\alpha\sin\alpha = 2Ukl/K^2$. For this case, $k = 6.66 \times 10^{-7}\,\mathrm{m}^{-1}$ and $K_s^2 = \beta/U = 1.08 \times 10^{-12}\,\mathrm{m}^{-2}$, so that $l = (K_s^2 - k^2)^{1/2} = 7.97 \times 10^{-7}\,\mathrm{m}^{-1}$. Hence, $c_{gy} = 14.7\,\mathrm{m\,s}^{-1}$, and the time taken to cover 20° of latitude, i.e. $2.2 \times 10^6\,\mathrm{m}$, is $1.5 \times 10^5\,\mathrm{s}$ or 1.7 days.

6.6 The *speed* of a wave packet was given by Eq. (6.19): this is simply $d_g s/dt$ (where d_g/dt is the rate of change following the packet):

$$\frac{d_g s}{dt} = 2U\cos\alpha.$$

Since $\tan\alpha = l/k$, it follows that:

$$\frac{d_g\alpha}{dt} = \frac{1}{(1 + l^2/k^2)}\frac{1}{k}\frac{d_g l}{dt} = \frac{k}{K_s^2}\frac{d_g l}{dt}.$$

Noting that $l = (K_s^2 - k^2)^{1/2}$ and that k is conserved by the packet, we find that:

$$\frac{d_g l}{dt} = \frac{k}{2K_s^2 l}\frac{d(K_s^2)}{dy}\frac{d_g y}{dt} = \frac{k}{2K_s^2 l}\frac{d(K_s^2)}{dy}c_{gy}.$$

Finally, noting that $\cos \alpha = k/K_s$, $\sin \alpha = l/K_s$ and that $c_{gy} = 2U \cos \alpha \sin \alpha$, we obtain the required second equation:

$$\frac{d_g \alpha}{dt} = \frac{U \cos^4 \alpha}{k^2} \frac{d}{dy}(K_s^2).$$

This formulation has certain practical advantages, principally that it avoids the difficulty of integrating Eq. (6.24b) near the poleward turning point of the ray, where $l \to 0$ and $k \to K_s$

6.7 Take a value of $[\overline{v'\theta'}]$ of 13 K m s^{-1} as representative of the lower troposphere. Take f as 10^{-4} s^{-1} and $s^2/h = 5.7 \times 10^{-4}$ K Pa^{-1} (see problem 5.6). Then the vertical component of the Eliassen–Palm flux is $F_2 = -hf[\overline{v'\theta'}]/s^2$, i.e. 2.3 m s^{-2}Pa^{-1}. The zonal flow acceleration is $\partial F_2/\partial p$; assuming F_2 is small in the upper troposphere, with $\Delta p \simeq 70$ kPa, this means that the acceleration is 3.3×10^{-5} m s^{-2} or 2.8 m s^{-1} day^{-1}. From Fig. 5.5, a typical $[\overline{v'\theta'}]$ is around 20 m^2 s^{-2} which varies on a typical horizontal scale of 2×10^6 m. Hence the acceleration due to eddy momentum flux convergence is around 10^{-5}m s^{-2} or 0.9 m s^{-1} day^{-1}. This is not negligible, but we must conclude that the eddy acceleration is dominated by the vertical component of the Eliassen–Palm flux.

6.8 From the definition of the Eliassen–Palm flux in pressure coordinates, Eq. (6.73), it follows that:

$$\mathbf{F} \cdot \nabla[\bar{u}] = -[\bar{u}]_y \overline{[u^*v^*]} - \frac{hf}{s^2} \overline{[v^*\theta^*]}[\bar{u}]_p.$$

The zonal mean thermal wind relationship, Eq. (1.53), can be written in the same notation as $[\bar{u}]_p = h[\bar{\theta}]_y/f$, and so:

$$\mathbf{F} \cdot \nabla[\bar{u}] = -[\bar{u}]_y \overline{[u^*v^*]} - \frac{h^2}{s^2} \overline{[v^*\theta^*]}[\bar{\theta}]_y.$$

Hence, taking the global average, we find the required result, that

$$\frac{d\langle E \rangle}{dt} = \langle \mathbf{F} \cdot \nabla[\bar{u}] \rangle.$$

(see Eq. (5.37)).

A.7 Solutions to Chapter 7

7.1 Note that a pure Kelvin wave has zero $[u^*v^*]$ since $v^* = 0$, and the planetary and Rossby gravity waves all have zero $[u^*v^*]$ since they have no meridional phase tilt. However, the combination of a Kelvin wave and a planetary wave does have strong phase tilts in the vicinity of the heating region, such as to generate an equatorward flux of westerly momentum (or, if you prefer, a poleward flux of easterly momentum). Analysis of the linear numerical inegrations used to generate Fig. 7.4 shows that the maximum momentum fluxes are indeed of this sign, with largest values just north and south of the equator, falling to small values on the meridional scale $(c_0/2\beta)^{1/2}$.

7.2 From Fig. 7.3, we note that the dispersion curves for the $n = -1$ and $n = 1$ modes pass through the origin and are virtually linear in the vicinity of the origin, i.e. as $\omega \to 0$ and $k \to 0$. In this limit, Eq. (7.10) gives:

$$\frac{(2n+1)}{c_0}\omega = -k, \text{ or } c_g = -\frac{c_0}{(2n+1)}.$$

For an equivalent depth of 400 m, $c_0 = 63 \text{ m s}^{-1}$. For $n = -1$ (the Kelvin wave) $c_g = 63 \text{ m s}^{-1}$ and for $n = 1$, the planetary waves, $c_g = -21 \text{ m s}^{-1}$. Then with $\tau_D = 5 \text{ days} = 4.3 \times 10^5 \text{ s}$, the length scales $c_g\tau_D$ are 27 000 km to the east of the source region and 9 000 km to the west of the source region.

7.3 Since $\mathbf{E}$ is predominantly zonal, $\nabla \cdot \mathbf{E}$ will be dominated by $-2M_x$. From Fig. 7.17 at 45 °N in the Atlantic storm track, the following values of $-2M$ can be extracted:

—	4.7×10^6 m	—	3.5×10^6 m	—
$17\text{m}^2\text{s}^{-2}$		$57\text{m}^2\text{s}^{-2}$		$23\text{m}^2\text{s}^{-2}$
105 °W		45 °W		0 °E

From these figures, we estimate $\nabla \cdot \mathbf{E} = 8.5 \times 10^{-6} \text{ m s}^{-2}$ or 0.74 m s^{-1} day^{-1} at the western end of the storm track and $-9.7 \times 10^{-6} \text{ m s}^{-2}$ or -0.84 m s^{-1} day^{-1} at the eastern end.

7.4 From Fig. 7.18, we note that a typical vertical component of $\mathbf{E}$ in the Atlantic storm track is 2.5 Pa m s^{-2} and a typical x-component at 25 kPa is 60 m^2 s^{-2}. Hence the slope in the x-direction is 4.2×10^{-2} Pa m^{-1}, i.e. 42 kPa per 1000 km. This is probably an underestimate, since the x-component is large in the upper troposphere where the p-component is becoming small. Using the hydrostatic relation, the slope is therefore at least 3.5×10^{-3}. The typical y-component of the $\mathbf{E}$-vector is smaller, not more than about 20 m^2 s^{-2} (see Fig. 7.8(d)), and so the slope is around 180 kPa per 1000 km or 1.5×10^{-2}. From Fig. 4.2(a), note that the typical slope of the midlatitude isentropes is around 20 kPa per 1000 km or 1.6×10^{-3}.

7.5 For simplicity, suppose that the temperature wave is a 'square wave', with temperature $T_0 + \Delta T$ for half its wavelength and temperature $T_0 - \Delta T$

for the other half. The zonal mean temperature is therefore T_0. Now the humidity mixing ratio will be

$$r = \frac{R\epsilon e_{s0}}{p} \exp\left[\frac{L}{R_v T_0}\left(\frac{1}{T_0} - \frac{1}{T}\right)\right] \simeq \frac{R\epsilon e_{s0}}{p} \exp\left[\frac{\pm L\Delta T}{R_v T_0^2}\right]$$

for $|\Delta T| \ll T_0$, where R is the (constant) relative humidity. It follows that the zonal mean humidity mixing ratio is

$$[r] = \frac{R\epsilon e_{s0}}{p} \cosh\left(\frac{L\Delta T}{R_v T_0^2}\right).$$

However, the apparent zonal mean saturated humidity mixing ratio, based upon the zonal mean temperature, will simply be $\epsilon e_{s0}/p$, so that the apparent zonal mean relative humidity R_a is

$$R_a = R\cosh\left(\frac{L\Delta T}{R_v T_0^2}\right).$$

Now $R_v T_0^2/L$ is 13.8 K. If R is 0.9, it follows that if ΔT exceeds about 6 K, R_a will be greater than 1.

7.6 This problem makes use of a result proved in problem 2.2. If the motions of air parcels were purely horizontal and q were conserved by fluid parcels, then the rms meridional displacement would be $\overline{\eta'^2}^{1/2} = \overline{q'^2}/\bar{q}_y$. Analogously, if the motions were purely vertical and q was conserved, then $\overline{p'^2} = \overline{q'^2}/\bar{q}_p$. From Fig. 7.20(a) in the vicinity of 30°N in the lower troposphere, $\partial q/\partial y \simeq 2.8 \times 10^{-9}$ m^{-1}, while $\partial q/\partial p \simeq 2.5 \times 10^{-7}$ Pa^{-1}. Hence the meridional displacement would be around 360 km, while the vertical displacement would be around 4 kPa. In reality, the midlatitude motions are at a small angle to the horizontal (see Section 5.4) and the actual dispersal of fluid elements will be in both the vertical and the horizontal.

7.7 From

$$e_s = e_{s0} \exp\left\{\frac{L}{R_v}\left(\frac{1}{T_0} - \frac{1}{T}\right)\right\},$$

we find that $e_s(200\,\text{K})$ is 0.44 Pa. If the tropopause is at a pressure of 15 kPa, it follows that $r = \epsilon e_s/p$ is 1.8×10^{-5}, i.e. 18 parts per million by mass.

A.8 Solutions to Chapter 8

8.1 Calculate variances and co-variances from formulae such as:

$$\overline{U'^2} = \overline{U^2} - \overline{U}^2, \quad \overline{U'B'} = \overline{UB} - \overline{U}\,\overline{B}, \quad \text{and so on.}$$

The symmetric matrix of variances and covariances is:

$$\begin{pmatrix} \overline{U'^2} & \overline{U'A'} & -\overline{U'B'} \\ \overline{U'A'} & \overline{A'^2} & \overline{A'B'} \\ \overline{U'B'} & \overline{A'B'} & \overline{B'^2} \end{pmatrix} = \begin{pmatrix} 0.2880 & 0.0732 & -0.0843 \\ 0.0732 & 0.7581 & -0.0507 \\ -0.0843 & -0.0507 & 0.8213 \end{pmatrix}$$

The eigenvalues and eigenvectors of this matrix can be determined by standard methods (see, for example, Chapter 11 of Press *et. al.*, (1992)). The eigenvalues are 0.87, 0.73 and 0.27, so that they account for 47%, 39% and 14% of the variance of the time series respectively. The corresponding eigenvectors are $(1.20, 0.46, 1.18)$, $(-0.13, 0.85, -0.50)$ and $(0.98, 0.04, -0.19)$. The first two are more or less parallel to the (A, B) plane and the last is roughly parallel to the U axis. It is worth remarking that the covariances, and hence the eigenvectors, are quite sensitive to details such as the length of the calculation and the integration scheme for this chaotic system.

8.2 For the configuration suggested, $k = 2/(a \cos 45) = 4.44 \times 10^{-7}\,\text{m}^{-1}$ and $l = \pi/5 \times 10^6 = 6.28 \times 10^{-7}\,\text{m}^{-1}$. Thus $\beta/K^2 = 27.4\,\text{m s}^{-1}$ (which is the wind speed at which resonance occurs) and $(\tau_D k)^{-1} = 3.0\,\text{m s}^{-1}$. Assuming a value of $H = 7\,\text{km}$, then $\mathscr{D} = 15.1\,\text{m s}^{-1}\,\text{day}^{-1}$ at resonance. At this wind speed, the friction drag is $4.5\,\text{m s}^{-1}\,\text{day}^{-1}$, smaller than $\mathscr{D}$, which is a necessary (but not sufficient) condition for resonance. Drawing the graph of $\mathscr{D}$ vs U and of $(U_E - U)/\tau_D$ vs U, multiple equilibria will be found, with a 'blocked' state for $U = 24\,\text{m s}^{-1}$ and a zonal state at $U = 46\,\text{m s}^{-1}$, with an unstable equilibrium at $U = 34\,\text{m s}^{-1}$.

8.3 For stationary Rossby waves, $c_{gy} = U \sin(2\alpha)$. For the DJF flow at 25°N and 25 kPa, $[u]$ is about $25\,\text{m s}^{-1}$ and $\beta = 2\Omega \cos(\phi)/a$ is $2 \times 10^{-11}\,\text{m}^{-1}\,\text{s}^{-1}$. For simplicity in this order of magnitude calculation, assume that $[q]_y$ is dominated by β. It follows that $K_s = (\beta/[u])^{1/2} = 9.1 \times 10^{-7}\,\text{m}^{-1}$. For a zonal wavenumber m disturbance, $k = m/a \cos(\phi)$. Hence, for $m = 3$, $k = 5.2 \times 10^{-7}\,\text{m}^{-1}$, $l = (K_s^2 - k^2)^{1/2} = 7.5 \times 10^{-7}\,\text{m}^{-1}$. Hence $\alpha = \tan^{-1}(l/k) = 55°$ and $c_{gy} = 23\,\text{m s}^{-1}$. The time for the packet to travel $2 \times 10^6\,\text{m}$ in the meridional direction is therefore $2 \times 10^6/23$ s or about one day. For a wavenumber 1 disturbance, $c_{gy} = 9.4\,\text{m s}^{-1}$ and the time taken is about 2.5 days.

A.9 Solutions to Chapter 9

9.1 The scale height $H = RT/g$ is about 6.4 km for the stratosphere. For an isothermal atmosphere, $z = -H \ln(p/p_s)$. Then, the height of the 3 kPa surface is 22 km, of the 1 kPa surface is 29.5 km and that of the 0.1 kPa surface is 44.2 km.

9.2 The Brunt–Väisälä frequency is given by:

$$N^2 = \frac{g}{\theta}\frac{\partial\theta}{\partial z} = \frac{g}{T}(\Gamma_a - \Gamma),$$

where Γ_a is the adiabatic lapse rate g/c_p and Γ is the actual lapse rate. In the stratosphere, assume that $\Gamma = 0$. Then $N^2 = g^2/(c_p T)$. For a temperature of 220 K, N^2 is $4.4 \times 10^{-4}\,\mathrm{s}^{-2}$. The period of buoyancy oscillation is typically $2\pi/N$, which is around 5 minutes, compared to around 10 minutes in the troposphere.

9.3 Use the thermal wind equation in the form:

$$[u]_z = (g/fT)[T]_y.$$

A typical temperature difference between 70 °S and 50 °S is around 20 K. Assuming a mean temperature of around 200 K and that the wind at 15 km is close to zero, it follows that the wind at 30 km is $58.5\,\mathrm{m\,s}^{-1}$, in reasonable accord with Fig. 9.6(b). Now elementary synoptic meteorology gives the relationship between the gradient wind u_r and the geostrophic wind u_g to be:

$$u_r^2 + (fr)u_r - (fr)u_g = 0.$$

Only the positive root is relevant, the negative root being inertially unstable. Assuming that the vortex is circular and is centred on the pole, $r = 3.33 \times 10^6\,\mathrm{m}$ and so (fr) is $420\,\mathrm{m\,s}^{-1}$. It follows that for $u_g = 58.5\,\mathrm{m\,s}^{-1}$, $u_r = 52\,\mathrm{m\,s}^{-1}$. We conclude that the gradient wind makes a correction of not more than about 10% to the strength of the polar night jet.

9.4 Evanescent Rossby waves depend upon height as $\exp\{-(\mu - 1/2H)z\}$, where μ is given by Eq. (6.47):

$$\mu = \frac{N}{f}\left(\frac{f^2}{4N^2H^2} + k^2 + l^2 - \frac{\beta}{U}\right).$$

Taking $k = 5/(a\cos\phi)$, $l = 0$, $H = 6.4\,\mathrm{km}$ and $\beta = 5.88 \times 10^{-11}\,\mathrm{m}^{-1}\,\mathrm{s}^{-1}$, we find

$$\mu = 179 \times (1.91 \times 10^{-13} + 1.49 \times 10^{-12} - 5.88 \times 10^{-13})^{1/2} = 1.87 \times 10^{-4}\,\mathrm{m}^{-1}.$$

The e-folding height for amplitude, $(\mu - 1/2H)^{-1}$ is therefore 9.2 km, while the e-folding height for energy per unit volume, μ^{-1} is 5.3 km.

9.5 The condition for vertical propagation is

$$U < \frac{\beta}{k^2 + l^2 + f^2/(4N^2H^2)}.$$

Taking $k = 2/(a \cos \phi)$, $l = 0$, and the usual values for other quantities gives $U < 18 \, \mathrm{m \, s^{-1}}$ for propagation.

9.6 Following the theory of Section 6.4, the perturbation streamfunction for vertically evanescent Rossby waves may be written:

$$\psi^* = A e^{ikx} e^{(1/2H - \mu)z} = A e^{ikx} e^{\mu' z}.$$

Then the eddy meridional wind, $\partial \psi^* / \partial x$, is:

$$v^* = \frac{A k e^{\mu' z}}{2} \left(i e^{ikx} - i e^{-ikx} \right),$$

where we have explicitly taken the real part of the streamfunction. Similarly, the eddy temperature is:

$$T^* = \frac{\mu' f T_0 A}{2g} \left(e^{ikx} + e^{-ikx} \right).$$

The product $v^* T^*$ is therefore:

$$v^* T^* = \frac{\mu' f T_0 A^2 e^{-2\mu' z}}{2g} \left(i e^{2ikx} - i e^{-2ikx} \right),$$

which is purely wavelike. Consequently, $[v^* T^*]$ is identically zero.

9.7 Consider the levels around $1 \, \mathrm{kPa}$ (around $30 \, \mathrm{km}$ above the ground) at $60 \, °\mathrm{S}$. At this latitude, $\beta = 2\Omega \cos \phi / a = 1.14 \times 10^{-11} \, \mathrm{m^{-1} \, s^{-1}}$, which is the planetary contribution to $[q]_y$. Now, the zonal winds at $70 \, °\mathrm{S}$, $60 \, °\mathrm{S}$ and $50 \, °\mathrm{S}$ are $50 \, \mathrm{m \, s^{-1}}$, $70 \, \mathrm{m \, s^{-1}}$ and $68 \, \mathrm{m \, s^{-1}}$ respectively. Using simple second-order finite differences, $[u]_{yy} = (50 - 2 \times 70 + 68)/(1.23 \times 10^{12}) = -1.79 \times 10^{-11} \, \mathrm{m^{-1} \, s^{-1}}$. The vertical term may, for present order of magnitude purposes, be taken as $(f^2/N^2)[u]_{zz}$. Reading off values of $[u]$ at $37.4 \, \mathrm{km}$, $30 \, \mathrm{km}$ and $22.6 \, \mathrm{km}$ (75, 70 and $50 \, \mathrm{m \, s^{-1}}$ respectively) and using the same finite difference formula yields a value of $(f^2/N^2)[u]_{zz}$ of $-1.1 \times 10^{-11} \, \mathrm{m^{-1} \, s^{-1}}$. We conclude that all three terms are comparable, with $[u]_{yy}$ being somewhat the largest. This contrasts with the troposphere, where the smaller value of N^2 and the smaller vertical scale of the jet means that $(f^2/N^2)[u]_{zz}$ is generally the dominant term in $[q]_y$ in the vicinity of the major jets.

9.8 Start from the dispersion relation for Rossby waves (Eq. (6.61)):

$$\omega = U k - \frac{\beta k}{K_T^2}, \quad \text{where } K_T^2 = k^2 + l^2 + \frac{f^2}{N^2} \left(m^2 + \frac{1}{4H^2} \right).$$

For steady ($\omega = 0$) waves, $K_T^2 = \beta/U$. Differentiating the dispersion relation with respect to m gives the vertical component of the group velocity, c_{gz} (Eq. (6.63)):

$$c_{gz} = \frac{2\beta f^2 km}{N^2 K_T^2}.$$

Substituting for K_T, we find:

$$c_{gz} = \frac{2U^2(f^2/N^2)km}{\beta}, \text{ as required.}$$

For a zonal wavenumber 1 disturbance at 60°N, assume $l = 0$, $N^2 = 4 \times 10^{-4}\,\mathrm{s}^{-2}$ (see problem 9.2). Then $\beta = 1.14 \times 10^{-11}\,\mathrm{m}^{-1}\mathrm{s}^{-1}$, $(f/N)^2 = 4.0 \times 10^{-5}$. Then $K_T^2 = 3.8 \times 10^{-13}\,\mathrm{m}^{-2}$, $k^2 = 1/(a\cos\phi)^2 = 3.14 \times 10^{-7}\,\mathrm{m}^{-2}$ and so $m = 3.09 \times 10^{-5}\,\mathrm{m}^{-1}$. Hence $c_{gz} = 6.13 \times 10^{-2}\,\mathrm{m\,s}^{-1}$ and so the time taken for a packet of waves to propagate from the tropopause, at 15 km, to 50 km is about $35 \times 10^3/c_{gz} = 6.6$ days.

9.9 In the lower troposphere, the poleward temperature flux reaches a maximum some 3 km above the ground. Ignoring, for order of magnitude purposes, the vertical variation of ρ_R, the condition for the residual circulation in the lower troposphere to vanish is:

$$[v] = \frac{1}{\theta_{Rz}}\frac{[v^*\theta^*]}{\Delta z} = \frac{g}{\theta_R N^2}\frac{[v^*\theta^*]}{\Delta z}.$$

Substituting in values, we find a value of $[v^*\theta^*]$ of around $4.5\,\mathrm{K\,m\,s}^{-1}$. The observed values are larger than this, so we conclude that the residual circulation is certainly thermally direct, reversing the thermally indirect Ferrel cell, in the mid-latitudes.

A.10 Solutions to Chapter 10

10.1 Assume that the streamlines have a sinusoidal shape; since at the maximum amplitude of the waves, $v^* = \Delta U$ while the background flow $[u] = \Delta U$, it follows that they cross the $y = 0$ axis at an angle of 45°. The maximum displacement L is to be identified with the 'mixing length', which is related to the wavelength of the waves by $\pi L = \lambda/2$. Now, $\lambda = 2\pi/k$, and so $L = k^{-1}$. The most unstable Eady wave has $k = 1.61/L_R$ (see Eq. (5.50) et. seq.). Hence the required result: $L = 0.61L_R$.

10.2 Let us assume that the horizontal eddy scale L is half the most unstable wavelength, i.e. $L = \pi NH/(1.61f) = 1.95NH/f$. Using the data in Tables 10.1 and 10.2 yields $f = 1.65 \times 10^{-4}\mathrm{s}^{-1}$ and the atmospheric scale height $H = 20.6$ km. Then $N = Lf/(1.95H) = 4.93 \times 10^{-3}\,\mathrm{s}^{-1}$. In terms

of the adiabatic and actual lapse rates, the Brunt–Väisälä frequency can be
written

$$N^2 = \frac{g}{T} (\Gamma_a - \Gamma)$$

so that

$$\Gamma_a - \Gamma = 0.133 \, \text{K km}^{-1}.$$

10.3 Express the cyclostrophic thermal wind equation, Eq. (10.23) in height
coordinates:

$$\frac{\partial [u^2]}{\partial z} = \frac{ga}{T \tan \phi} \frac{\partial [T]}{\partial y}.$$

For this estimate, assume that $\partial [T]/\partial y$ and $[u]$ are zero below 50 km. For
the slab between 50 km and 70 km, write $\partial [T]/\partial y \simeq \Delta T/a$. Then

$$\Delta [u^2] = \frac{g \Delta z}{\tan \phi} \frac{\Delta T}{T}.$$

Substituting values from Tables 10.1 and 10.2 gives a value of $[u]$ at 70 km
of $105 \, \text{m s}^{-1}$.

10.4 If the number of jets is n, and the pole-equator distance is $\pi a/2$,
then the jet width is $\pi a/(2n)$. But from Eq. (10.25) *et. seq.*, the jet width
is $\pi (U/\beta)^{1/2}$. Hence, $U = \beta a^2/(4n^2)$. For Neptune's midlatitudes, $\beta = 6.31 \times 10^{-12} \, \text{m}^{-1} \text{s}^{-1}$, and so for $n = 4$, we find $U = 60 \, \text{m s}^{-1}$.

10.5 In the limit of $\alpha \gg 1$, Eq. (10.17) reduces to $\gamma^2 B^3 = B_E^3$ and so
$B = B_E/\gamma^{2/3}$, $V = 0.55 B_E/(\gamma^{1/3} \delta)$. The slope of the isentropes is B/V, i.e.
$1.82 \delta/\gamma^{1/3}$. The slope of the eddy heat flux vector is $[w^* \theta^*]/[v^* \theta^*]$. From
Eqs. (10.10) and (10.11), this is found to be $0.62 \delta/\gamma^{1/3}$, which is about 1/3
that of the isentropes, as we were asked to show.

10.6 From Eq. (4.18), the depth of the Ekman boundary layer is $D = (2K/f)^{1/2}$ and so the eddy diffusion parameter is $f D^2/2$. For midlatitudes,
this gives a K of $50 \, \text{m}^2 \, \text{s}^{-1}$. Take the width of the baroclinic zone as $3 \times 10^6 \, \text{m}$;
then the Taylor number is around 2×10^{17}. The thermal Rossby number,
$\Delta U/(f L)$ is 6.7×10^{-2}, assuming an average value of ΔU of $20 \, \text{m s}^{-1}$. These
values put us well outside the range of the experimental parameters shown in
Fig. 10.17 and we might speculate that they indicate a highly irregular flow,
as observed. But this is an unjustifiable extrapolation to extreme parameters
and a different geometry.

10.7 From Fig. 4.1, the surface winds over the southern oceans have a
maximum of about $7 \, \text{m s}^{-1}$ at 50 °S and fall to zero at 70 °S. The maximum

wind stress is therefore $\rho C_D [u]^2 = 1.25 \times 10^{-3} \times 50 = 6.3 \times 10^{-2}$ Pa. The wind stress curl is $\tau_x/L = 6.3 \times 10^{-2}/(2 \times 10^6) = 3.1 \times 10^{-8}$ Pa m^{-1}. The vorticity forcing is therefore $3.1 \times 10^{-8}/(10^3 \times 1.5 \times 10^3) = 2.1 \times 10^{-14}$ s^{-2}. Hence, the poleward current is $2.1 \times 10^{-14}/\beta = 1.8 \times 10^{-3}$ m s^{-1}.

10.8 For an incompressible fluid such as sea-water, the Brunt–Väisälä frequency may be written $N^2 = -(g/\rho)\partial\rho/\partial z$. In this problem,

$$\rho = \rho_0\{1 - \alpha(T - T_0)\}$$

(the real equation of state for sea water is a good deal more complicated than this, and includes quadratic terms). Hence, $N^2 = \alpha g \partial T/\partial z \simeq \alpha g \Delta T/\Delta z$. From the given values, N^2 is 7.2×10^{-6} s^{-1}. The Rossby radius NH/f (take H as 1.5 km) is therefore around 40 km.

Bibliography

The purpose of this bibliography is to suggest fuller and more detailed reading on the topics covered for the interested student. No attempt has been made to give an exhaustive cover of all the available literature, though references in the papers and articles cited should lead the student into this. I have generally tried to give an accessible and readable account of topics rather than be pedantic about priority. Inevitably, the references are biased towards papers by my colleagues. This is not intended as a slight to the many scientists who I do not know so well, so I hope they will accept my apologies if they feel unjustly omitted. But I am persuaded that my approach is more appropriate for a text that is intended as a teaching tool rather than a research monograph. References for each chapter are given when possible, together with sources for the specific topics covered in each section.

For the student who wishes to research the literature on a particular topic, I hope that these references will serve as a starting point. Many of the texts and reviews I have cited give thorough lists of references to earlier work. Careful research with the Science Citations Index will bring the student up to date.

Global circulation data

A number of Atlases of global circulation data exist. Where ever possible, I have re-plotted my figures from the data acquired for the Atlas of Hoskins *et al.* (1989), which is based on the European Centre for Medium Range Weather Forecasts operational analyses. This is not because it is necessarily superior to other compilations, but for the sake of consistency between the different fields, and also because I have easier access to the fields to compute other diagnostics. Other useful compilations are due to Oort (1983) and to Lau (1984).

Chapter 1: The governing physical laws

This chapter is a summary of the major results which will be derived in any good text on dynamical meteorology. The book by Holton (1992) is a comprehensive and clear account. More detail will be found in Pedlosky (1987) and Gill (1982), though these latter authors are perhaps more motivated by ocean than by atmospheric applications. The basic fluid dynamics, especially the relationship between the large scale flow and the subsynoptic scale and small scale turbulence, is treated thoroughly by Brown (1991).

Section 1.1: Thermodynamics is dealt with more fully in Wallace and Hobbs (1977). An alternative, somewhat more advanced account, is given in the opening chapters of Rogers and Lau (1989).

Section 1.8: The 'Q-vector' form of the omega equation is discussed by Hoskins *et. al.* (1978), and compared with the more traditional versions. Section 1.9: Hoskins *et. al.* (1985) review the modern renewal of interest in potential vorticity thinking, which follows on the improvements of the global observing system and improved analysis systems, so that potential vorticity can be estimated with acceptable accuracy.

Chapter 2: Observing and modelling global circulations

Section 2.3: Haltiner & Williams (1980) is a well-known text dealing with numerical weather prediction in some detail. The relevant chapters of Holton (1992) are brief but helpful.

Section 2.4: Daley (1991) has written a recent text book on the analysis of meteorological data.

Section 2.5: A good survey of global circulation modelling is given by Washington and Parkinson (1986). A more recent collection of articles is contained in Trenberth (1992). The 'SGCM' is introduced in James & Gray (1986).

Chapter 3: The atmospheric heat engine

The text by Piexóto & Oort (1992) gives a more advanced account of these important matters.

Chapter 4: The zonal mean meridional circulation

Section 4.2: This argument was given in its simplest form by Held & Hou (1980) and had been anticipated to some extent by Schneider & Lindzen

(1977). The asymmetric Hadley circulation was discussed by Lindzen & Hou (1988).

Chapter 5: Transient disturbances in the mid-latitudes

Section 5.3: This section is similar to the development in Holton (1992), chapter 10. See also Peixóto & Oort (1992). A more sophisticated formulation of the available potential energy, which removes any assumption that static stability is constant on pressure surfaces was given by Pearce (1978). Observed levels of energy and conversions are based on data in Oort & Piexóto (1983).

Section 5.4: Gill (1982) gives a valuable and thorough account of the Eady model. Pedlosky (1987) gives details of the Charney and two-level models, and derives the general conditions for baroclinic instability of an arbitrary zonal flow. The original papers are Eady (1949) and Charney (1947).

Section 5.5: Simmons & Hoskins (1978) is a classic account of baroclinic lifecycles. Pedlosky (1971) describes weakly nonlinear theory applied to the two-level model, while Drazin (1970) does the same for the Eady model.

Chapter 6: Wave propagation and the steady eddies

Section 6.2: Hoskins & Karoly (1981) describe the application of ray tracing to Rossby waves.

Section 6.3: Grose & Hoskins (1979) show how a linear barotropic model can simulate the observed northern hemisphere winter steady waves. Figs. 6.16 and 6.17 were prepared by reproducing their calculations with more recent zonal wind data. Held (1983) gives a review of theories of steady waves in terms of Rossby wave propagation.

Section 6.5: Edmon *et. al.* (1980) give a helpful account of Eliassen–Palm diagnostics for tropospheric flows. Andrews *et. al.* (1987), chapter 3, derive the transformed Eulerian mean equations in a stratospheric context.

Section 6.6: Again, see Edmon *et. al.* (1980).

Chapter 7: Three-dimensional aspects of global circulation

Section 7.1: Gill (1980) provides a readable and illuminating account of tropical wave motions, and uses them to discuss the response of the flow to tropical heating anomalies. See also Chapter 11 of the text book by Gill (1982).

Section 7.2: Fein & Stephens (1987) contains several chapters which discuss various aspects of the physics and dynamics of monsoon circulations, generally at a nonspecialist level.

Section 7.3: The present study of storm tracks was initiated by Blackmon (1976), though it seems that Klein (1957) and Sawyer (1970) were the first authors to note the relationship between transient eddy variance and the tracks of individual weather systems. Wallace *et. al.* (1988) gives a more recent review of storm track observations. Hoskins *et. al.* (1983) contains some diagnostics of the storm tracks, while James & Anderson (1984) and Trenberth (1991) discuss the Southern Hemisphere storm track.

Section 7.4: The theory of the E–vector was given in Hoskins *et. al.* (1983). Other alternative formulations of the interaction between the eddies and the zonally varying mean flow are possible and have different advantages. See, for example, Plumb (1985). The forcing of storm tracks by observed heating distributions was demonstrated by Hoskins & Valdes (1990).

Section 7.5: Because of the difficulties in obtaining high quality moisture analyses, not many detailed descriptions of the global circulation of moisture have been published. That by Piexóto & Oort (1983) is one of the best. Chapter 12 of Piexóto & Oort (1992) gives a good introduction to the global water cycle.

Chapter 8: Low frequency variability of the circulation

Section 8.2: Wallace & Gutzler (1981) give a comprehensive discussion of northern hemisphere teleconnection patterns. See also Wallace & Blackmon (1983). The theoretical basis of EOF analysis is discussed by Mo & Ghil (1987); see also Piexóto & Oort (1992), appendix B for a full derivation of the mathematics. The theory and examples of the Rossby source function is given by Sardeshmukh & Hoskins (1988). Some southern hemisphere teleconnection patterns are discussed by Mo & White (1985).

Section 8.3: A fuller discussion of observations and theory of stratospheric oscillations, together with relevant references, is given in Andrews *et. al.* (1987). Figure 8.8 was taken from Naujokat (1986) and the laboratory analogue is given in Plumb & McEwan (1978). See also Holton (1983) for a discussion of the QBO.

Section 8.4: Intraseasonal oscillation was identified by Madden & Julian (1971) and its spatial structure was elucidated in Madden & Julian (1972). A recent review of both observations and theory is given in Slingo & Madden (1991).

Section 8.5: Philander (1990) gives a recent and thorough account of the observations and theory of ENSO and ENSO related phenomena.

Section 8.7: Lorenz (1984) introduced this simple model. The method of determining the dimensionality of the solution sub–space is essentially the algorithm of Grassberger & Procaccia (1983). James & James (1992) described the ultra low frequency variability of a simple atmospheric circulation model.

Chapter 9: The stratosphere

The topics covered in this chapter are treated much more thoroughly and fully in the texts by Andrews *et. al.* (1987) which deals with the dynamics, and by Wayne (1991), which devotes considerable space to stratospheric chemistry, especially the chemistry of ozone. A more concise review of dynamical issues is given by Holton (1983).

Chapter 10: The circulation of the atmospheres of the planets

Section 10.1: Beatty & Chaikin (1990) give a useful overview of modern planetary studies in a series of popular articles written by leading researchers.

Section 10.2: Williams (1988a, b) gives a most thorough and illuminating account of the regimes of flow of terrestrial-like atmospheres. The parametrization of baroclinic waves in an atmosphere is a simplified version of calculations due to Stone (1972). The reduced drag calculations were given by James & Gray (1986), and the central role of horizontal shear was stated by James (1987).

Section 10.3: See Read (1986) for a discussion of the problems encountered in generating intense superrotation. Schubert *et. al.* (1980) give a post-Pioneer review of the circulation of the Venus atmosphere.

Section 10.4: The generation of zonal jets by two-dimensional turbulence on a β–plane was described by Rhines (1975). The application to the Jovian circulation was discussed by Williams (1978).

Section 10.5: See Gill (1982) for a thorough treatment of ocean circulations. A more concise review of large scale ocean circulations is given by Anderson (1983).

Section 10.6: Hide & Mason (1975) is a classic review of work on the rotating fluid annulus. Hart *et. al.* (1986) describe experiments in a spherical system.

References

Anderson, D.L.T. (1983): The oceanic general circulation and its interaction with the atmosphere. In *Large Scale Dynamical Processes in the Atmosphere*, eds. Hoskins, B.J. and R.P. Pearce. Academic Press (London), pp. 305–36.

Andrews, D.G., J.R. Holton & C.B. Leovy (1987): *Middle Atmosphere Dynamics*. Academic Press (Orlando, Florida), 489pp.

Barnes, J.R. (1981): Midlatitude disturbances in the Martian atmosphere: a second year. *J. Atmos. Sci.*, **38**, 225–34.

Batchelor, G. (1967): *An Introduction to Fluid Dynamics*. Cambridge University Press, 615pp.

Beatty, J.K. & A. Chaikin (1990): *The New Solar System* (third edition). Cambridge University Press, 326pp.

Blackmon (1976): A climatological spectral study of the 500 mb geopotential height of the Northern Hemisphere. *J. Atmos. Sci.*, **33**, 1607–23.

Brown, R.A. (1991): *Fluid Mechanics of the Atmosphere*. Academic Press (San Diego), 486pp.

Charney, J.G. (1947): The dynamics of long waves in a baroclinic westerly current, *J. Meteorol.*, **4**, 135–63.

Daley, R. (1991): *Atmospheric Data Analysis*. Cambridge University Press, 457pp.

Drazin, P.G. (1970): Non-linear baroclinic instability of a continuous zonal flow. *Q. J. Roy. Met. Soc.*, **96**, 667–76.

Eady, E.T. (1949): Long waves and cyclone waves, *Tellus*, **1**, 33–52.

Edmon, H.J., B.J. Hoskins & M.E. McIntyre (1980): Eliassen-Palm cross sections for the troposphere, *J. Atm. Sci.*, **37**, 2600–16 (see also corrigendum *ibid*, **38**, 1115).

Fein, J.S. & P.L. Stephens, eds. (1987): *Monsoons*. J. Wiley (New York), 632pp.

Fleming, E.L., S. Chandra, J.J. Barnett & M. Corney (1990): Zonal mean temperature, pressure, zonal wind and geopotential height as functions of latitude. *Adv. Space Res.*, **10**, 11-62.

Gill, A.E. (1980): Some simple solutions for heat-induced tropical circulations. *Q. J. Roy. Met. Soc.*, **106**, 447-462.

Gill, A.E. (1982): *Atmosphere Ocean Dynamics*. Academic Press (New York), 662pp.

Gille, J.C. & L.V. Lyjak (1986): Radiative heating and cooling rates in the middle atmosphere. *J. Atmos. Sci.*, **43**, 2215-29.

Grassberger, P. & I. Procaccia (1983): Measuring the strangeness of strange attractors. *Physica*, **9D**, 189–208.

Grose, W.L. & B.J. Hoskins (1979): On the influence of orography on large scale atmospheric flow. *J. Atmos. Sci.*, **36**, 223–34.

Haigh, J.D. (1985): A fast method for calculating scale-dependent photochemical acceleration in dynamical models. *Q. J. Roy. Met. Soc.*, **111**, 1027–38.

Haltiner, G.J. & R.T. Williams (1980): *Numerical Prediction and Dynamic Meteorology.* J. Wiley (New York), 477pp.

Hart, J.E., G.A. Glatzmaier & J. Toomre (1986): Space laboratory and numerical simulations of thermal convection in a rotating hemispherical shell with radial gravity. *J. Fluid Mech.*, **173**, 519–44.

Held, I.M. (1983): Stationary and quasi-stationary eddies in the extratropical troposphere: theory. In *'Large Scale Dynamical Processes in the Atmosphere'* eds. Hoskins, B.J. and R.P. Pearce. Academic Press (London), 127–68.

Held, I.M. & A.Y. Hou (1980): Nonlinear axially symmetric circulations in a nearly inviscid atmosphere. *J. Atm. Sci.*, **37**, 515–33.

Hide, R. & P.J. Mason (1975): Sloping convection in a rotating fluid. *Advances in physics*, **24**, 47–100.

Holton, J.R. (1983): The stratosphere and its links with the troposphere. In *'Large Scale Dynamical Processes in the Atmosphere'* eds. Hoskins, B.J. and R.P. Pearce. Academic Press (London), 207–304.

Holton, J.R. (1992): *An Introduction to Dynamic Meteorology*, 3rd edition. Academic Press (San Diego), 507pp.

Hoskins, B.J., I. Draghici & H.C. Davies (1978): A new look at the ω–equation. *Q. J. Roy. Met. Soc.*, **104**, 31-8.

Hoskins, B.J., H.H. Hsu, I.N. James, M. Masutani, P.D. Sardeshmukh & G.H. White (1989): *Diagnostics of the Global Atmospheric Circulation, Based on ECMWF Analyses 1979-1989.* WCRP-27, World Meteorological Organisation (Geneva), 217pp.

Hoskins, B.J., I.N. James & G.H. White (1983): The shape, propagation and mean-flow interaction of large scale weather systems. *J. Atm. Sci.*, **40**, 1595–1612.

Hoskins, B.J. & D. Karoly (1981): The steady linear response of a spherical atmosphere to thermal and orographic forcing. *J. Atm. Sci.*, **38**, 1179–96.

Hoskins, B.J., M.E. McIntyre and A.W. Robertson (1985): On the use and significance of isentropic potential vorticity maps. *Q. J. Roy. Met. Soc.*, **111**, 877-946.

Hoskins, B.J. & P.J. Valdes (1990): On the existence of storm tracks. *J. Atm. Sci.*, **47**, 1854–64.

Ingersoll, A.P. (1990): Atmospheres of the giant planets. *The New Solar System*, eds. J.K. Beatty & A. Chaikin, 139-52.

James, I.N. (1987): Suppression of baroclinic instability in horizontally sheared flows. *J. Atmos. Sci.*, **44**, 3710-20.

James, I.N. (1988): On the forcing of planetary–scale Rossby waves by Antarctica. *Q. J. Roy. Met. Soc.*, **114**, 619–637.

James, I.N. & D.L.T. Anderson (1984): The seasonal mean flow and distribution of weather systems in the southern hemisphere: the effects of moisture transports. *Q. J. Roy. Met. Soc.*, **110**, 943-66.

James, I.N. & L.J. Gray (1986): Concerning the effect of surface drag on the circulation of a planetary atmosphere. *Q. J. Roy. Met. Soc.*, **112**, 1231-50.

James, I.N. & P.M. James (1992): Ultra low frequency variability of the flow in a simple atmospheric circulation model. *Q. J. Roy. Met. Soc.*, **118**, 1211-33.

Klein, W.H. (1957): Principal tracks and mean frequencies of cyclones and anticyclones in the Northern Hemisphere. *Research Paper No 40*, US Weather Bureau, Washington DC, 60pp.

Lau, N-C. (1984): Circulation statistics based on FGGE level IIIb analyses produced by GFDL. *NOAA Data Report ERL GFDL-5*, US Dept of Commerce. 427pp.

Lindzen, R.S. & A.Y. Hou (1988): Hadley circulations for zonally averaged heating centred off the equator. *J. Atmos. Sci.*, **45**, 2416-27.

Lorenz, E. (1984): Irregularity: a fundamental property of the atmosphere. *Tellus*, **36A**, 98–110.

Madden, R.A. & P.R. Julian (1971): Detection of a 40–50 day oscillation in the zonal wind in the tropical Pacific. *J. Atmos. Sci.*, **28**, 702–8.

Madden, R.A. & P.R. Julian (1972): Description of global scale circulation cells in the tropics with a 40–50 day period. *J. Atmos. Sci.*, **29**, 1109–23.

Mo, K.C. & M. Ghil (1987): Statistics and dynamics of persistent anomalies. *J. Atmos. Sci.*, **44**, 877–901.

Mo, K.C. & G.H. White (1985): Teleconnections in the southern hemisphere. *Mon. Wea. Rev.*, **113**, 22–37.

Naujokat, B. (1986): An update of the observed quasi-biennial oscillation of the stratospheric winds over the tropics. *J. Atmos. Sci.*, **43**, 1873–7.

Oort, A.H. (1983): Global atmospheric circulation statistics 1958–1973. *NOAA Professional Paper 14*, US Dept of Commerce, 180pp + 47 microfiches.

Oort, A.H. & J.P. Piexóto (1983): Global angular momentum and energy balance requirements from observations. *Advances in Geophysics*, **25**, 355–490.

Pearce, R. P. (1978): On the concept of available potential energy, *Q. J. Roy. Met. Soc.*, **104**, 737-55.

Pedlosky, J. (1971): Finite amplitude baroclinic waves with small dissipation. *J. Atmos. Sci.*, **28**, 587–97.

Pedlosky, J. (1987): *Geophysical Fluid Dynamics*. Springer-Verlag (New York), 710pp.

Peixóto, J.P. & A.H. Oort (1992): *Physics of Climate*. American Physical Society (New York), 520pp.

Peixóto, J.P. & A.H. Oort (1983): The atmospheric branch of the hydrological cycle and climate. In *Variations of the Global Water Budget*, eds., Reidel, London, 5–65.

Philander, S.G.H. (1990): *El Niño, La Niña and the Southern Oscillation*. Academic Press (San Diego), 293pp.

Plumb, R.A. (1985): On the three dimensional propagation of stationary waves. *J. Atmos. Sci.*, **42**, 217–29.

Plumb, R.A. & A.D. McEwan (1978): The instability of a forced standing wave in a viscous stratified fluid: a laboratory analogue of the quasi–biennial oscillation. *J. Atmos. Sci.*, **35**, 1827–39.

Press, W.H., B.P. Flannery, S.A. Teukolsky & W.T. Vetterling (1992): *Numerical Recipes: the Art of Scientific Computing* (2nd edition). Cambridge University Press, 992pp.

Read, P.L. (1986): Super-rotation and diffusion of axial angular momentum: II. A review of quasi-axisymmetric models of planetary atmospheres. *Q. J. Roy. Met. Soc.*, **112**, 253–72.

Rhines, P.B. (1975): Waves and turbulence on a beta plane. *J. Fluid Mech.*, **69**, 417–43.

Rogers, R.R. & M.K. Yau (1989): *A Short Course in Cloud Physics*, third edition. Pergamon (Oxford), 293pp.

Sardeshmukh, P.D. & B.J. Hoskins (1988): The generation of global rotational flow by steady idealized tropical divergence. *J. Atmos. Sci.*, **45**, 1228-51.

Sawyer, J.S. (1970): Observational characteristics of atmospheric fluctuations with a timescale of a month. *Q. J. Roy. Met. Soc.*, **96**, 610–25.

Schneider, E.K. & R.S. Lindzen (1977): Axially symmetric models of the basic state for instability and climate studies. Part II: Nonlinear calculations. *J. Atmos. Sci.*, **34**, 280–96.

Schofield, J.T. & F.W. Taylor (1983): Measurements of a mean, solar-fixed temperature and cloud structure of the middle atmosphere of Venus. *Q. J. Roy. Met. Soc.*, **109**, 57-80.

Schubert,G., C. Covey, A. Del Genio, L.S. Elson, G. Keating, A. Seiff, R.E. Young, J. Apt, C.C. Counselman, A.J. Kliore, S.S. Limaye, H.E. Revercomb, L.A. Sromovsky, V.E. Suomi, F. Taylor, R. Woo & U. von Zahn (1980): Structure and circulation of the Venus atmosphere. *J. Geophys. Res.*, **85**, 8007–25.

Shapiro, M.A. (1980): Turbulent mixing within tropopause folds as a mechanism for the exchanges of chemical constituents between the stratosphere and troposphere. *J. Atm. Sci.*, **37**, 994–1004.

Shine, K.P. (1987): The middle atmosphere in the absence of dynamical heat fluxes. *Q. J. Roy. Met. Soc.*, **113**, 603–33.

Simmons, A.J. & B.J. Hoskins (1978): The lifecycles of some nonlinear baroclinic waves. *J. Atm. Sci.*, **35**, 414–32.

Slingo, J.M. & R.A. Madden (1991): Characteristics of the tropical intra-seasonal oscillation in the NCAR community climate model. *Q. J. Roy. Met. Soc.*, **117**, 1129–70.

Solomon, S., J.T. Kiehl, R.R. Garcia & W. Grose (1986): Tracer transport by teh diabatic circulation deduced from satellite observations. *J. Atmos. Sci.*, **43**, 1603–17.

Stone, P.H. (1972): A simplified radiative-dynamical model for the static stability of rotating atmospheres. *J. Atm. Sci.*, **29**, 405–18.

Trenberth, K.E. (1991): Storm tracks in the Southern Hemisphere. *J. Atmos. Sci.*, **48**, 2179–94.

Trenberth, K.E. (ed.) (1992): *Climate System Modelling*. Cambridge University Press (Cambridge), 788pp.

Townsend, R.D. & D.R. Johnson (1985): A diagnostic study of the isentropically and zonally averaged mass flux during the first GARP global experiment. *J. Atmos. Sci.*, **42**, 1565–79.

Wallace, J.M. & M.L. Blackmon (1983): Observations of low-frequency atmospheric variability. In '*Large Scale Dynamical Processes in the Atmosphere*' eds. Hoskins, B.J. and R.P. Pearce. Academic Press (London), 397pp.

Wallace, J.M. & P.V. Hobbs (1977): *Atmospheric Science: An introductory survey*. Academic Press (New York), 467pp.

Wallace, J.M. & D.S. Gutzler (1981): Teleconnections in the geopotential height field during the Northern Hemisphere winter. *Mon. Wea. Rev.*, **109**, 784–812.

Wallace, J.M., G-H. Lim & M.L. Blackmon (1988): Relationship between cyclone tracks, anticyclone tracks and baroclinic waveguides. *J. Atm. Sci.*, **45**, 439–62.

Washington, W.M. & C.L. Parkinson (1986): *An Introduction to Three Dimensional Climate Modelling*. Oxford University Press, 422pp.

Wayne, R.P. (1991): *Chemistry of Atmospheres*, (2nd edition). Clarendon Press (Oxford), 447pp.

Williams, G.P. (1978): Planetary circulations: 1. Barotropic representation of Jovian and terrestrial turbulence. *J. Atmos. Sci.*, **35**, 1399–1426.

Williams, G.P. (1988a): The dynamical range of global circulations – I. *Climate Dynamics*, **2**, 205–60.

Williams, G.P. (1988b): The dynamical range of global circulations – II. *Climate Dynamics*, **3**, 45-84.

Index

417

Printed in the United States
By Bookmasters